Moshe R. Heller (Editor)
CONTROL DATA

Nuclear Simulation

Proceedings of an
International Symposium and Workshop
October 1987, Schliersee, West Germany

With 107 Figures

Springer-Verlag Berlin Heidelberg New York
London Paris Tokyo

The International Nuclear Simulation Symposium and
Mathematical Modelling Workshop

Initiated, sponsored and oranized by:
Control Data GmbH, West Germany

Co-Sponsors:
Gesellschaft für Reaktorsicherheit (GRS)
The Society for Computer Simulation (SCS)

Editor:
Moshe R. Heller, Control Data GmbH, FRG

ISBN-13: 978-3-642-83223-9   e-ISBN-13: 978-3-642-83221-5
DOI: 10.1007/978-3-642-83221-5   ·

Library of Congress Cataloging-in-Publication Data
Nuclear simulation.
Proceedings of the International Nuclear Simulation Symposium
and Mathematical Modelling Workshop;
sponsored by Control Data GmbH, Gesellschaft für Reaktorsicherheit,
and the Society for Computer Simulation.
1. Nuclear engineering – Simulation methods – Congresses.
2. Nuclear engineering – Mathematical models – Congresses.
I. International Nuclear Simulation Symposium
    and Mathematical Modelling Workshop (1987: Schliersee, Germany)
II. Heller, Moshe R.
III. Control Data GmbH.
IV. Society for Computer Simulation.
V. Gesellschaft für Reaktorsicherheit.
TK9006.N827 1987 621.48 87-23461

2161/3020 543210

"For the best and safest method of
philosophizing we should say seems
to be, first to inquire deligently into
the properties of things, and of
establishing these properties by
experiment, and then to proceed more
slowly to hypothesis for the
explanation of them."

ISAAC NEWTON, 1670

Preface

Welcome to Bavaria - Germany and to the INTERNATIONAL NUCLEAR SIMULATION SYMPOSIUM AND MATHEMATICAL MODELLING WORKSHOP. A triennial international conference jointly promoted by Control.Data, GRS and SCS, which takes place at Schliersee, a small town near the Alps.

The aim of the Symposium is to cover most of the aspects of nuclear modelling and simulation in theory and practice, to promote the exchange of knowledge and experience between different international research groups in this field, and to strengthen the international contact between developers and users of modelling and simulation techniques.

On the occasion of the Symposium people of scientific and engineering disciplines will meet to discuss the state-of-the-art and future activities and developments.

A large number of contributed papers has been strictly examined and selected by the papers committee to guarantee a high international standard.

The book contains the accepted papers which will be presented at the Symposium. The papers have been classified according to the following topics:

1. HARDWARE TOOLS
2. SIMULATION-SOFTWARE-TOOLS
3. PLANT ANALYSER
4. REACTOR CORE
5. NUCLEAR WASTE

Authors from 9 countries will meet at the Symposium. They work for Industrial Companies, Universities and the Research and Development Institutes so that a broad spectrum of simulation activities is covered: Theory and application, hardware and software, research and operations.

The editor is greatful to the authors for making possible the publication of this book, and especially to WOLFGANG F. WERNER, for the selection of the papers and the contribution to the success of the Symposium.

Mr. VON HAGEN and Ms. RAUFELDER of Springer-Verlag for the excellent publication of the proceedings to whom I would like to extend my thanks.

My thanks also go to all of the Control Data people and specially to Ms. JULIE ESCH who have been involved beyond everyday's work in the promotion of the Symposium.

Munich, October 1987   M. R. Heller

Table of Contents

Introduction
Simulation "In the Core"

M. R. Heller

Control Data GmbH, München, W-Germany

Before the widespread use of the digital machines, relationships and
activities in the scientific-engineering approach were largely determined by the human mind, its formalization and analytic powers. The
scientist or engineer who approaches a real world process tries to
gain insight or an understanding of the phenomena on the process under
study. Today after Chernobyl, which exploded in the world's worst nuclear
accident on 26 April 1986. The desire to prove that the errors which led
to it will never be repeated is understandable, but the main impression
left by the shuttered villages and abandoned fields in which bushes are
begining to sprout is the length of the time necessary to eradicate the
consequences of the original accident.

One of the very powerful methods consists in trying to obtain an abstract
or formal model or representation of the accident (process). The activity
is defined as   m o d e l   b u i l d i n g   and   f o r m a l i z a -
t i o n .   In essence the procedure requires abstraction and simplification. Simplification is necessary to restrict the complexity of the
representation. One only chooses those properties within given boundaries
of space or time which are believed to be connected with each other but
unconnected with other properties or other parts of the world. Basically
the model builder proceeds by hypothesis, induction and deduction.
The complete body of methods is called modelling methodology.

Modelling itself is for a part still considered as an art. There are a
large number of factors that come into play. The representation process
involves almost always "inter- or extrapolation". Certainly, a large
body of methodology is objective and mathematically sound. Many tools
are well-defined and clearly stated. Statistical techniques and para-

meter estimation procedures have a firm logical basis, but standard
techniques cannot solve all problems. The final product is the result
of careful trade-off's between existing facts, decision on the choice
of representative details, careful experimental work and its interpretation. Examples of such issues relate to the choice of formalisms, the
evaluation of the validity of à priori facts, the required level of
descriptive details etc.

It has sometimes been said that systems with automatic support devices
are far better off than a system with a crew that has to evaluate threat
and then take precautions; the automation is thought to be faster and
without having THE TROUBLE WITH HESITATION BETWEEN STEPS to be taken.

M o d e l   i n t e g r a t i o n is introduced; different descriptions
of the same real world process are compared, screened for consistency
and integrated into a whole. This activity is a basic step in scientific
work, especially in the process of developing a theory and working out
general principles. The formal model, though a simplified representation
of reality, always summarizes a vast amount of information, comprising
facts, axioms and hypotheses. If its validity is high or in other words
if its descriptive quality is good, it can be used to obtain useful knowledge on the system under study. In this sense a model can be seen as an
extremely compact and useful extension of a data base containing loose
pieces of data. With modern computerized receivers and possibly computer-
controlled system arrays A HIGHLY EDUCATED OPERATOR CAN LOOK THROUGH
DENSE ENVIRONMENT and give decision-makers correct information.

The   s i m u l a t i o n   a c t i v i t y , being the experimentation
with models, supports not only the model utilization but the model
building as well. The introduction of computational devices requires
provisions for suitable man-machine communication. The nature of this
communication has its impact on the modelling methodology. Man is
especially apt in reasoning and in recognizing patterns; his computat-
tional powers however are limited. A process modelbase is built up,
composed of several candidate models, "primitives" and combined ones,
with their own features. The creation and testing of an extensive variety
of such models is of key importance and relies upon an efficient use of
advanced parallel processing and supercomputing. At present there are

arguments that, with time, modelling methodology will be incorporated for its major part in "intelligent", "self-organizing" machines, so that human intervention could be brought to a bare small level. In that case simulation will in a sense supplant modelling. More over it can be stated that SIMULATION is a COST EFFECTIVE solution to enginerring effectivness.

More probably the reason for using Nuclear Simulation is the fear of the damage to the prestige and credibility of this industry if Chernobyl were completly abondoned.

Nowadays in advanced information processing a fundamental transition is taking place from data-processing to k n o w l e d g e - p r o c e s - s i n g , which will be key of the next-generation computers. Advanced information processing uses the automation of data acquisition and data processing as well as the automation of the reasoning process in order to combine these functions to create a system capable of showing an "intelligent" behaviour. This will allow simulators to rely on flexible, human-like thought processes to diagnose problems, rather than on rigid procedures expressed in flowcharts or "decision trees". To incorporate this approach, the central knowledge-base contains rules generated from discussions eith the specialist of the process under study. This added flexibility will greatly improve the diagnostic ability of current simulators. New simulator architectures enable these developments in expert system design.

The ultimate goal in process-studies consists in the integration of knowledge from the model level to the à priori level. Therefore, existing models must be generalized. Extensive experimentation, results in the validation of the generalized models; and last but not least these general models have to be interpreted in laws and theories. Future developments should reveal these frontiers of simulation.

REFERENCES

(1) Cockburn, P.: "Cleaning up Soviet nuclear act", F. T. June 22, 1987

(2) Heller, M.R.: Simulation "in the Boat", First Intercontinental Maritime Simulation Symposium, Proceedings by Springer-Verlag, June 3-5, 1985.

(3) Van Steenkiste, G.C.: Simulation "On the Road", The first European Cars/Trucks Simulation Symposium, Schliersee, May 2-4, 1984.

The Role of Symbolic Processing in Supercomputing

Robert M. White

Control Data Corporation
Minneapolis, Minnesota, 55440, U.S.A.

Most of you are dealing with what I would call a "megaproblem" -- a problem that requires the highest performance computer available. Such problems occur mainly in applications where physical phenomena are simulated, such as weather, earthquakes, and ocean currents. Such problems also arise in the design of systems that depend on physical processes but where experimental study is uneconomical, difficult or hazardous, such as the behavior of airfoils in a wind stream, the vibration of buildings and other structures, and the detonation of nuclear weapons. In addition to simulations of physical phenomena, the simulation of economic activities, such as with linear programming, also can use high performance computers. Cryptography is another megaproblem area. Control Data has traditionally provided the tools to solve such problems--both hardware and software.

Megaproblem solvers have an insatiable appetite for raw computer power. And one can always imagine problems just beyond the range of our existing capabilities. Figure 1, for example, shows the capabilities of a few supercomputers relative to the size of aerodynamic computations. The traditional mode has been to "chase the problem." That is, to increase the raw computer power largely through the introduction of more advanced electronics. However, over the past few years there have been a number of developments that should at least make one question this approach.

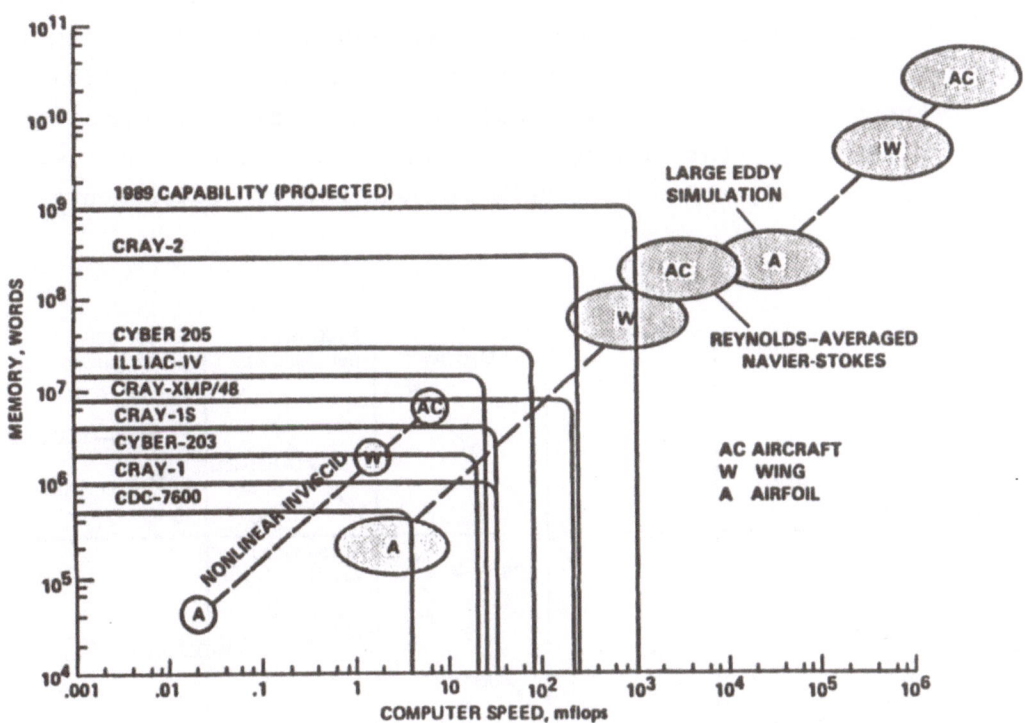

Figure 1   Computer speed and memory requirements compared with computer capabilities for 15 minute runs with 1985 algorithms. (Courtesy K. G. Stevens, NASA Ames Research Center)

The first development is the appearance of minisupercomputers. The performance of some of these machines is plotted in Figure 2 as a function of price. Gordon Bell, in a manuscript entitled "Preparing for Changing Scientific Computer Environments," has pointed out that since the performance/price ratio is essentially

the same for all these machines, there will be a hierarchy
of use where some users will trade off the ability to work
interactively for ten hours on what might be an one hour
job on a supercomputer since the supercomputer turn-around
time in batch-mode may in fact also be ten hours.
However, the situation is much more complex than Figure 2
would imply. For example, as we saw in Figure 1, memory
size is also an important consideration. In addition,
there is I/O bandwidth, software availability, etc. The
point is that there are many variables, and each
particular problem will have its own set of requirements.

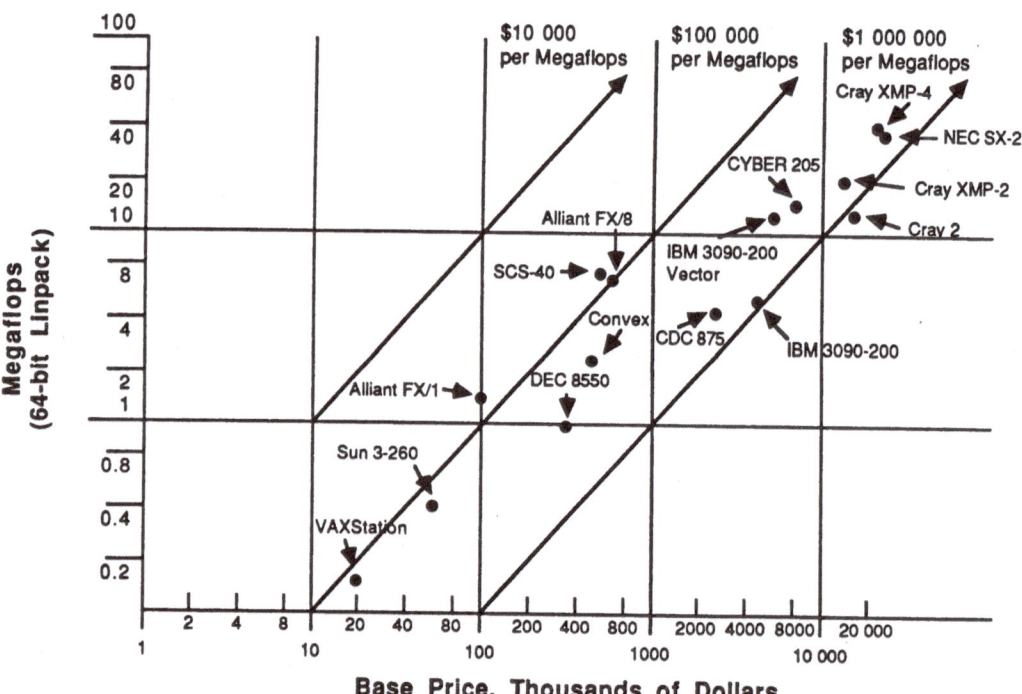

Figure 2   Ranking of various computers by their
            performance on the Linpack benchmark.[1]

The second major development is the explosion in parallel architectures. Dongarra lists several reasons for this including:

- Use of off-the-shelf processors
- Standard bus interfaces
- Custom gate arrays
- Availability of venture capital as well as government funding to university researchers

Control Data presaged parallel processing when it introduced multiple functional units in the 6600 (1962) and vector processing in the STAR 100 (1970), the predecessor of the CYBER 205.

Another factor that has promoted parallel processing has been its ready application to problems in the area of artificial intelligence. The fact that the human brain processes symbols and images so effectively with such slow devices (the neurons) is due to its massive parallelism.

The extent to which parallelism is employed depends upon the AI application. These applications range from expert systems to image processing. Expert systems are based on symbolic processing. The characteristics of symbolic processing are fundamentally different from those of numerical processing. Rule-based systems employ forward and/or backward chaining; semantic networks use marker propagation; and frames require procedural attachments. These capabilities require symbolic knowledge representations, search-intensive operations, a large memory requirement without the locality-of-reference property, variable sizes of messages, the use of nondeterministic algorithms, interactive I/O, and knowledge database requirements, features not found on the supercomputers of today.

The locality-of-memory-reference problem must be faced by all AI architectures. AI applications tend to spray memory requests randomly and conditionally over the entire active address space. This means that distributed addressing and anticipatory fetches are of little help in masking inherent memory cycle times. Large, fast memories seem to be the only solution -- the ICOT PSI 2 machine has 320 megabytes of main memory!

LISP and PROLOG are the dominant languages of symbolic computing. A LISP program may be thought of as a set of functions among which data is passed; a PROLOG program may be thought of as a set of assertions and relations over which inferences are made. Most commercial LISP machines, such as Symbolics, have a uniprocessor architecture but with "tagged" memory to help manage the amorphous data structures. The concurrent execution of these functions forms the basis for parallelism, and parallel symbolic machines are beginning to appear.

Parallel architectures have had a major impact on the intelligent interface aspects of AI such as image processing. Examples include CMU's WARP Machine, and BBN's Butterfly. The topology of these architectures is also specific to the application. Image processing, for example, is best served by a hypercube or other topology with tight local coupling. These applications also typically involve a global memory that is shared by the many processors.

Parallel processing has also entered supercomputer architectures. However, unlike artificial intelligence applications, numeric processing has a long history built on the von Neuman architecture. Figure 3 illustrates the solution heirarchy and indicates the parts that play a role in parallelism. The "special" processor part is where the parallel architecture is defined. This might be the vector part of a vector machine, with the scalar part being the "general" processor, or it might be the multiprocessors in a concurrent architecture. So much code for scientific calculation has been developed in FORTRAN, and use of the language is so widespread, that it is expected to be an important language for decades. There also seems to be a strong preference for the UNIX operating system.

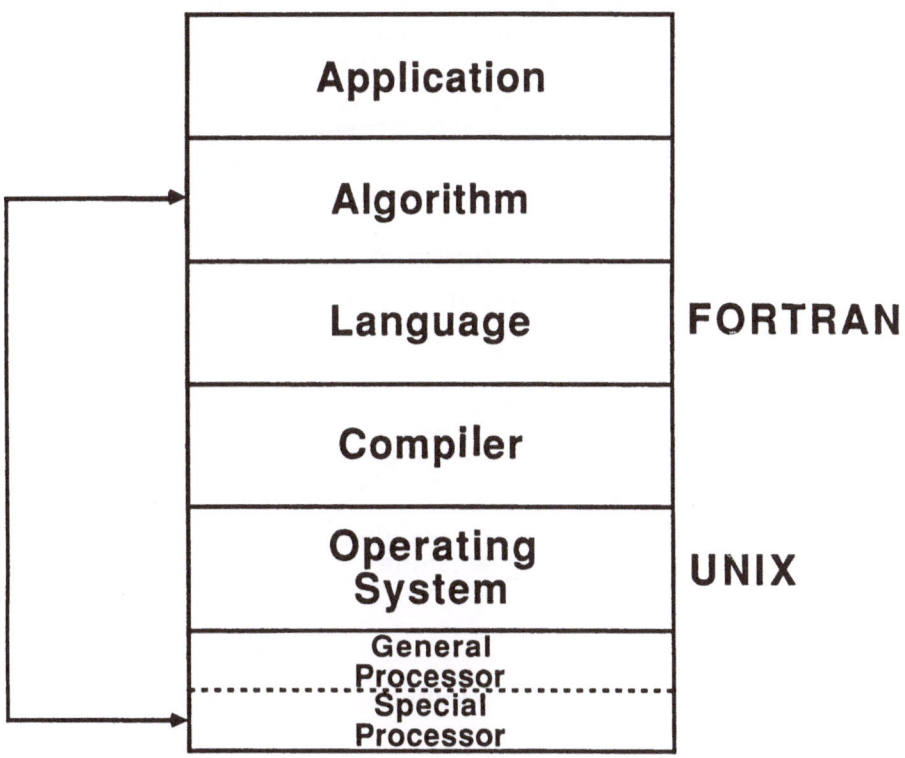

Figure 3 Solution Heirarchy

The transition to a vector machine can be accomplished by various means: the establishment of an assembly code

library that can be called from FORTRAN, building a
compiler that vectorizes the code, or extensions of the
FORTRAN. In the case of a multiprocessor architecture, as
we noted above, one should also develop new algorithms
that utilize this hardware configuration. The compiler
must support these changes. This is very difficult in the
case of massively parallel architectures. As a result,
numeric parallel architectures are starting at the
coarse-grained level with four or eight processors.
Figure 4 illustrates how the three general application
domains we have discussed map onto parallel architectures.

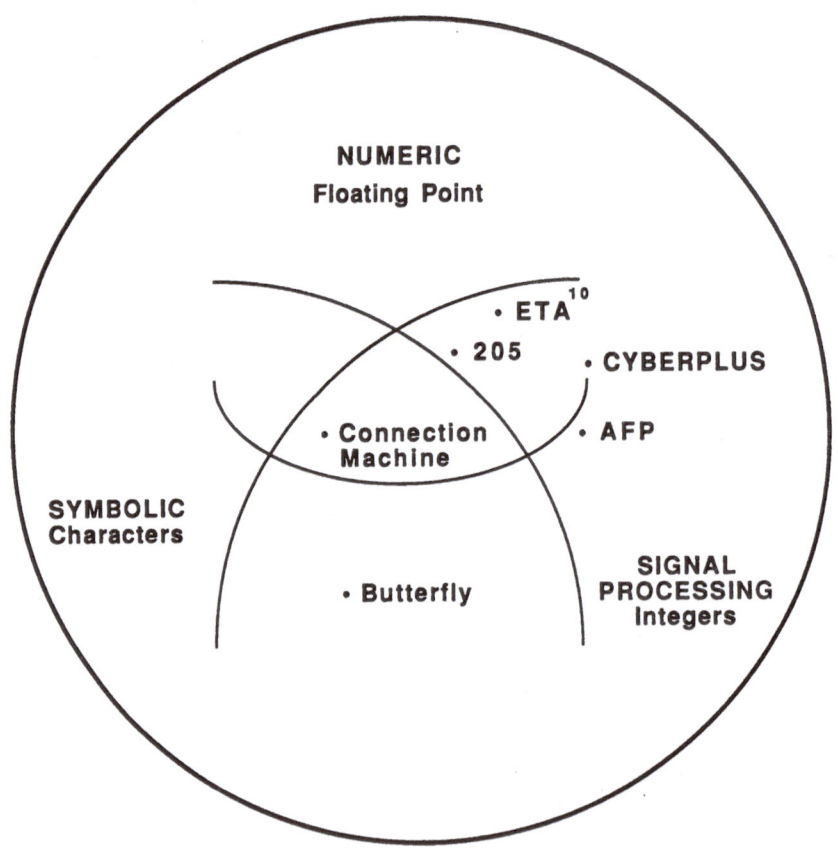

Figure 4 Schematic mapping of application domains onto
parallel architectures.

This is not to say that numeric applications run most efficiently on coarse-grained architectures. On the contrary. Fox[2] has demonstrated that many scientific problems can be explicitly mapped onto certain local memory gridlike architectures so they make efficient utilization of the hardware. For example, the finite difference equations, finite element equations, and partial differential equations encountered in geophysical or aerodynamical problems map onto a 3D mesh. The fast Fourier transforms used in fluid dynamics map onto a hypercube. This "coupling" between algorithms and architecture is evident even in the present coarse-grained machines. The Alliant FX/8, for example, has 8 "special" 64-bit processors, each with a vector instruction set, and 12 "general" processors for handling I/O and running the operating system. Were the programs that were run in the Linpack benchmark shown in Figure 1 hand-coded to exploit this particular architecture of the Alliant, its performance would increase from the 8 megaflops indicated up to 100 megaflops. Since large sectors of the supercomputer market are associated with very specific problems, it may be more efficient to design minisupers with architectures specific to those problems. For example, both the U.S. Weather Service and the European Weather Service use the s a m e program for weather prediction. It would be interesting to compare the accuracy of the predictions run on a "general" supercomputer with those run on a specially-architected minisuper.

While artificial intelligence applications face the locality-of-memory-reference problem, for a given algorithm and a given heuristic strategy, the faster the processor, the larger the portion of the search space that can be explored. Thus, it should not be surprising that some AI researchers are exploiting existing supercomputers. LISP has been implemented on a CRAY machine where it runs 5 to 15 times the speed of a Symbolics 3600.

While I have no doubt that supercomputers do offer opportunities for AI research, I believe the benefits t o supercomputing f r o m AI can be even greater. We are already beginning to see the appearance of "intelligent tools" for supercomputer users such as versions of FORTRAN that autovectorize or parallelize, interactive debuggers, run-time analyzers, simulators, etc. But the technology of expert systems should prove even more powerful to the hardware designers and the developers of algorithms, providing unprecedented opportunities for _adaptive_ megaproblem simulations and solutions.

REFERENCES

1.   J. Dongarra, J. L. Martin, and J. Worlton, "Computer Benchmarking = Paths and Pitfalls," IEEE SPECTRUM, July 1987, p. 38.

2.   G. C. Fox and S. W. Otto, "Algorithms for Concurrent Processors," Physics Today 37(5): 50-59 (1984); G. Fox, "The Performance of the Caltech Hypercube in Scientific Calculations," Supercomputers -- Algorithms, Architectures and Scientific Computation, ed. F. A. Masten, T. Tajima, Texas: University of Texas, (1985)

The ETA^{10} Supercomputer System

CHARLES D. SWANSON

ETA Systems, Incorporated
1450 Energy Park Drive
St. Paul, MN 55108

ABSTRACT

The ETA Systems, Incorporated ETA^{10}Ⓡ is a next-generation supercomputer featuring multiprocessing, a large hierarchical memory system, high performance input/output, and network support for both batch and interactive processing. Advanced technology used in the ETA^{10} includes liquid nitrogen cooled CMOS logic with 20,000 gates per chip, a single printed circuit board for each CPU, and high density static and dynamic MOS memory chips. Software for the ETA^{10} includes an underlying kernel that supports multiple user environments, a new ETA FORTRAN compiler with an advanced automatic vectorizer, a multitasking library and debugging tools. Possible developments for future supercomputers from ETA Systems are discussed.

I. INTRODUCTION

ETA Systems, Incorporated was established in August, 1983, to produce a next-generation supercomputer, the ETA^{10}. By the summer of 1986, a prototype ETA^{10} CPU was running at ETA Systems headquarters in St. Paul, Minnesota. The first shipment of an ETA^{10} occurred in December, 1986, to Florida State University.

A functional digram of the ETA^{10} is shown in Figure 1. Up to eight Central Processing Units (CPUs), each with 4 million words (32 million bytes) of Central Processor Memory, can be configured. The Central Processor Memory is complemented by up to 256 million words (2 billion bytes) of Shared Memory and up to 1 million words (8 million bytes) of Communication Buffer memory. Input/Output Units (IOUs) provide access to peripheral devices and networks.

This paper will present an overview of the technology, architecture and software of the ETA^{10}, and conclude with a brief discussion of possible future supercomputers.

II. TECHNOLOGY

The technology chosen for supercomputer implementation has a significant impact. It defines the robustness and speed available to the overall system. The technology selected for ETA Systems' supercomputers is a CMOS based VHSIC (Very High Speed Integrated Circuit) logic chip designed by ETA Systems. CMOS was the desired logic technology because its operating characteristics are well understood and it has low power requirements.

Each VHSIC chip contains 20,000 gates (circuit building blocks). An entire ETA^{10} central processor consists of only 240 of these hips. This reduces the processor component count and interconnections, increasing system reliability.

The circuit density of the chip provides two additional benefits. First, the processing speed is increased by minimizing signal distances. Also, there is sufficient circuitry available to implement a patented, built-in evaluation and self testing feature for circuit monitoring. This advanced feature allows complete functional testing of the individual chips and chip-to-chip interconnect testing. It increases reliability while enhancing maintainability.

An important characteristic of CMOS is that its switching speed is directly related to its temperature, getting faster as temperature goes down. Another is that a cool environment reduces the possibility of heat build-up damage in its active elements. Therefore, the ETA^{10} central processor is immersed in a cryostat containing liquid nitrogen at 77° K (-196° C). Cryogenic technology, although new to the computer industry, is widely used throughout the world.

Another significant technology was employed to allow the VHSIC chips to be effectively used. ETA Systems developed an extremely dense, state-of-the-art circuit board (44 layers). This allows an entire central processing unit to fit on a single board. It also increases the central processor speed and reliability by further reducing signal lengths and external wiring requirements.

III. ETA^{10} HARDWARE ARCHITECTURE

Five major hardware components provide the processing capability of the ETA^{10}: the Central Processing Unit, Shared Memory, Communication Buffer, Input/Output Unit, and Service Unit. These components are shown in the ETA^{10} Functional Diagram, Figure 1.

In addition, an ETA^{10} system includes peripheral devices and connections to data networks. The following subsections describe each of these hardware elements.

A. Central Processing Unit

Up to eight identical central processors can be configured into a single system. The ETA^{10} supercomputer can be upgraded up to a maximum configuration on the customer's site.

The Central Processing Units (CPUs) provide the computational power for the ETA^{10} supercomputer. Features provided in each CPU include:

* Scalar processor - Traditional sequential processing is provided by the scalar processor. It is capable of issuing an instruction every clock cycle. These instructions can be bit, byte, half-word or full 64-bit word operations. It also includes a high speed 256 word register file and a 64 word instruction stack to minimize memory accesses.

* Vector Processor - The ETA^{10} features memory-to-memory vector processing. The independent vector processor has two floating point pipelines so that two 64-bit results are produced each clock cycle following vector startup. The ETA^{10} also supports 32-bit (half precision) arithmetic, providing four 32-bit results per clock cycle. A vector shortstop path in the hardware permits vector results to be sent to the vector pipeline for subsequent vector operations rather than forcing them back to memory first. This feature reduces the vector startup time relative to the predecessor CYBER 205.

* Central Processor Memory - Each central processor contains 4 million words (32 MBytes) of memory dedicated to that processor. CP memory utilizes 64K bit Static Random Access Memory (SRAM) MOS memory chips.

* Virtual Addressing - The CPU provides a virtual addressing capability that allows a program address space of up to 2 trillion words. The virtual memory size being larger than the physical memory size is transparent to the user.

B. Shared Memory

Shared Memory (SM) is a large high-speed memory which is shared by all CPUs, Input/Output Units and Service Units. Shared Memory sizes range from 32 million to 256 million 64-bit words. Dynamic Random Access Memory (DRAM) MOS chips with 256K bits per chip are utilized. Like all memory on the ETA^{10}, Shared Memory is air cooled using a special technique which blows dry, chilled air on each individual memory chip.

Shared Memory provides one port per CPU which can sustain transfer rates of greater than 9 billion bits per second (9,000 Mbps). It also provides up to twenty ports for connecting Input/Output Units and Service Units, each of which can sustain 280 million bits per second (280 Mbps). All CPU and IOU ports can operate at their rated speed concurrently.

Data flow through the Shared Memory is controlled by the Shared Memory Interface consisting of a 44 layer printed circuit board and 20,000 gate CMOS chips, cooled by air.

C. Communication Buffer

The Communication Buffer provides rapid access by all system processors to small amounts of shared data and synchronizing functions, used for queuing and multitasking operations. The Communication Buffer consists of a one million 64-bit word memory and an interface processor consisting of a 44 layer PC board populated with 20,000 gate CMOS logic chips.

The Communication Buffer high-speed memory is shared between all of the system elements. Its bandwidth is approximately 2,000 Mbps per port. All ports can operate at their rated speed concurrently. Using the Communication Buffer for these interprocessor communication functions reduces the potential for contention in the Shared Memory.

D. Input/Output Unit

The Input/Output Unit (IOU) provides a powerful, flexible interface between the other system elements and peripherals and networks attached to the ETA^{10}. The IOU is interfaced to the Shared Memory by a Data Pipe Channel. The Data Pipe Channel is a fiber optic connection to a low-speed Shared Memory port.

The general philosophy for I/O is that the resources and traffic will be distributed between the CPUs and the IOUs. In general, the CPUs will manage logical I/O and IOUs will manage physical I/O. This is done transparently to the using process.

The IOU consists of up to five Functional Units. Each of the Functional Units has a Motorola 68020® microcomputer, up to five megabytes of memory, a Data Pipe Buffer and a controller interface for a disk channel, tape channel or network channel. The Functional Unit combinations within an IOU are determined by customer requirements. Up to 18 IOUs may be configured on an ETA^{10} system.

E. Service Unit

The Service Unit is an independent system which provides all system maintenance interfaces and performs power and cooling supervisory functions. It also acts as the interface for the operator and service console(s).

F. Peripheral Devices

Peripheral devices supported on the ETA^{10} include disk storage units and magnetic tapes.

The disk subsystem includes storage devices (disks), a Functional Unit which implements the file management capabilities assigned to the IOU and at least one Functional

Unit which controls the flow of information to and from the disk(s). Each disk storage unit has a capacity of 1250 million bytes, a burst transfer rate of 96 million bits per sec and a sustained transfer rate of 80 million bits per second.

The magnetic tape subsystem includes the storage devices (tape drives), storage controllers and a Functional Unit which includes a FIPS-60 standard interface for the tape unit(s). Tapes supported are 9-track, 1600 CPI Phase Encoded and 9-track, 6250 CPI Group Code Recorded. The tape speed is 200 inches per second.

G. Networks

The ETA^{10} interfaces with both the Open Interconnection Network (OIN) and the Multi-Host Network (MHN).

The OIN is based on the IEEE 802.3 standard for Ethernet. OIN supports the U.S. Department of Defense standard TCP/IP protocol. The supported standard application protocols include File Transfer Protocol (FTP) and TELNET. The OIN is rated at 10 million bits per second and is capable of connecting 6,000 devices with 5,000 concurrently active. The OIN will be primarily used to provide interactive access to the ETA^{10} via workstations and/or terminals.

The MHN can be either the Control Data Corporation Loosely Coupled Network (LCN®) or the Network Systems Corporation HYPERChannel®. Rated at 50 million bits per second, the MHN is primarily for high speed file transfers between mainframes and for batch processing on the supercomputer.

IV. ETA^{10} SOFTWARE

A. Operating System

The ETA^{10} operating system consists of a single underlying kernel which supports multiple user environments. The kernel provides process management, file and record management, file resource management, input/output, operator interfaces, accounting, and user validation. The user environments provide the command languages, file systems and utility packages. The two user environments ETA Systems has chosen to support are ETA System V, based on AT&T UNIX® System V, and VSOS®, the user interface of the CYBER 205®.

The System V user environment is based on AT&T UNIX System V, Release 2, with some Berkeley extensions, primarily for networking. It has both the Bourne and C shells. The System V user environment has been enhanced to support multiple CPUs, the ETA^{10} memory hierarchy and high performance distributed I/O, thousands of processes, and additional security features.

The VSOS programming environment is provided for compatibility with the CYBER 205 and is based on the CYBER 205 VSOS Release 2.2. It has been enhanced to provide richer interactive features along with traditional batch processing. In addition to the other ETA^{10} compilers, the CYBER 205 FORTRAN 200 compiler is provided for compatibility.

B. Compilers

Several software products are designed to be used with either user environment. These products include the standard programming languages (ETA FORTRAN, C, and Pascal), multi-tasking capabilities, and some program development tools.

1. ETA FORTRAN

The ETA FORTRAN compiler supports the ANSI 77 standard language and the anticipated array notation of the next ANSI standard. A state-of-the-art automatic vectorizer also is included in the ETA FORTRAN compiler, with the following features:

    DO and IF loop vectorization
    Scalar promotion
    IF THEN ELSE construct vectorization
    User feedback and directives
    Automatic strip mining
    Array bounds checking
    Round-off error control
    Nested loop collapse
    Alternate complex storage

The ETA FORTRAN compiler also includes performance analysis capability and is compatible with CYBER 205 FORTRAN 200.

2. C

The C language, as defined in "The C Programming Language"[1], will be fully implemented on the ETA^{10} system. The C compiler emphasizes compatibility with other portable C compilers, fast compilation speed, and diagnostics.

3. Pascal

A full implementation of Pascal, as defined in the "Pascal User Manual and Report"[2], will also be available. The language will be extended to include a vector notation and will emphasize fast compilation speed and good diagnostics.

C. Multitasking Library

The ETA^{10} hardware and system software provide the communication, synchronization and CPU allocation mechanisms to support multitasking such that parts (tasks) of a job which can run in parallel with other parts of the same job can simultaneously execute on two or more CPU's with shared memory.

A multitasking library is provided to permit users to explicitly describe the parallelism in their applications. Users can access the multitasking library from either programming environment.

The multitasking library is an extendable set of basic routines that supports several models for multitasking. It acts as an extension to ETA^{10} compilers to allow explicit use of the parallel architecture of the ETA^{10} system. The multitasking models define and manage the following three data types to implement the multitasking models:

1. A task is an independent, executable environment that is a subset of a user's program.

2. A counter is a shared semaphore with an integer value. It is used to provide mutual exclusion and to signal global events.

3. A shared array is used to share data between cooperating tasks of a user's program.

D. Applications

The ETA Systems' Applications Group is arranging with vendors in several diciplines to provide applications for the ETA^{10}. In addition, ETA Systems has established and staffed an applications research laboratory in cooperation with the University of Georgia, with primary focus on computational chemistry, math libraries and graphics.

Applications currently operational on the CYBER 205 will be directly transferable by virture of preserving FORTRAN 77 and the VSOS user environment from the CYBER 205.

V. FUTURE DEVELOPMENTS

A discussion of future supercomputers will involve larger memories, faster and denser logic devices, and continuation of multiprocessor architectures.

A. Memory

Within the next few years, memory devices with four times the current bit densities will be available. This means that the 64K static RAM chips used in the ETA^{10}'s Central Processor Memory can be replaced with 256K SRAM chips, so that each CPU can have 16 million 64-bit words of local memory. Similarly, the 256K dynamic RAM chips used in the Shared Memory (SM) can be replaced with 1 million bit DRAM chips providing up to 1 billion words of SM. The ETA^{10} is designed to handle these higher density chips; upgrading the ETA^{10} memory can be done in the field.

In the early 1990s, another four-fold increase in memory chip densities is expected, leading to still larger memories (or smaller volumes for current memory sizes).

B. Logic

Logic choices in the early 1990s may include CMOS, HEMT GaAs and ECL. CMOS may be available with faster switching speeds and another order of magnitude increase in gate density to 200,000 gates per chip, so that a supercomputer CPU can be made with less than 40 chips. HEMT GaAs chips with about 20,000 gates/chip are considered a possibility. This GaAs technology could achieve extremely fast switching speeds utilizing liquid nitrogen cooling. Finally, faster silicon ECL logic may also be available at gate densities of 20,000 gates/chip.

ETA Systems is investigating all of these logic choices for implementation in future supercomputers.

C. Architecture

Since the CMOS logic technology used in the ETA^{10} does not require liquid nitrogen for functionality but for speed, a supercomputer based on an air-cooled ETA^{10} CPU board is possible. Such a supercomputer would have performance in the range of current generation supercomputers, would require very little power, would not require special cooling equipment, and would run all of the ETA^{10} software.

In the early 1990s, the availability of larger memories and faster CPUs in smaller packages means that extending multiprocessor architectures to more than 8 CPUs is likely.

ETA^{10} is a registered trademark of ETA Systems, Inc.

Motorola 68020 is is a registered trademark of Motorola Corp.

LCN, VSOS and CYBER 205 are registered trademarks of Control Data Corporation.

HYPERChannel is a registered trademark of Network Systems Corp.

UNIX is a registered trademark of AT&T Bell Laboratories.

23

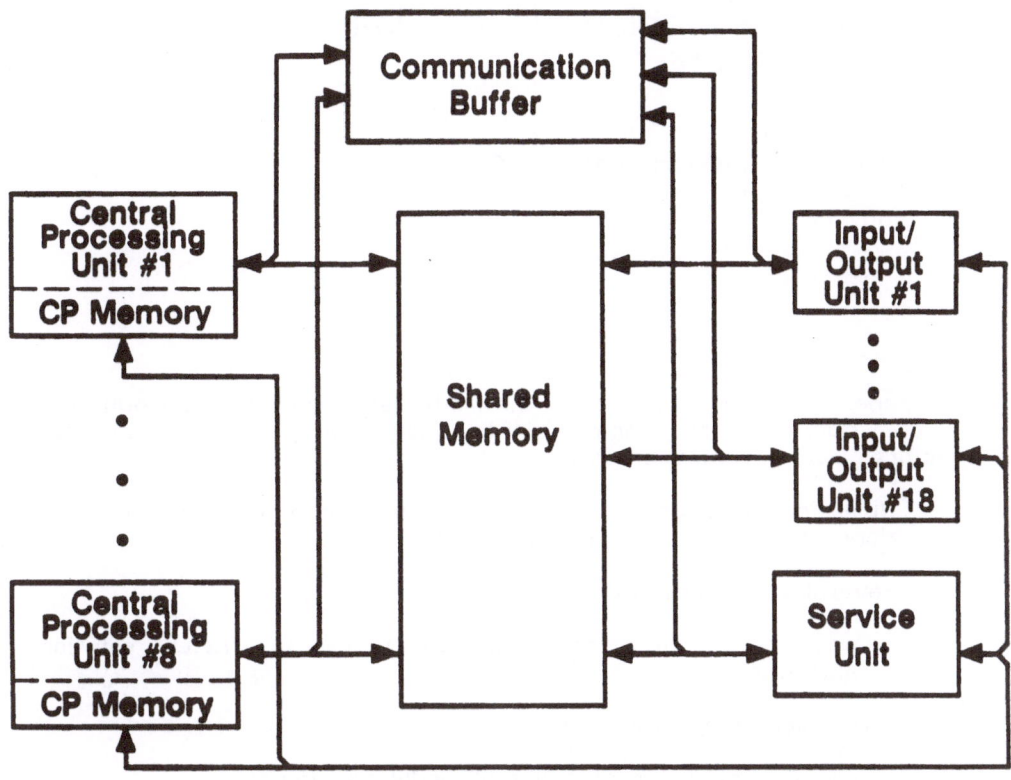

Figure 1. ETA[10] Functional Diagram

REFERENCES

1. Brian W. Kernighan and Dennis M. Ritchie, "The C Program-
    ming   Language" (Englewood Cliffs, NJ: Prentice-Hall,
    1978).

2. Kathleen Jensen and Niklaus Wirth, "Pascal User Manual
    and Report" (New York: Springer-Verlag, 1985).

Mathematical Problems in the Simulation of Reactor Plants

W.F. WERNER

Gesellschaft für Reaktorsicherheit (GRS) mbH
Forschungsgelände
Garching, FRG

Abstract
The paper gives an overview of the mathematical problems encountered in the simulation of reactor plants, particularly with light water reactors. Major problem areas are:

i)   Numerical methods for the solution of partial differential equations, describing the process models.

   Several topics are addressed:

   - ways of spatial discretisation which lead to efficient solution algorithms

   - efficient time integration

   - coupling of different types of model equations

   - algorithms suitable for use on multi-parallel processors.

ii)   Principals and methods for the quantification of uncertainties which influence the results of reactor simulation calculations.

   Several of the commonly used methods are summarized. Criterions for their use are discussed.

1.   Introduction

Design and operation of large and complex engineering systems require the utilization of mathematical simulation models in many instances and for a large variety of purposes.

A few examples are

- mathematical tools for effective, efficient and optimal design of components of complex systems requiring the modelling of the governing physical processes

- mathematical models to simulate the interactions of components of complex systems

- mathematical models for efficient and economic operations of complex systems

- mathematical models of control and safety systems.

This short list already indicates that a wide range of mathematical disciplines is involved in such simulation problems:

- process models are mostly described in terms of systems of ordinary and partial differential equations. In general, analytical solutions of such systems are unavailable; therefore the modelling of physical processes has to take resort to numerical solution methods in most cases.

- The simulation of the interaction of systems components, and of control and safety systems additionally utilizes methods of control theory.

- Optimization under constraints requires the use of linear and non-linear programming techniques.

- In many cases, the mathematical effort is substantially increased by the necessity to perform sensitivity analyses and quantification of uncertainties with regard to key parameters and key phenomena, in order to arrive at a thorough understanding of the behaviour of the system under important normal and abnormal conditions. The mathematically rigorous treatment of this aspect of simulation problems is based on methods of probability theory.

All the mentioned aspects of simulation problems have to be considered in the design and in the economic and safe operation of nuclear systems, particularly nuclear power reactors.

2. Object of the Simulation

In order to be able to discuss the involved mathematical methods in some depth the following presentation will be restricted to nuclear power plants with pressurized water reactors as steam supply system.

The principal components of nuclear power plants with pressurized water reactors are

- reactor core
- coolant loops, consisting of reactor pressure vessel, steam generators, pumps, pipes
- heat sink (steam generators, turbine)
- auxiliary systems
- balance of plant systems.

Simulation models must describe the behaviour of the plant under a wide range of conditions and constraints; they are used for many purposes, e.g.

- design of nuclear and thermohydraulic components
- design and evaluation of safety systems
- development and optimization of procedures for normal operation
- optimization of fuel utilization
- development of predictive capabilities for the containment and termination of abnormal occurrences
- development of capabilities to predict the course of severe accidents and to derive accident management strategies from such predictions.

3. Mathematical Modelling

In the discussion of mathematical modelling techniques it is useful to distinguish models of physical-technical processes in the plant (process model) from models for control and operation of the plant (balance of plant models).

3.1 Process Models

Process models are used to describe the physical and technical processes occurring in the plant. Mathematically they are forumlated in terms of partial and ordinary differential equations derived form first principles. In general, analytical solutions obeying the prescribed side conditions are not available. Therefore, numerical techniques are needed to obtain solutions. Such numerical solutions have to be provided for a large variety of initial and boundary conditions.

Process models are related to several problem areas

- Neutron Kinetics Models, describing the nuclear power generation in the reactor core. The mathematical model consists of a system of partial and ordinary differential equations. For whole-core analysis, time dependent, three-dimensional parabolic (diffusion theory) differential equations are used.

  Some analyses also require the use of transport equations in various degrees of approximation. The theoretical background of both types of calculations is well understood /1/.

  Numerical methods for the calculation of whole-core power distributions are well established /2-4/. The introduction of subdomain weighted residual techniques /5-11/ has greatly increased their efficiency. Three-dimensional time dependent analyses on the basis of homogenized cross-sections have become routine taks. Many mathematical benchmark problems have demonstrated the high accuracy and reliability of these methods /15-17/. However, great care is required in the derivation of homogenized cross-sections from the heterogenous structure of the fuel elements, and in the reconstruction of neutron fluxes from solutions obtained with homogenized cross-sections /12-14/.

- Thermo-Fluiddynamic Models describing the coolant flow and transport of energy inside the pressure vessel and in the loops of the primary and secondary systems.
  In full generality the governing equations are time dependent,

three-dimensional Navier-Stokes equations for multiphase flows /18/. For practical use in reactor simulations these equations have to be greatly simplified /19-21/. In many applications, the modelling of viscous forces and of turbulence is disregarded. Friction forces often are described by simple resistance laws.

Unfortunately, the use of multi component-, or multi-phase flow equations may lead to mathematically ill-posed problems, if the pressure difference between different components or phases of the fluid is ignored. Though these forces normally are very small, relative to other forces appearing in the balance equations, the consequences of their neglect are drastic: the eigenvalues of the convective (Euler) part of the differential operator can become complex for certain ranges of flow parameters, making the problem ill-posed.

Though there is general agreement that eigenvalues have to the real for a correctly formulated problem of the considered type there have been many controversial arguments about the practical implications of the occurence of non-real eigenvalues /22-24/. Recent research results /25-27/ indicate that correct modeling of interfacial and drag forces leads to well posed problems with hyperbolic convective terms, at least in regions where the flow is stable.

The consequences for numerical techniques of the occurrence of non-real eigenvalues are severe. In order to dampen the unphysical oscillations caused by complex eigenvalues, numerical schemes need a substantial amount of numerical diffusion, which is introduced through the use of staggered grids /21/, in conjunction with first order Finite Difference approximations. Such schemes possess enough robustness to brush over the ill-posedness of the problem, however, at the expense of unusually high costs for accurate calculations. Computing costs can be significantly reduced, if the eigenvalues are real. Then, staggering of the computational grid is not necessary, and modern techniques, like vector flux splitting /28/ or coefficient matrix splitting /29/ can be utilized for the evaluation of the convective terms. Also, second order subdomain WR-methods have been shown to be very accurate and efficient for the solution of well-posed flow problems /31/.

The differential equation systems modeling neutron kinetics and thermo-fluiddynamics are coupled through physical processes. The whole process model consists of a large system of hyperbolic, parabolic, and ordinary differential equations, which, additionally is coupled to the balance of plant models. The number of variables in the process models may be as high as 10^4 - 10^5.

Two properties of the system require particular consideration in the numerical solution procedure:

i)   The overall system of ordinary differential (obtained after semi-discretisation of the partial differential equations) is "stiff", which means that
  - the ratio of the modulus of the largest eigenvalue to the modulus of the smallest eigenvalue is large; typically 10^6:1 or larger
  - the behaviour of the solution is dominated by the eigenvalues of small modulus (large time constant).

ii) The evaluation of the right hand side of a large portion of the differential equation system is very costly, mainly due to the complexity of equations of state and algebraic correlations of the equations of thermo-fluiddynamics.

The combination of properties (i) and (ii) require implicit time integration for the majority of analyses.

The requirement on the accuracy of the calculation is very high in some cases, while not being severe in others. Accuracy requirement may also vary for different solution components.

3.2   Balance of Plant Models

Balance of Plant Models describe the actions of

- auxiliary and front line systems

- control systems, limitation systems, reactor protection systems

They are modelled by sets of ordinary and algebraic equations.

The number of variables in the balance of plant models is between 10^2 and 10^3.

- The system of differential equations is not or at most moderately stiff.

- The evaluation of the right hand side requires little computing time.

The last two properties suggest explicit time integration.

3.3   Coupling of Process Models and Balance of Plant Models

The simulation of the reactor plant requires the time integration of the coupled system of process models and balance of plant models.

The coupling between the two types of models is asymmetric in the sense that balance of plant variables react much more sensitively to variations of process variables, then vice versa.

Consequently, the time integration of the process-model equations can be carried out with significantly larger time sleps than the time integration of the balance of plant models.

4.   Desirable Properties of Numerical Methods Used in Reactor Simulation Analyses

From the above discussion it is possible to derive a list of desirable properties of numerical methods used in the framework of reactor simulation problems.

- Flexible, mathematically sound control of the error must be available. The specification of parameters controlling the error must be a user option.

  Error control is standard in many of the modern ordinary differential equation (ODE) solver packages /32,33/. However, their use in this context requires that a semi-discrete approach /34/ (continuous in time, discrete in space) is taken for the solution of the partial differential equations (PDE's) of the process models. Then the influence of the temporal discretisation error can be kept at any desired level. The control of the spatial discretisation error is much more cumbersome. Though there are encouraging attempts to control the spatial discretisation errors by adaptive or moving grids /35,36/, such techniques are still far away from general usuability for multidimensional problems. In this situation the validation of reactor simulation analyses requires careful convergence analyses. Presently, Gesellschaft für Reaktorsicherheit (GRS) is performing research contract work directed at catalogueing reactor simulation problems with regard to required spatial discretisations.

The items discussed above suggest that <u>one</u> ODE-solver package should be used for the time integration of all differential equations, and that the simulation problem should be formulated in terms of <u>one</u> large non-linear system. However, this desire meets great resistance for practical reasons: many codes incorparated in reactor simulation packages have originated as stand-alone programs; in order to be able to use them for simulation purposes they are coupled into the simulation package in an in-tandem fashion, operating on grids which are staggered in time. Unfortunately, straightforward use of this coupling strategy defeats any control of the time-discretisation error performed within subprograms. Careful analyses of the truncation error occurring through coupling of originally stand-alone codes is therefore required in order to preserve overall error control.

- It should be possible to integrate some parts of the whole system implicitly, and other parts explicitly, depending on stability requirements, evolution of solution components, and computational effort for the evaluation of the right hand side.

- Since such partition into implicity, resp., explicity integrated equations is not known a priori, an algorithm should be available which can perform such partitioning automatically.

- It should be possible to integrate different parts of the system with different time-step sizes; for example significantly smaller time-step sizes for balance of plant models than for the process models.

- The Jacobian of the system must be updated from time to time. It should be possible to perform this update locally, according to requirements dictated by the evolution of the solution, and an automatic algorithm should be able to detct when and where up-dating is required.

5. Algorithms for Multiparallel Processing

Numerical algorithms are not independent of the structure of the computers which are used for their execution.

In recent years computer have come into use, which permit some degree of parallelism in carrying out computational tasks. A significant mark of destinction of computer architecture is the level on which parallel execution is performed. Two types of parallelisms are found in present-day computers:

- pipeline (vector) processing on the level of micro instructions. A vector processor is capable of applying one operation to many sequentally queued data.

- multiparallel processing on the level of subroutines. Here, several complete processors perform relatively complex calculations inde-pendent of each other in an asynchronious manner. After completion of the individual processors tasks, data are exchanged and the calculation may contiunue.

The first of the two alternatives can, for example, conveniently be exploited for all explicit time integration problems.

The second of the two alternatives - multiparallel processing - is particular attractive for the solution of multidimensional problems by means of implicit fractional step methods.

5.1 Parallelism in Fractional Step Methods

The basic idea in fractional step methods is that an operator Ω is split into suboperators $\Omega_1, \ldots, \Omega_m$. Frequently, $\Omega_1, \ldots, \Omega_m$ is related to spatial coordinates $X_1, \ldots, X_m$. For a given difference scheme each of the set of the equations $\frac{\partial u}{\partial t} + \Omega_i = 0$, $i = 1, \ldots, m$ is solved for a fraction $\frac{\Delta t}{m}$ of the time step Δt.

Let the linearized ODE-System (obtained after semidiscretisation) be written as

$$\frac{du}{dt} + \Omega u = \frac{du}{dt} + \Omega_x \, u + \Omega_y \, u + \Omega_z \, u = 0 \tag{1}$$

where the operator Ω is split into the space directions. Equation (1) is solved in three steps. In each step only one of the suboperators Ω_x, Ω_y or Ω_z is treated implicitly, while the others are treated explicitly:

First step:

$$\frac{u^{n+1/3} - u^n}{\Delta t} + \Omega_x \, u^{n+1/3} + \Omega_y \, u^n + \Omega_z \, u^n = 0$$

Second step:

$$\frac{u^{n+2/3} - u^{n+1/3}}{\Delta t} + \Omega_y \, u^{n+2/3} - \Omega_y \, u^n = 0$$

Third step:

$$\frac{u^{n+1} - u^{n+2/3}}{\Delta t} + \Omega_z \, u^{n+1} - \Omega_z \, u^n = 0$$

Thus the solution of the implicit 3-dimensional problem

$$\frac{u^{n+1} - u^n}{\Delta t} + \Omega \, u^{n+1} = 0$$

is replaced by successive solutions of 1-dimensional problems. Due to the explicit treatment of transverse directions, all implicit 1-dimensional problems related to a selected space coordinate can be executed in parallel.

By eliminating the intermediate values $u^{n+1/3}$ and $u^{n+2/3}$ the following combined formula is obtained:

$$\frac{u^{n+1} - u^n}{\Delta t} + \Omega \, u^{u+1}$$
$$+ \Delta t \, (\Omega_x \, \Omega_y + \Omega_x \, \Omega_z + \Omega_y \, \Omega_z) \, (u^{n+1} - u^n)$$
$$+ \Delta t^2 \, \Omega_x \, \Omega_y \, \Omega_z \, (u^{n+1} - u^n) = 0$$

The following observations can be made. The first line represents the approximation of the equation $\frac{du}{dt} + \Omega u = 0$. The error terms of the second and third line are introduced by the splitting of the operator Ω. For steady state flows (when $u^{n+1} = u^n$) these error terms vanish and the solution of the fractional step method is the solution of the unsplit equation.

Because of the split error terms one might expect time step reductions relative to a fully implicit method. Indeed, experiences with fractional step methods reported in the literature are mixed, and severe time step reductions have been observed in some applications /37/.

Essentially, two options are available to influence the splitting error:

- the products of discrete operators representing the splitting error are proportional to $\frac{1}{h}$, resp. $\frac{1}{h^2}$, h being the spatial mesh-width. Thus, if large meshes can be used (without impairing accuracy) the splitting error can be reduced.

- In the overall time integration problem the splitting errors appear as additional terms in the collection of all contributions to the second- and third order truncation errors. If the time-advancement algorithm is capable of eliminating these terms, i.e. if its order is higher then 3, then the splitting error influence can be kept small.

Experience with a recently published method indeed shows /38/ that through the combination of higher order methods for both temporal- and spatial discretisation, excellent performance of fractional step methods can be achieved.

5.2 Parallelism in ODE-Solvers based on Local Extrapolation

Variable-(high)order ODE-solvers based on local extrapolation are in practical use since many years /39, 32/. Independent of their computational details, all extrapolation methods are based on the following idea:

initial values U^n of the solution $u(t)$ of

$$\frac{du}{dt} = f(t, u(t)) \tag{2}$$

are prescribed at $t = t_u$.

In order to obtain an approximation U^{n+1} to the solution $u(t_n + \Delta t) = u(t_{n+1})$, the basic step length Δt is subdivided into k subintervals of the length $\frac{\Delta t}{k}$, $k \; \varepsilon \; \{1,2,3,4,5,6,8,12\}$

On each subinterval a discrete approximation of equation (2) is solved to yield the initial entries u_k^{n+1}, $k = 1,2,\ldots$ (which are different approximations of the solution $u(t_{n+1})$ of (2)) for the construction of an extrapolation tableau. The extrapolation tableau values are rational or polynomial extrapolations of the u_k^{n+1}, $k = 1,2,\ldots$ to zero mesh width).

For each k the solution of equation (2) with time step size $\frac{\Delta t}{k}$ is independent of the others. Therefore the values u_k^{n+1}, $k = 1,2,\ldots$ can be calculated in parallel. For instance if 3 parallel processors P_1, P_2 and P_3 are available the work can be distributed accordingly to the following table:

k	P_1	P_2	P_3	WORK	EFFICIENCY
4	1,2	3	4	10	0.833
6	2,3	1,4	6	16	0.888
8	1,3,4	2,6	8	24	1.0
12	1,2,3,6	4,8	12	36	1.0

The error and time step control algorithm of FEBE estimates for which order (i.e. for which k) and time step size the user specified error bounds are expected to be satisfied in the next step. Hence the work for the next step can be distributed to the 3 parallel processors according to the estimated number of k.

It can be seen that the 3 parallel processors can be utilized with efficiency > 0,8.

5.3 Parallelism in the computation of the Jacobian matrix of ODE systems

Implicit time integration of system (2) requires evaluation and updating of the Jacobian matrix $\frac{\partial f}{\partial u}$. This may be performed by numerical differentiation:

$$\frac{\partial f}{\partial u_i} \sim \frac{f(u_1, \ldots u_i + R, \ldots u_N) - f(u_1, \ldots, u_N)}{R}$$

where R is a small disturbance of the i-th solution component. Thus the numerical evaluation of the Jacobian matrix requires N calls of the subroutine which evaluates f (u,t). In two-phase flow models this calculation consumes a significant portion of the total computing time of the algorithm because of the necessary evaluations of water - and steam table routines. The evaluations of the columns of the Jacobian matrix are independent of each other and can be done in parallel with roughly equal workload on processors.

This few examples illustrate that there is enough potential for application of multi-parallel processing in reactor simulation problems to justify more research in this area.

6.   Quantification of Uncertainties in Simulation Problems

The computational results obtained with simulation code systems are subject to uncertainties. In general two different type of uncertainties occur in computational assessments

- uncertainty due to possible stochastic variations

- uncertainty due to inaccurate knowledge of deterministic items.

In the context of simulations of nuclear power plants, uncertainties of the first kind are responsible for the multitude of event sequences which may occur with some likelihood determined by availabilities or unavailabilities of system functions.

Uncertainty due to inaccurate knowledge means that along each event sequence path many alternative calculational results describing the physical processes have to be regarded as possibly correct. Thus the quantification of such inaccuracies requires great attention in the framework of reactor simulation. The first step in any analysis of uncertainties of the second type is to set up a compilation of the potentially important "uncertain parameters". This phrase stands for any inaccurately known deterministic item (constants or functional laws).

In the second step parameter uncertainties are quantified via indication of the smalles and largest possibly appropriate values of the constants and via compilation of the sets of possibly appropriate alternative functional laws. It would make litte sense to consider only the ranges between minimum and maximum as well as only the sets of alternatives if there are well justified preferences for subranges and subsets. These preferences are expressed via degrees of belief for the respective subrange or subset to contain the appropriate value or functional law. If the degrees of belief comply with the axioms of probability

theory the wealth of well-established methods and tools from probability calculus and statistics is at disposal to arrive from the degrees
of belief at parameter level in a traceable manner at the logically
resulting degrees of belief at the output level. This procedure is
called "probabilistic uncertainty analysis".

Probabilisticly modelled parameter uncertainties define a (subjective)
probability distribution of calculational results. It is obtained by
propagating the probabilisticly modelled uncertainties through the
computational model.

6.1   Methods for quantification of uncertainties

A number of methods are in practical use /40-45/:

- Direct Monte Carlo (Statistical Tolerance Limits)
- Discrete Variables (Scenarios)
- Method of Moments
- "Response Surface" Technqiues

Which one to choose depends on

- complexity of the computational model (number of uncertain parameters, size and structure of the code)

- accuracy requirements on the distribution function, resp. on selected fractiles

- available resources.

The conceptionally simplest approach is Direct Monte Carlo. However,
its use is limited to fairly simple models.

If certain fractiles interest more than the overall distribution function,
then order statistic permit to obtain p% confidence limits of distribution
fractiles on basis of relatively few computer runs. For example 95 %
confidence limits of the 95 % fractile can be obtained with 59 computer
runs of the model, independent of the number of uncertain parameters.

Simplification and reduced effort results from the use of Discrete Variables. Here, the set of possibly correct parameter values is restricted to a finite number of discrete values, instead of continuous distributions. The discrete distribution of the result is available, if one computer run has been performed for each admissible combination of parameter values. However, the number of required computer runs increases with the number of uncertain parameters and their discrete values. Through the definition of "Scenarios", the computational expenses can be reduced.

"Response Surface" Techniques approximate the hyperplane of model responses over the parameterspace by "response surface" functions which can be easily evaluated. Thus, an approximation of the distribution of the results and their fractiles is obtained. Since the expense of model evaluation is greatly reduced relative to the use of the full model, Direct Monte Carlo becomes feasible. However, checking the adequacy of the response surface may be difficult.

6.2   Results of Uncertainty Quantification

After propagation of the uncertainties through the computational models of the simulation programs, intervals which are supports of subjective probability density functions are obtained instead of point values for output parameters. Upper and lower confidence limits can be read off from such presentation of results.

In case of time histories of output variables, bounds representing p % confidence limits are obtained by connecting points representing p % fractiles. Mean values or best estimates can also be obtained from such representation of computational results.

It is obvious that in order to determine what a "best estimate result" is, it is necessary to perform an uncertainty quantification process. The use of "best estimates" of uncertain parameters does, in general, not lead to best estimate results.

The conduction of uncertainty quantification greatly increases the confidence in the computed results of simulation programs and can improve the understanding of a plants behaviour.

The discussion of methods of uncertainty quantification makes it clear than an enormous amount of work is involved in this task. One of the prerequisits for its successful conduction is the availability of efficient numerical techniques for the solution of process model equations and the availability of powerful computing resources.

References

/ 1/ Henry, A.F.: Nuclear Reactor Analysis, MIT Press, Cambridge, Mass., 1975.

/ 2/ "Nuclear Reactor Core Analysis Code: VENTURE", Oak Ridge National Lab., 1976.

/ 3/ Cadwell, W.R.: PDQ-7 Reference Manual, WAPD-TM-678, January 1967.

/ 4/ Henry, A.F.: Review of Computational Methods for Space-Dependent Kinetics, Dynamic of Nuclear Systems, University of Arizona Press, Tuscon, Ariz., 1972.

/ 5/ Birkhofer, A., Langenbuch, S., and Werner, W.: Coarse-Mesh Method for Space-Time Kinetics, Trans. Am. Nucl. Soc., 18, Page 153, 1974.

/ 6/ Langenbuch, S., Maurer, W., and Werner, W.: Simulation of Transients with Space-Dependent Feedback by Coarse Mesh Flux Expansion Method, MRR 145, Proc. of Joint NEACRP/CSNI Specialists' Meeting on New Development in Three-Dimensional Neutron Kinetics, pp. 173-188, 1975.

/ 7/ H. Finnemann: A Consistant Nodal Method for the Analysis of Space-Time Effects in Large LWR's, MRR 145, Proc. of Joint NEACRP/CSNI Specialists' Meeting on New Developments in Three-Dimensional Neutron Kinetics, pp. 145-172, 1975

/ 8/ R.A. Shober, A.F. Henry: Trans. Am. Nucl. Soc. 24, 193 (1976)

/ 9/ S. Langenbuch, W. Maurer, W. Werner: High-Order Schemes for Neutron Kinetics Calculations, Based on a Local Polynomial Approximation, Nuclear Science and Engineering 64, 508-516 (1977)

/10/ J. Dorning: Modern Coarse Mesh Methods - A Development of the 70's, Proc. Conf. on Comput. methods in Nucl. Eng., CONF-790402, Vol. 1, p. 3-1, Williamsburg, Virginia, 1979

/11/ R.A. Rydin, M.A. Robinson, P.J. Wantuck: Proc. Conf. on Comput. Methods in Nucl. Eng., CONF-790402, Vol. 1, p. 3-73, Williamsburg, Virginia, 1979

/12/ G. Greeman, K. Smith, A.F. Henry: Proc. Conf. on Comput. Methods in Nucl. Eng., CONF-790402, Vol. 1, p. 3-49, Williamsburg, Virginia, 1979

/13/ K. Koebke: Advances in Homogenization and Dehomogenization, Proceedings of the Int. Topical Meeting on Advances in Mathematical Methods for the Solution of Nuclear Engineering Problems, Hilton International München, April 27-29, 1981

/14/ K.S. Smith, K.R. Rempe: Testing and Application of the QPANDA Nodal Model, Int. Topical Meeting on Advances in Reactor Physics, Mathematics and Computation, Hotel Meridien Montparnasse, Paris, 27-30 Avril 1987, Volume 2

/15/ Argonne Code Center: Benchmark Problem Book, ANL-7416 Supplement 2, Mathematics and Computers (UC-32), problem 11

/16/ Argonne Code Center: Benchmark Problem Book, ANL-7416 Supplement 3, Mathematics and Computers (UC-32), problem 17

/17/ Argonne Code Center: Benchmark Problem Book, ANL-7416 Supplement 2, Mathematics and Computers (UC-32), problem 14

/18/ D. Drew: Mathematical Modeling of Two-Phase Flow, Ann. Rev. Fluid Mech., Vol. 15, 261-291

/19/ D.R.H. Beattie: Two-Phase Flow Structure and Mixing Length Theory, J. Nucl. Eng. Design, 21, 46-64

/20/ M. Ishii: Thermal-Fluid Dynamics of Two-Phase Flow, Eyrolles, Paris

/21/ TRAC-PF1/MOD1: An Advanced Best-Estimate Computer Program for Pressurized Water Reactor Thermalhydraulic Analysis, LA-10157-MS, NUREG/CR-3858

/22/ R.W. Lyczkowski, D. Gidaspow, Ch.W. Solbrig, E.D. Hughes: Characteristics and Stability Analyses of Transient One-Dimensional Two-Phase Flow Equations and Their Finite Difference Approximations, Nuclear Science and Engineering, 66, 378-396 (1978)

/23/ B. Wendroff: Two-Fluid Models: A Critical Survey, EPRI-Workshop on basic 2-Phase-Flow Modeling, Tampa, Florida, 27.2.-2.3.79

/24/ J. Kesting, M.Z. Podowski: International Workshop on Two-Phase Flow Fundamentals, National Bureau of Standards, Gaithersburg, Maryland 20899, Sept. 22-27, 1985

/25/ Pauchon, Banarjee: Interface Momentum Interaction Effects in the Average Multifield Model, Part I, Int. Journal of Multiphase Flow, Vol. 12, 1986

/26/ Pauchon, Banarjee: Interface Momentum Interaction Effects in the Average Multifiel Model, Part II, Proceedings of Conference on National Heat Transfer, Pittsburgh, USA, August 1987

/27/ S.J. Lee, R.T. Lahey, jr., O.C. Jones, jr.: The Prediction of Two-Phase Turbulence and Phase Distribution Phenomena using a K-ε Model, submitted to, Int. J. of Multiphase Flow

/28/ J.L. Steger, R.F. Warming: Flux Vector Splitting of the Inviscid Gasdynamic Equations with Application to Finite-Difference Methods, Journal of Computational Physics 40, 263-293 (1981)

/29/ S.R. Chakravarthy, D.A. Anderson, M.D. Salas: The Split-Coefficients Matrix Method for Hyperbolic Systems of Gas Dynamic Equations, AIAA Paper 80-0268, 1980

/30/ U. Graf: Survey of a numerical procedure for the solution of hyperbolic systems of three-dimensional fluid flow, Atomkernenergie-Kerntechnik Vol. 49 (1986), No. 1/2

/31/ U. Graf, W. Werner: Numerical Benchmark Problem "Transient without Loss of Coolant" in a PWR, Int. Topical Meeting on Advances in Reactor Physics, Mathematics and Computation, Hotel Meridien Montparnasse, Paris, 27-30 Avril 1987, Volume 3

/32/ E. Hofer: An A(α)-Stable Variable Order ODE-Solver and its Application as Advancement Procedure for Simulations in Thermo- and Fluiddynamics, Proc. Topical Meeting on Advances in Mathematical Methods for the Solution of Nuclear Engineering Problems, Munich 1981, pp. 529-544

/33/ P.W. Gaffney: A survey of FORTRAN subroutines suitable for solving stiff oscillatory ordinary differential equations, Oak Ridge National Laboratory, ORNL/CSD/TM-134, Febr. 1982

/34/ W. Werner: Weighted Residual methods for the Solution of Fluid Dynamic Problems, Proceedings 1981 ANS/ENS International Topical Meeting on Advances in Mathematical Methods for the Solution of Nuclear Engineering Problems, Munich, Germany, April 27-29, 1981, Vol. 1, p. 461, Kernforschungszentrum Karlsruhe (1981)

/35/ J.F. Thompson, F.C: Thames, C.W. Mastin: TOMCAT - a Code for Numerical Generation of Boundary-Fitted Curvilinear Co-ordinate Systems of Fields Containing any Number of Arbitrary Two-Dimensional Bodies, J. of Comp. Physics, Vol. 24, 1974, pp. 299-319

/36/ U. Graf, P. Romstedt, W. Werner: Use of a Dynamic Grid Adaptation in the Asymmetric Weighted Residual Method, Nuclear Science and Engineering, 92, 66-70 (1986)

/37/ J. Douglas, H.H. Rachford: On the Numerical Solution of the Heat Conduction Problems in two and three variables, Trans. Americ. Math. Soc. 82 (1956), p. 421

/38/ U. Graf, W. Werner: Solution of 2- and 3-Dimensional PDE Problems: "An Implicit Time-Integration Method for Parallel Processing", International Topical Meeting on Advances in Reactor Physics, Mathematics and Computation, Hotel Meridien Montparnasse, Paris, 27-30 Avril 1987, Volume 1

/39/ J. Stoer: Einführung in die Numerische Mathematik I, Heidelberger Taschenbücher, pp 114-118, Springer Verlag

/40/ Methodology for Probabilistic Risk Assessment of Nuclear Power Plants, Pickard, Lowe and Garrick Inc., PLG-0209, Washington, Juni 1981

/41/ P. Baybutt: CREED: A Computer Code for the Construction of 2nd Fractional Factorial Statistical Designs, Report from BCL to NRC, Nov. 1977

/42/ J.K. Vaurio: PROSA-2: A Probabilistic Response Surface Analysis and Simulation Code, Argonne National Laboratory, ANL-81-33, Argonne, Mai 1981

/43/ M. Mazumdar et al: Review of the Methodology for Statistical Evaluation of Reactor Safety Analyses, EPRI 309 (1975)

/44/ R.L. Iman, J.C. Helton, J.E. Campbell: An Approach to Sensitivity Analysis of Computer Models: Part I - Introduction, Input Variable Selection and Preliminary Variable Assessment, Journal of Quality Technology, Vol. 13, No. 3 (July 1981) - Part II - Ranking of Input Variables, Response Surface Validation, Distribution Effect and Technique Synopsis, Journal of Quality Technology, Vol. 13, No. 4 (Oct. 1981)

/45/ E. Hofer et al.: Uncertainty and Sensitivity Analysis of Accident Consequence Submodels, Proceedings of the Int. ANS/ENS Topical Meeting on Probabilistic Safety Methods and Applications, San Fracisco, Febr. 24 - March 1 1985, EPRI (1985)

Nuclear Power Plant Transient Analysis
Plant Data Compared to Simulation Results

G. BREILING

Brown Boveri Reaktor GmbH
Dudenstr. 44
6800 Mannheim 1

Abstract

For the 1308 MWe NPP Mülheim-Kärlich, with a two loop nuclear
steam supply system and once-through steam generators a computer code for the analysis of operational transients and
control system optimization has been developed. After completion of startup testing a comparison of code results against
field data was made. The comparison shows good agreement and
demonstrates that the scope of the simulation and the mathematical models meet the requirements of a best estimate analysis tool for operational transients.

Introduction

With the commissioning of the 1300 MWe Mülheim-Kärlich plant,
field data for a large variety of operational transients are
available for comparison against results of computer simulations for the first time for this type of PWR.

Mülheim-Kärlich is a turn-key power plant built by Brown,
Boveri & Cie. / Brown Boveri Reaktor GmbH in Germany. It contains a Babcock & Wilcox (USA) 205 FA raised loop NSS incorporating integral economizer once-through steam generators.
The turbine and secondary system were supplied by Brown,
Boveri & Cie. A six month power escalation test phase ended
in September 1986. During this period major transient tests
were performed at power levels of 15 %, 40 %, 75 %, and
100 %, i.e. transients such as load changes, RC-pump trips,
turbine trip, load rejection, scram, loss-of-offsite-power

etc. For these and a series of other operational transients, both contractual and licensing requirements had to be met. These transients therefore defined major system characteristics and required an extensive design verification program prior to startup testing.

The major verification tool was the plant simulation code POWER TRAIN. This code, developed by the Babcock & Wilcox Company, was originally implemented on a hybrid computer installation. Since 1985 a pure digital version has been available that simulates the Mülheim-Kärlich plant and runs on a CDC CYBER 855. POWER TRAIN was the primary code used during the design phase to ensure, that the plant control system would meet contract requirements for both normal and upset operational transients. During startup it was again used to tune control system parameters to optimize performance.

Code Description

Fig. 1 shows the scope of the simulated process components in POWER TRAIN. The code structure is specific for the Mülheim-Kärlich NPP, see /1/ as a reference for plant description. The scope of simulation comprises all primary and secondary systems which directly contribute to the dynamic performance of the plant during operational transients.

The core model uses point kinetics and one-dimensional thermal-hydraulics. Boron concentration can be changed to influence steady state rod position. Rod control is provided by the control system model.

The pressurizer is simulated using three non equilibrium control volumes. The pressurizer model contains spray and heater functions, relief valve and safety valve models and level control by means of a makeup function.

The simulation of the turbine generator set includes genera-

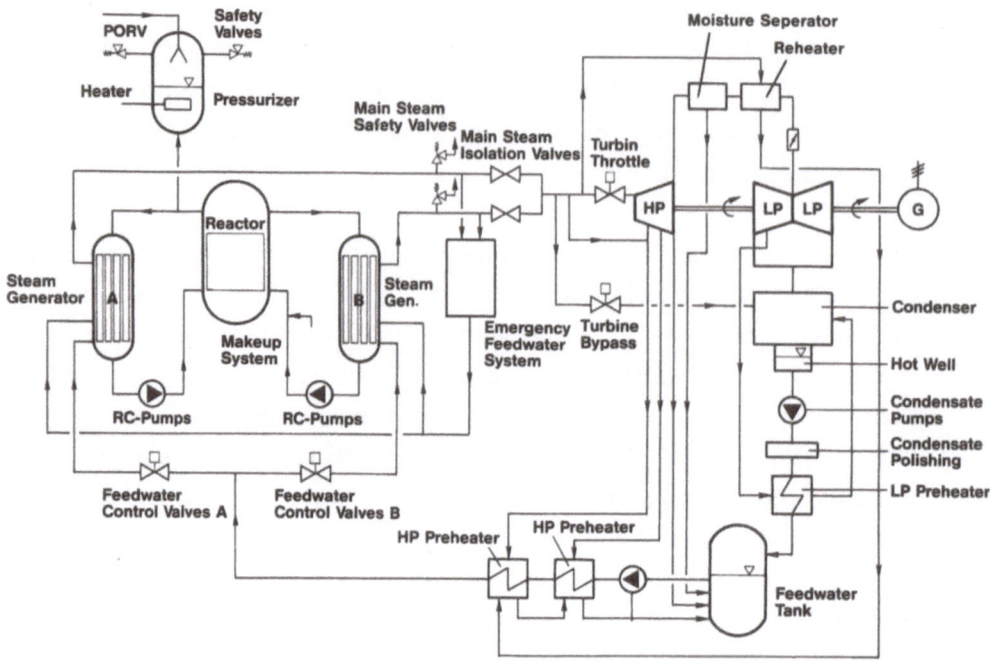

Fig. 1: POWER TRAIN, process components

tor torque, frequency and load, the main condenser and the turbine extractors to feedwater preheaters.

In addition to these plant models, the plant control system is provided. Fig. 2 shows the Integrated Control System (ICS) of the Mülheim-Kärlich plant. The load demand as induced by grid frequency or by the operator is fed in three parallel channels to the three process control stations: reactor, feedwater, turbine. A change in load demand is therefore simultaneously observed by these three subsystems. This keeps the "power train" balanced during demand changes. The nonlinearities are corrected by the temperature control and the steam pressure control. To support the change of steam generator inventory, a steam line pressure feedback to the reactor / steam generator demand is provided. This type of plant has a floating boiling length and holds steam line pressure and RC-temperature constant /1/.

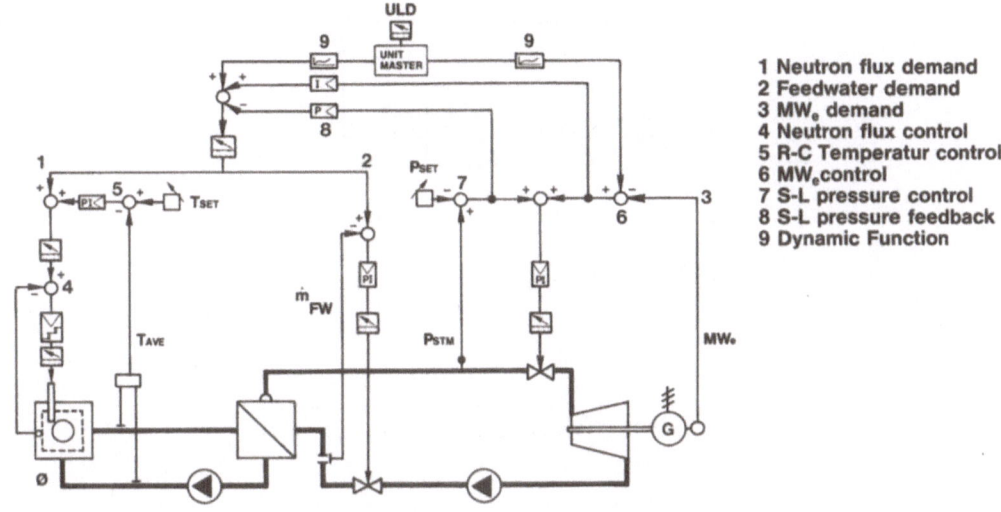

Fig. 2: Integrated Control System

The POWER TRAIN control system model is a highly detailed representation of the unit's control system, including major limiting functions. Together with the bypass control and the turbine control stations the models comprise about 1000 functional elements. The control system structure is defined in form of an input file. POWER TRAIN reads and interpretes this input file and thus determines the control functions. Changes to the control schematics, e.g. for optimization purposes, can easily be translated into POWER TRAIN input, which follows control system flow diagram terminology. As such, the code is very flexible with respect to control system modifications, see fig. 3 as an example for the input format.

In addition to the control functions, a selection of the reactor protection signals is provided. Load demand changes, operator interventions, system upsets and component malfunctions are introduced by means of input specifications.

As modeling principles, the fixed control volume – fixed time step – approach is applied. The partial differential equations for mass and energy transport as well as the control functions are solved by finite differencing. No momentum

48

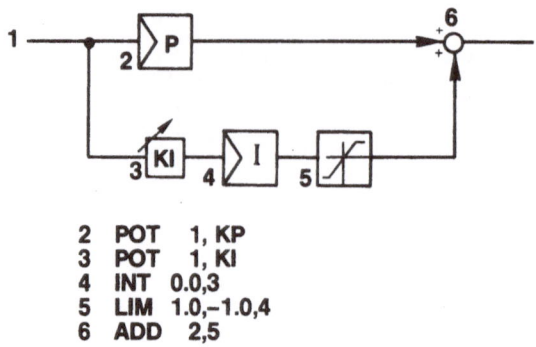

```
2  POT  1, KP
3  POT  1, KI
4  INT  0.0,3
5  LIM  1.0,-1.0,4
6  ADD  2,5
```

Fig. 3: PI-Controler input format

equation is applied, but pump coastdown is modeled by use of
input tables. Steam generator pressure loss is calculated by
a Martinelli-Nelson friction model. Metal mass heat capacity
is taken into account. A forced convection heat transfer
correlation is used for the calculation of the steam generator primary side heat transfer. Secondary side heat transfer
correlations comprise forced and natural convection, subcooled boiling, nucleate boiling, film boiling, and forced
convection to superheated steam.

In the two phase region, phase drift is taken into account by
means of a drift flux corrector to the energy equation, see
/2/:

$$\varrho \cdot A \cdot \frac{\partial h}{\partial t} + \dot{m} \frac{\partial h}{\partial z} = A \cdot \frac{\partial p}{\partial t} + \dot{q} + D$$

$$D = A \cdot \varrho_g \cdot (h_g - h_f) \cdot \frac{\partial}{\partial z} (\alpha \cdot v_{gj})$$

A	= cross section
h	= enthalpy
ϱ	= density
$\dot{m}$	= mass flow rate
z	= axial coordinate
p	= pressure

q   = transferred heat rate

D   = drift flux corrector

α = void fraction

v_{gj} = steam velocity minus mixture velocity

The drift velocity was modeled by

$$v_{gj} = \frac{\varrho_f}{\varrho} \cdot 0{,}457 \, \frac{m}{sec} \cdot (1 - \alpha)$$

A typical calculation time step is 0.25 sec. This time step is doubled for the steam generators. For models with quick acting valves such as feedwater valves or steam line safety valves the time step is reduced by 1/32. The number of steam generator control volumes is 275. The code runs in real time.

Comparison Of Code Results To Field Data

In the following, operational transients are presented and their course is explained. Plant data as recorded during startup testing are compared to POWER TRAIN results. The POWER TRAIN calculations were made using only the proper input specifications defining the initiating event to calculate the transient.

Load step of + 120 MWe (9.2 %)

The control principle applied to achieve a quick turbine response before the reactor can follow consists of a momentary stepwise reduction of steam pressure. This instantaneously increases the temperature difference for heat transfer in the boiling region of the steam generators, thereby producing the necessary steam flow. Until the slower responding reactor balances the energy demand, the energy mismatch is compensated for by the heat capacity of the primary coolant.

As can be seen from the control schematic, the desired reduc-

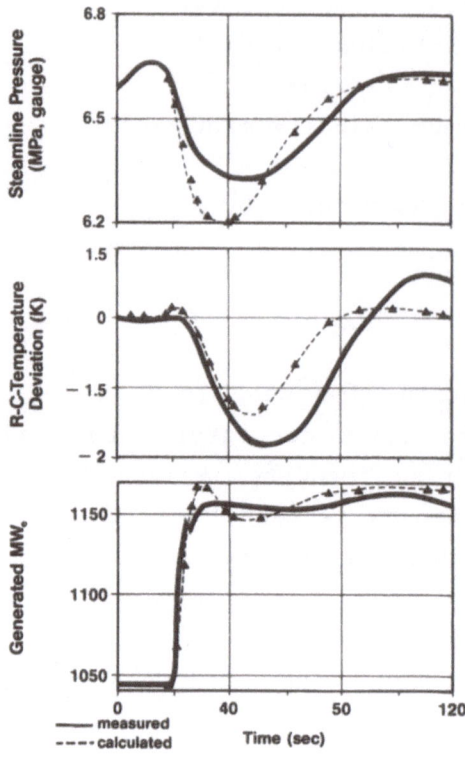

Fig. 4: + 120 MWe Step Load Change

tion of the pressure control setpoint is only achieved when a
deviation of generated power versus power demand exists. To
this end, a dynamic device (kicker) "exaggerates" the load
demand step in a way that the necessary pressure deviation
and turbine load response is achieved. The influence of this
kicker function dies out corresponding to the reactor control
response time and steamline pressure recovers. The opposite
process occures for a load reduction.

The comparison of plant data to existing POWER TRAIN design
calculations shows good agreement (Fig. 4) and demonstrates
the code's capability to predict the dynamic behaviour of the
process. Using the design setpoints for the kicker functions,
the plant was able to supply 80 % of the 120 MWe demand step
within 4 s. The full + 120 MWe change was provided in 14 s.

Load Rejection

Fig. 5 shows data from a load rejection test with runback to houseload. In this transient, the grid circuit breakers open, the turbine starts to accelerate and the control system immediately closes the turbine throttle completely. This appears very similiar to a turbine trip. Of primary interest is the magnitude of the steamline pressure peak which occurs before the 100 % turbine bypass fully controls pressure. In Mülheim-Kärlich the turbine throttle and the bypass control valves are physically identical. For 100 % steam flow, the necessary valve stroke is 66 % for the turbine and 94 % for the bypass. This difference is due to the fact that the two identical sets of control valves operate with quite different pressure differences. However, at full power, the experienced turbine throttle position is below its design value, because the turbine capacity is slightly greater than design. In case of a load rejection (or a turbine trip), the throttle demand signal prior to the event (which is continously stored) is transfered to the bypass demand, thereby causing the bypass valves to open to a position which is appropriate to take the actual steam flow without producing large pressure deviations. Simultanously the pressure setpoint for the bypass control is set to the nominal operational value.

As fig. 5 shows, the design analysis for the control concept and the bypass valve servo system layout quite closly predicted plant behaviour (the initial R-C-pressure peak shows no deviation at all).

This analysis was essential to assure that the influence of the secondary pressure peak on primary coolant temperature is tolerable and that opening of the pressurizer relieve valve (PORV) does not occur. This design goal has been achieved with a satisfactory margin. It should be noted, however, that the design bypass valve characteristic used in the analysis deviates from plant data. This can be seen from the plotted valve position on fig. 5.

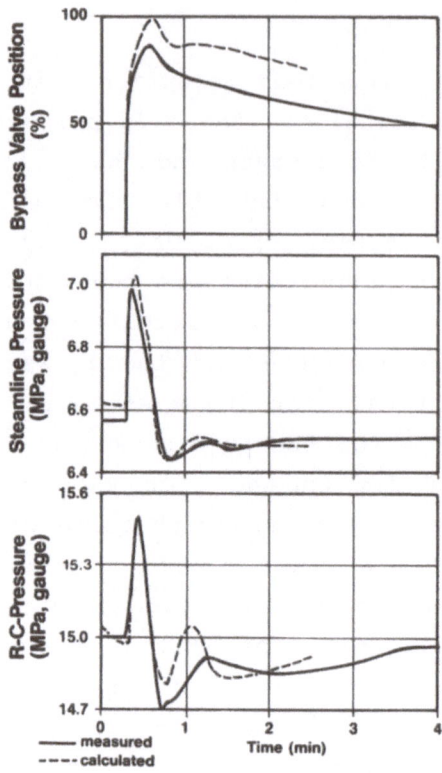

Fig. 5: Load Rejection

The key simulation model for this transient is the bypass
control system. The control process, i.e. transfer of turbine
throttle position, P-I-characteristics, pressure setpoint
change, electric-hydraulic valve servo characteristics, has
to be accurately modeled. The steamline pressure peak results
from the fact that the initial bypass opening is only to the
turbine throttle position before the load rejection (66 %).
It then requires a (slow) P-I-controlled valve movement to
obtain full capacity. It is mandatory to know the actual
steady state turbine throttle positions at load for a correct
simulation of this transient. With correct input data, POWER
TRAIN can be used to further optimize system response.

The turbine and the turbine control models in POWER TRAIN were developed for fast operational load changes, and cannot quantitatively predict the frequency peak in the first few seconds after a load rejection. Therefore the models have been tuned to reproduce frequency data furnished by the turbine manufacturer. With this, the effect of the frequency peak on neutron flux (due to accelerating R-C pumps and coolant flow) could be studied. It has been shown, that no reactor trip occurs due to high neutron flux, even at the end of a fuel cycle.

After initiation of the event as shown in fig. 4, the turbine throttle opens to control houseload. The reactor and the steam generators perform a runback to 15 % load with steam pressure controlled by the turbine bypass. Precalculations and tests both showed that load rejection transients, as well as failure of one out of two main transformers can be handled and runback to houseload operation without difficulty.

Trip of one RC-pump

The final operational transient presented in this paper is the trip of one reactor coolant pump. In such an event, the control system overrides the unit load demand and performs a runback of the three demand signals to 75 % load at a rate of 30 %/min.

This is the maximum load change possible at design rod velocity. The modified demand signals are input into the control channels via min-select gates, located below the "kicker" devices (dynamic functions, fig. 2). Thus the "exaggerated" increase in steamline pressure to follow load is avoided (see explanation of load step control functions above).

If the reactor cannot follow the load ramp, a limit-signal reduces the feedwater ramprate accordingly. In order to balance cold leg temperatures, the total feedwater demand, is

54

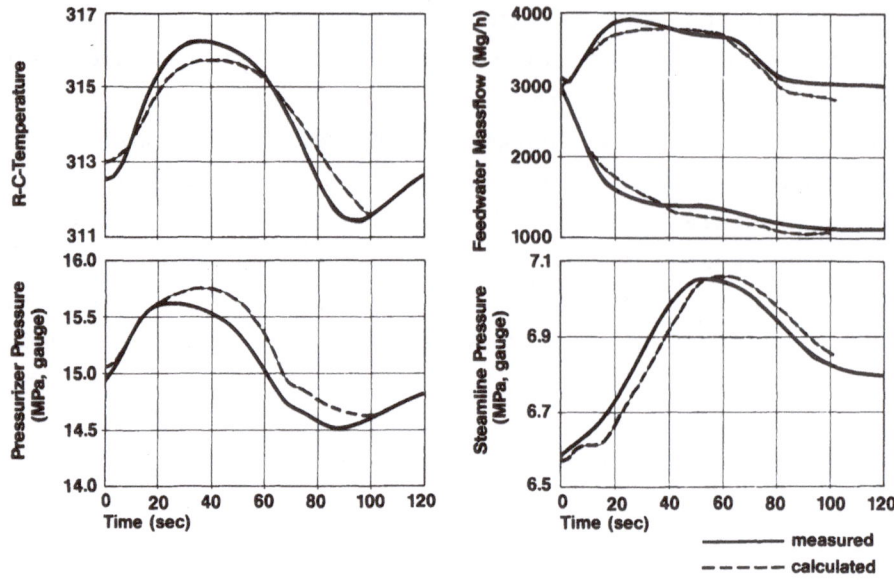

Fig. 6: Trip of one R-C-pump at 90 % power

ratioed corresponding to the primary coolant flowrates during pump coastdown. Reactor coolant loop and pump characteristics are such that within about 30 s, 75 % primary coolant flow is reached.

For the startup test series performed at Mülheim-Kärlich the maximum load change capability of the reactor was only about 20 %/min due to the rod worth at beginning-of-life for the first fuel cycle. The necessary 25 % load reduction would take about 75 s as compared to the 30 s for the pump coast-down. Until the generic cycle load change capability of 30 %/min is reached, R-C-pump trip testing was restricted to 90 % initial core power. The runback was therefore extended to 65 % load in order to test the full design span of load change.

As shown in fig. 6, the dynamics of the process lead to an increase of reactor coolant average temperature and to a corresponding pressure peak. This is a result of the slower reduction of reactor power (20 %/min) as compared to the run-back rate induced by the control system, which, when the deviation exceeded a preset value actuated a limitation of the feedwater ramprate. The second contributor to the coolant temperature increase is the increasing steamline pressure (power reduction always leads to a controlled pressure increase, see fig. 2).

The POWER TRAIN results for this transient (fig. 6) compare very well to the plant data. The difference in primary pressure peak is only about 0.1 MPa (15 psi), which is mainly due to the 0.1 MPa difference in initial values.

Conclusion

The application of POWER TRAIN as presented in this paper has demonstrated the code's capability as a best estimate analysis tool for operational transients and control analysis problems. The scope of the simulation and the mathematical models have proved to be sufficient to calculate a great variety of transients. POWER TRAIN is also easy to handle and to modify for the purpose of optimization studies. For high accuracy, however, very precise component characteristics data and core physics data must be provided.

References

1. R. Bieselt, P. Wirtz:
   "Der BBR-Druckwasserreaktor mit Geradrohrdampferzeuger"
   Sonderdruck aus VGB-Kernkraftwerks-Seminar 1978

2. R. T. Lahey, F. J. Moody:
   "The Thermal-Hydraulics of a Boiling Water Nuclear Reactor"
   American Nuclear Society 1977

Development of an Advanced Thermal Hydraulics Model for Nuclear Power Plant Simulation

Richard Moffett
CAE Electronics Ltd
St-Laurent, Québec H4T 1G6

Abstract

This paper summarizes the development of an advanced digital computer thermal hydraulics model for nuclear power plant simulation which has been carried out at CAE Electronics Ltd. A review of thermal hydraulics code design options is presented together with a review of existing engineering models.

CAE has developed an unequal temperatures–unequal velocities five equation model based on the drift flux formalism. CAE has selected the model on the basis that phase separation and thermal non-equilibrium are required to simulate complex and important phenomena occurring in systems such as reactor cooling systems (RCS) and steam generators (SG) of nuclear power plants. The drift flux approach to phase separation and countercurrent flow was selected because extensive testing and validation data supports full-range drift flux parameters correlations [1]. The five equation model was also chosen because it conserves important quantities, i.e. mass and energy of each phase, and because of numerical advantages provided by the ease of coupling phasic mass conservation equations with phasic energy conservation equations.

The basis of CAE's five equation thermal hydraulics model as well as supporting models for convection and conduction heat transfer, break flow, interphase mass and heat transfer are described. Comparison of code calculations with experimental measurements taken during a small break LOCA test with the OTIS facility are presented.

The use of such advanced thermal hydraulics model as plant analyzer considerably improves simulation capabilities of severe transient as well as of normal operation of two phase systems in nuclear power plants.

1 Introduction

There is a broad variety of thermal hydraulics models to choose from. Before investing in developing a thermal hydraulics code, a brief survey of the available coding design options is indicated. These options are not the essential of the code but can make development and usage much easier. Coding design characteristics give the code more or less flexibility depending on the requirement of the application.

Furthermore, an enlightened study of the implications of model's engineering design basis/assumptions on scope and numerical/physical performance should be made with the requirements of the application in mind. In the domain of full scope plant analyzers, a wide spectrum of conditions must be covered and the model must have the intrinsic capabilities to simulate all of them in a continuous manner.

Full scope plant analyzers must have the capabilities to realistically address normal operation as well as severe transients and accident conditions of two-phase systems. Even though not directly perceived by operators, insight provided by an advanced model has a considerable impact on visible overall performance of simulators:

- non equilibrium effects impact on predicted heat transfer rates, surface temperature of fuel rods and steam generator tubes, vapour generation and condensation and pressure transients

- unequal velocities impact on the distribution of transported quantities (mass and energy) in the system

- the combination of the above influence

  - the location of levels and heat transfer regime boundaries, important for prediction of cladding rupture,

  - the determination of flow interruption in RCS as due to accumulation of void in the hot-leg U-bend,

  - the prediction of void distribution in reactor which in turn impact on reactivity feedback.

2 Coding design options

The development, validation and use of an advanced thermal hydraulics code requires considerable investment of time and effort and this, at all the phases of its "life":

- the initial development,

- the maturation period, during which the code is validated against data taken from various test facilities, constitutive equations are corrected or new ones selected and numerics improved,

- the production period, during which the code is used to model a variety of plants.

A common requirement for efficiently going through the two last phases is the ability for the code to represent the test facility or plant. To achieve this goal, several code design options are available [7]:

- fixed, flexible or free topology

- fixed, flexible or free nodalization

- one, two or three dimensional calculations

- fixed or selectable coordinate system

- fixed, moving or adaptive grid

- component-by-component modeling or unified approach

CAE has developed a code that provides some flexibility in the topology and nodalization of the thermal hydraulics network. This is found to be useful and efficient, among other purposes, to match the design of experimental apparatus against which the code is validated and turn the code into production.

CAE has limited the scope of the model to one dimensional analysis because the code was developed for real-time simulation. This implies a linear coordinate system. A fixed size grid is used to divide the thermal hydraulics network into constant size control volumes.

As will be seen later, even though CAE's model is essentially one dimensional, some two dimensional effects, such as velocity and void fraction profiles in pipes cross-section are accounted for via the drift flux formalism. Further, even though the hydro and thermo dynamics modeling is implemented in a fixed grid environment, the heat transfer package models the effects of moving of collapsed or froth levels on heat transfer rates. This improves the quality of the feedback to the thermal hydraulics calculations because the model recognizes the existence of the movement and the relative importance of distinct heat transfer regimes along the relatively large heat transfer surfaces that are typically used on real-time simulators.

The unified approach was chosen because it provides a flexible frame in which peculiarities can be implemented as sources to a single general equation set. This way, only one version of an equation has to be written, debugged, validated and documented.

3 Engineering design options

The engineering basis of all thermal hydraulics model is a set three types of equations:

1. Conservation (or field) equations for mass, momentum and energy. These are partial differential equations.

2. Constitutive equations (or correlations) necessary to support the conservation equations. These include functions for heat transfer coefficients, friction coefficients, etc. related to the interface between the fluid and its immediate environment.

3. Fluid properties functions.

The conservation equations represent the physics of the model and as a consequence, the more conservation equations are used, the more physics is intrinsic to the model. Oppositely, the fewer conservation equations, the more constraints (assumptions on the state of the model) limit the scope of the model.

Since the physics of thermal hydraulics models is characterized by the conservation equations employed, models are generally classified according to number and type of conservation equations they are based on. This section presents some of the thermal hydraulics models together with their advantages and disadvantages.

Model identification	Constraints description	Constraints			Constitutive models					
		T_k	v_k	Total	F_w	F_i	Q_w	Q_i	w_i	Total
3C:1M.1K.1E	$T_l = T_{sat}$ $T_v = T_{sat}$ $v_l = v_v$	2	1	3	1	0	1	0	0	2
4C:2M.1K.1E	$T_l = f(T_v)$ $v_l = v_v$	1	1	2	1	0	1	0	1	3
4C:1M.2K.1E	$T_l = T_{sat}$ $T_v = T_{sat}$	2	0	2	2	1	1	0	0	4
4C:1M.1K.2E	$T_l = f(T_v)$ $v_l = v_v$	1	1	2	1	0	2	1	1	5
5C:1M.2K.2E	$T_l = f(T_v)$	1	0	1	2	1	2	1	1	7
5C:2M.1K.2E	$v_l = v_v$	0	1	1	1	0	2	1	1	5
5C:2M.2K.1E	$T_l = f(T_v)$	1	0	1	2	1	1	0	1	5
6C:2M.2K.2E	–	0	0	0	2	1	2	1	1	7

Table 1: Two phase thermal hydraulics models

Table 1 shows all the combinations of phasic or mixture conservation equations models are constructed from. The convention used to represent the models is cC:mM.kK.eE, where c is the total number of conservation equations, and m, k and e are respectively the number of mass, momentum and energy equations. In this table, it is easy to identify the limits of the model as well as the requirements for constitutive equations. In general, models based on one momentum equation need one wall to fluid friction term, F_w, while models based on two momentum equations need two in addition to an interphase friction term, F_i. Models based on two mass or two energy equations require one interphase mass flow term (w_i) for phase change rate. Models based on two energy equations require an interphase heat transfer model (Q_i) and the w_i model. Models with one momentum equation are subject to the homogeneous flow constraint ($v_l = v_v$) unless they are complemented by a model, such as the drift flux one, that relate the phase velocities to each other ($v_l = f(v_v)$). Models based on one mass and two energy or two mass and one energy equations are not true thermal non-equilibrium models because they are subject to a constraint of the form $T_l = f(T_v)$ on the phase temperatures. This constraint results from the fact that both phases mass and energy are necessary to independently define the phasic temperatures.

As seen in table 1, the most sophisticated model is based on six conservation equations: phasic mass, momentum and energy. As a consequence, the six independent variables necessary to independently define the state of each phase of a two phase mixture can be solved for without constraints. Because each phase is defined independently, this model is referred to as a "two-fluid" model; that is two one phase (three equations) model exchanging with each other and with the immediate environment.

The main disadvantage of the two-fluid model is the number and nature of the constitutive equations required to support it (interphase shear stress, wall/phase friction, momentum exchange, transient forces – e.g. virtual mass force) and the uncertainty, due to experimental difficulties with detailed measurements in two phase flow, related to them [8].

Fortunately, simpler models based on fewer conservation equations exist.

The least sophisticated thermal hydraulics model is based on three mixture conservation equations for mass, momentum and energy. As a consequence, the corresponding number of independent variables can be solved for and using them, the state of the fluid has to be determined. When one phase is present, the pressure, the phase velocity and the phase specific enthalpy are all that is known and required. However, when two phases are present, the pressure, the mixture velocity and the mixture specific enthalpy are also all that is known. As a consequence, assumptions must be used to determine the other variables of interest, namely phasic enthalpies, phasic velocities and mass fractions. Closure of the problem requires the following assumptions:

- liquid and vapour are at saturation when two phases are present (thermal equilibrium assumption),

- the two phases have the same velocity (homogeneous fluid assumption),

- the mass fraction of each phase is given by thermodynamic ratio.

The principal advantage of this model is its simplicity and the few constitutive equations required. In fact, only well known wall to mixture heat transfer, Q_w, and friction, F_w, are required. However, the restrictions imposed by the model greatly limit the scope of simulation of phenomena occurring in two phase systems such as in steam generators or reactor cooling systems i.e.:

- thermal disequilibrium, i.e. the fact that phasic state deviate from saturation in a two phase mixture

- phase separation, i.e. the fact that the phases flow with different velocities – vapour tends to flow upward faster than liquid or under some conditions, liquid flows downward while vapour flows upward

- countercurrent flow limitation (CCFL), i.e. a condition under which vapour upward flow limits downward liquid flow

- the impact of the above in overall performance as perceived by plant analyzer users

Intermediate models are constructed by adding conservation and related constitutive equations (and as a consequence simultaneously removing constraints) to the three equation model or, oppositely, by removing equations from the six equation model (and therefore adding constraints).

The four equation model is the next step of refinement after the three equation model. Essentially, it is a three equation model to which is added an extra mass, momentum or energy equation. There are three variants of the four equation model, each of which being distinguished by the type of equation added. As seen in table 1, the result is a decrease in the number of restrictions accompanied by an increase in the number of constitutive equations for phase to wall or interphase exchanges. None of these models cater with true thermal non-equilibrium and the ones which handle some thermal disequilibrium have disadvantages from the numerical point of view: it is not possible to couple the (one or two) mass equation(s) with the (two or one) energy equation(s).

Among the five equation models, the 5C:2M.1K.2E is the only true thermal non-equilibrium model because both phases mass and energy are calculated. By itself, this model cannot cater with non-homogeneous flow. However, the drift flux concept may be superimposed to it and make the model non-homogeneous. As a consequence, the mixture flow dynamics is governed by momentum effects while the more detailed individual phase dynamics depends on the drift flux parameters as is described in the next section. Therefore, this model offers all the physics required for simulation of complex two phase flow phenomena without the uncertainty generally associated to six equation models.

4   CAE's advanced thermal hydraulics model

CAE has developed a five equation non-homogeneous thermal non-equilibrium model based on two mass, two energy and one momentum conservation equations and the drift flux concept for phase separation. This is a coherent equation set in which important quantities, phasic mass and energy, are calculated from their corresponding conservation equations and the phase flows are derived from a mixture momentum equation and drift flux. The five equation model was also chosen because no had hoc restrictions are imposed on phase temperature and because of numerical advantages provided by the ease of coupling corresponding phasic mass and energy equations. In the following subsections, the conservation equations as well as some of the supporting constitutive models are presented.

4.1   Conservation equations

The engineering basis of CAE's thermal hydraulics model is a set of five conservation equations for phase mass, phase energy and mixture momentum.
Liquid phase continuity equation:

$$\frac{\partial(m_l)}{\partial t} = -\frac{\partial(w_l)}{\partial x}\Delta x - w_i + w_{sl}$$

Vapour phase continuity equation:

$$\frac{\partial(m_v)}{\partial t} = -\frac{\partial(w_v)}{\partial x}\Delta x + w_i + w_{sv}$$

Liquid phase energy equation:

$$\frac{\partial(m_l h_l)}{\partial t} = -\frac{\partial(w_l h_l)}{\partial x}\Delta x - w_i h_f + Q_{wl} + Q_{il} + w_{sl}h_{sl} + V(1-\alpha)\frac{\partial(P)}{\partial t}$$

Vapour phase energy equation:

$$\frac{\partial(m_v h_v)}{\partial t} = -\frac{\partial(w_v h_v)}{\partial x}\Delta x + w_i h_g + Q_{wv} + Q_{iv} + w_{sv}h_{sv} + V\alpha\frac{\partial(P)}{\partial t}$$

The mixture momentum equation is simply the sum of phasic momentum equations:

$$\frac{\partial(m_l \mathrm{v}_l + m_v \mathrm{v}_v)}{\partial t} = -\frac{\partial(w_l \mathrm{v}_l + w_v \mathrm{v}_v)}{\partial x} \Delta x - A \frac{\partial(P)}{\partial x} \Delta x - F_{gm} - F_{wm} + F_{sm}$$

Additional conservation equations are used to track non-condensible gases, boron and various species of radioactivity. Each product is transported throughout the thermal hydraulics network and to/from other systems based on the flow of the phase which carry it and on its concentration.

Phase separation was introduced in the model using the drift flux concept [5]. The drift flux approach is probably the most reliable and accurate analytical tool for full range advanced two phase flow modeling [8]. The correlation used has been extensively tested and successfully validated under numerous conditions[1]:

- up and down flows

- countercurrent and co-current two phase flows

- full range of pressures, flows and void fraction

- countercurrent flodding limitation (CCFL)

- continuous function

- validated against FRIGG, TLTA BWR, FLETCH SEASET, ORNL...

In the drift flux model, phasic velocities are function of a mixture flow quantity, say the volumetric flux (j), which is derived from solution of the momentum equation and the drift flux parameters (C_0 and V_{gj}). Practically, this means that the conservation equation set is not used to directly obtain each velocity or flow but instead is used to obtain the mixture volumetric flux which is then used, in conjunction with drift flux parameters to obtain the phasic velocities:

$$\mathrm{v}_l = \frac{(1 - C_0 \alpha)}{(1 - \alpha)} j - \frac{\alpha V_{gj}}{(1 - \alpha)}$$

$$\mathrm{v}_v = C_0 j + V_{gj}$$

and phasic mass flows:

$$w_l = \rho_l (1 - \alpha) A \mathrm{v}_l$$

$$w_v = \rho_v \alpha A \mathrm{v}_v$$

The C_0 parameter synthesizes the two dimensional effects of the potentially non-uniform radial void and volumetric flux profiles across the flow section. The V_{gj} parameter quantifies the effects of vapour drift in the mixture and is closely related to the terminal rise velocity of vapour through the liquid [2].

Beside its simplicity of application, the drift flux method has the advantage of not requiring two phasic momentum equations to achieve phase separation effects. This is a major benefit compared to models using two momentum equations because the latter must be supported by a number of uncertain correlations for phasic friction, interphase momentum exchange, interphase friction, and transient forces.

4.2 Constitutive equations

The constitutive equations are required to support and complement the physics provided by the conservation equations. For consistency reasons, the advanced five equation thermal hydraulics model requires an equally advanced set of constitutive equations for not only the essential (see table 1) wall/phase convection heat transfer, interphase heat and mass transfer, wall/phase friction but also for peripheral models such as conduction heat transfer in tubes and fuel rods, break flow and non-condensibles evolution. These are described in the following subsections.

4.2.1 Convection heat transfer

A single convective heat transfer package is used for all major heat transfer surfaces such as the surface of the fuel rods and on primary and secondary sides of steam generator tubes. The package calculates wall/phase heat rates, Q_{wl} and Q_{wv}, and wall/interphase heat rate, Q_{wi}, which are continuous with respect to flows, pressure, phasic and wall surface temperatures, mass or volume fraction and collapsed or froth level. The Q_{wi} heat transfer component represents that part of the wall heat flow that causes phase change (boiling or condensation) and is incorporated in the interphase mass and energy balance to predict phase change rate. The package considers natural and forced convection to liquid, subcooled and saturated nucleate boiling, transition boiling, film boiling as well as natural and forced convection to vapour. Critical heat flux (CHF) and the corresponding temperature are calculated for low and high flow and used to discriminate pre-CHF and post-CHF heat transfer regimes. A condensation heat transfer correlation completes the package. When the condensation heat transfer coefficient is significantly degraded by the presence of non-condensibles, a degradation factor is used to correct the "all water" condensation heat transfer coefficient [6].

4.2.2 Conduction heat transfer

A discrete parameter model is used to solve the conduction equations (in the radial direction) applied to the fuel rods and steam generator tubes. Rods and tubes are axially divided into a number of nodes corresponding to thermal hydraulics nodes. The rods and the tubes are further divided in the radial direction. Typically, two or three radial nodes are used in tubes while three or four are used in the rods. The effects of change in gap size, as outcoming from thermal expansion/contraction of fuel elements and cladding, is accounted for.

Heat storage and release from pipes and vessels is modeled using a lumped parameter model. The temperature of the surface in contact with the fluid is calculated and used in conjunction with a simplified convective heat transfer function for wall/fluid heat transfer.

4.2.3 Interphase heat/mass transfer

A phenomenological model as used by several design codes such as COBRA-NC [12], TRAC-BD1 [6] and RELAP5/MOD2 [13] was chosen. As shown in figure 1, the intephase heat transfer model considers heat transfer on both sides of a volumeless interface between the liquid and vapour phases and therefore cater with complete thermal non-equilibrium. The temperature gradient between the interface and each phase is a driving potential that directs the phases to an equilibrium condition, governed by the interphase temperature:

$$Q_{il} = h_{il}A_i(T_i - T_l)$$
$$Q_{iv} = h_{iv}A_i(T_i - T_v)$$

Interphase heat transfer coefficient – interphase area products are dynamically calculated based on a model inspired by Jones and Saha [9]. The form of the equation for interface area is applicable to various flow regimes. It was adapted so that it recognizes the existence of nucleation and condensation sites. As a result, both flashing and condensing of one phase fluid in an adiabatic environment can be simulated. When non-condensible gases are present and a driving potential for condensation exists, a degradation factor is applied to the interphase heat transfer coefficients.

Interphase mass transfer rates are pegged to the interphase heat transfer rates. In fact, application of the steady-state mass and energy equations to the interphase interface (often referred to as "interface jump conditions") gives the rate of mass transfer at the interface:

$$w_i = \frac{Q_{wi} - Q_{il} - Q_{iv}}{h_{fg}}$$

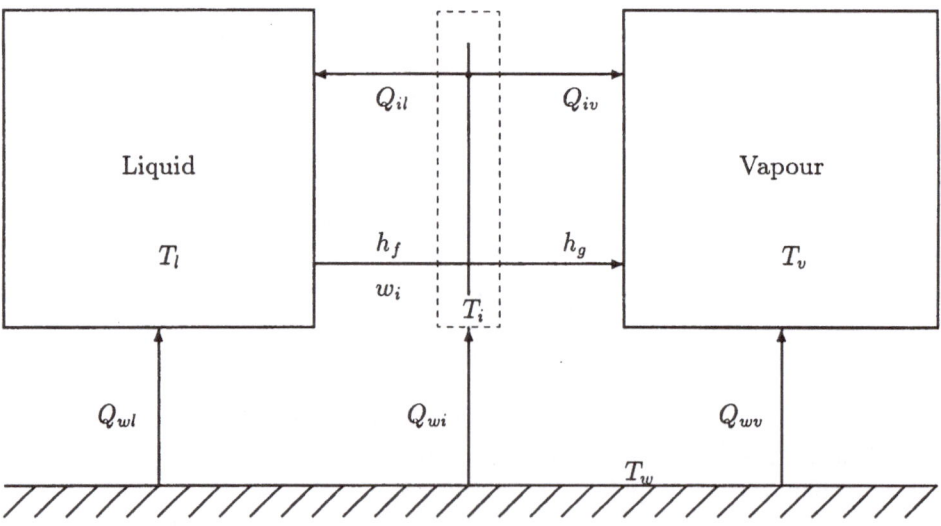

Figure 1: Interphase heat and mass transfer (positive directions shown) and wall/phase heat transfer

The current approach cater with phasic disequilibrium and handles all four sources of interphase heat and mass transfer:

- flashing of liquid alone ($T_l > T_i$)

- spontaneous condensation of vapour alone ($T_v < T_i$)

- vaporization of liquid due to contact with superheated vapour

- condensation of vapour due to contact with subcooled liquid

4.2.4   Break flow

The correct determination of break flow is particularly important when the break is small. Under this condition, the system's behaviour depends on its rate of change of mass over a relatively long period of time. The simulated performance becomes a direct function of the accuracy of the break flow calculation.

A package for break flow calculation was developed. The package accounts for:

- unchoked and choked flow

- one phase and two phase flow

- presence of non-condensibles

Break flows are implemented as sources to the thermal hydraulics conservation equations.

5   Code validation

The code was compared to the measurements from the OTIS (Once-Through Integral System) facility. OTIS, illustrated in figure 2, is a single loop facility with a plant to model scale factor of 1686. OTIS maintained the key elevations, approximate component volumes and loop flow resistances of a B&W 255 raised loop nuclear plant. OTIS simulated the key plant boundary conditions: high pressure injection, the pressurizer power operated relief valve, primary system vents and steam generator secondary steam and feed controls [10].

5.1   Small break LOCA (OTIS test #2202AA)

Plots comparing the OTIS data (solid lines) and the simulation results (dash lines) are presented in figures 3 to 8. The key events for this simulation are i) the cold leg suction (CLS) leak actuation at time 11 minutes, followed by ii) the pressurizer isolation 1.3 minute later and, another minute later, by iii) the core power decrease actuation, the high pressure injection (HPI) actuation and the steam generator secondary side level set-point increase from 5.7 ft to 38 ft at a rate of 3 ft/min.

On the primary side, the one phase liquid flow (1.42 lbm/s) is driven by natural circulation. On the secondary side, 113 F high auxiliary feed water (AFW) flow is injected at the top of the OTSG to maintain the collapsed level to 5.7 ft; the cold spray flows downward and condenses some of the upward flowing steam. When the cold leg leak is actuated, the primary side depressurizes to saturation at two rates representing the period before and after the pressurizer is isolated. When the primary pressure is such that the saturation temperature gets equal or below the liquid temperature, flashing occurs and greatly reduce the pressure decrease rate. Vapour starts to collect in the system's high points, such as in the upper part of the downcomer, reactor and hot leg. The formation of a vapour bubble in the hot leg U-bend causes a flow interruption on the primary side. The flow is briefly restored (spillover) due to the "push" created by the sudden arrival of vapour produced or accumulated in the reactor region into the hot leg. On the secondary side, the AFW flow responds to rise the level to its set-point. The increase of AFW flow depressurizes the secondary side, causing the pressure controller to close the steam valve, and thus reduce steam flow, to try to maintain the pressure. When the spillover occur, more heat is transferred to the secondary side and causes pressure peaks. The primary pressure suddenly decrease when the SG primary level uncovers the top part of the SG tubes which are exposed to the still active cold AFW flow on the secondary side. When the secondary level reaches its set-point, the AFW flow is reduced and the level stabilizes. Then, the primary side depressurizes more slowly. Because the leak flow exceeds the HPI flow, the hot leg and SG primary side levels decrease as the liquid inventory is lost.

This test exercised many aspects of the code and showed its ability to handle complex thermal hydraulics phenomena:

- flow dynamics and phase separation: flows in the primary loop and void out of upper bend of the hot leg, accumulation of vapour in reactor and downcomer high points, countercurrent flow

- thermal disequilibrium: subcooled liquid and superheated vapour near the injection point of the auxiliary feed water in the steam generator

- convective heat transfer package: pressure in the steam generator is governed by heat it exchanges with the primary side

- interphase heat/mass transfer: condensation of the secondary vapour on the auxiliary feed water flow, flashing of primary liquid when the saturation temperature gets equal or below the liquid temperature, flashing and recondensing of secondary side steam generator fluid when pressure changes

6 Conclusion

In this paper, two major aspects of the development of an advanced thermal hydraulics code were presented: coding design options and engineering options.

Coding design options such as flexible topology (network configuration) and nodalization (distribution of control volumes in network) are desirable because it allows thermal hydraulics code developers to more quickly validate their model against data taken from a variety of test facilities. The unified approach to components modelling (as opposed to component-by-component) is also a desirable option because only one general model has to be written, debugged and validated. Component/plant peculiarities can be easily implemented at the level of inputs to the general model/equations.

Thermal hydraulics models can be classified by the conservation equations they are based on. The conservation equations represent the physics of the model. The most sophisticated model is based on six conservation equations and requires uncertain correlations for phase flow dynamic. At the other end of the spectrum of thermal hydraulics models, the simpler three equation model does not have the scope to cater with phase separation and thermal disequilibrium. Intermediate models, such as the four equation model based on two mass conservation equations and the five equation model based on one mixture momentum equation were investigated. The latter was selected because it is conservative in the important quantities; the mass and energy of each phase and because the drift flux formalism on which the phase dynamics is based on is accurate.

CAE's model, including both conservation equations and constitutive models were described and some results presented. Comparison with the integral OTIS test #2202AA shows the abilities of the model to handle complex phenomena such as phase separation, void collection in high points and flow interruption, countercurrent flow and thermal disequilibrium which contribute to the overall response, timing and occurence of events.

Acknowledgment

The author would like to thank the contributors to the development of CAE's advanced thermal hydraulics code, especially Dr. J. Wasson, Mr. P.L. Rose and P. Vivier.

Nomenclature

Roman symbols

A	Area
h	Specific enthalpy
j	Volumetric flux
m	Mass
P	Pressure
Q	Heat flow
v	Velocity
V	Volume
w	Mass flow

Greek symbols

α	Void fraction
ρ	Density

Subscripts

f	Saturated liquid
g	Saturated vapour
i	Interphase
il	Interphase on liquid side
iv	Interphase on vapour side
k	$= l, v$ for liquid and vapour
l	Liquid
sl	Source of liquid
sv	Source of vapour
v	Vapour
w	Wall
wl	Wall to liquid
wv	Wall to vapour

References

[1] Chexal B. and Lellouche G., "A Full Range Drift Flux Correlation for Vertical Flows", EPRI NP-3989-SR, June 1985.

[2] Lahey R.T. and Moody J.F., "The Thermal Hydraulics of a Boiling Water Reactor", ANS, 1977.

[3] Schor A.L., Kazimi M.S. and Toderas N.E., "Advances in Two-Phase Flow Modeling for LMFBR Applications", Nuclear Design and Engineering, 82, pp. 127-155, 1984.

[4] Tran-Duc H., Morin P., "A Method of Calculation of Coupling Coefficients for a Reactor Core Nodal Model", Proceedings of the 1986 Summer Computer Simulation Conference, Reno, Nevada, 1986.

[5] Zuber N. and Findlay J.A., "Average Volumetric Concentration in Two-Phase Flow Systems", ASME Journal of Heat Transfer, November 1974.

[6] Taylor D. et al., "TRAC-BD1/MOD1: An Advanced Best Estimate Computer program for Boiling Water Reactor Transient Analysis", NUREG/CR-3633, Volume 1, April 1984.

[7] Wulff W., "Major System Codes, Capabilities and Limitations".

[8] Ishii M., Mishima K., "Two-fluid Model and Hydraudynamic Constitutive Relations", Nuclear Engineering and Design, 82, pp. 107-126, 1984.

[9] Jones O.C. Jr. et al., "Non-Equilibrium Vapor Generation ", Fifth Water Reactor Safety Information Meeting, November 7-11, 1977.

[10] Gloudemans J.R. et al., "OTIS Final Report", BAW-1905, January 1986.

[11] Judd R.A. et al., "ASSERT-4 User's Manual (Version 1)", Atomic Energy of Canada Limited, Chalk River Nuclear Laboratories, AECL-8573, September 1984.

[12] Wheeler C.L. et al., "COBRA-NC: A Thermal-Hydraulic Code for Transient Analysis of Nuclear Reactor Components – Equations and Constitutive Models", NUREG/CR-3262, Vol. 1.

[13] RELAP5/MOD2

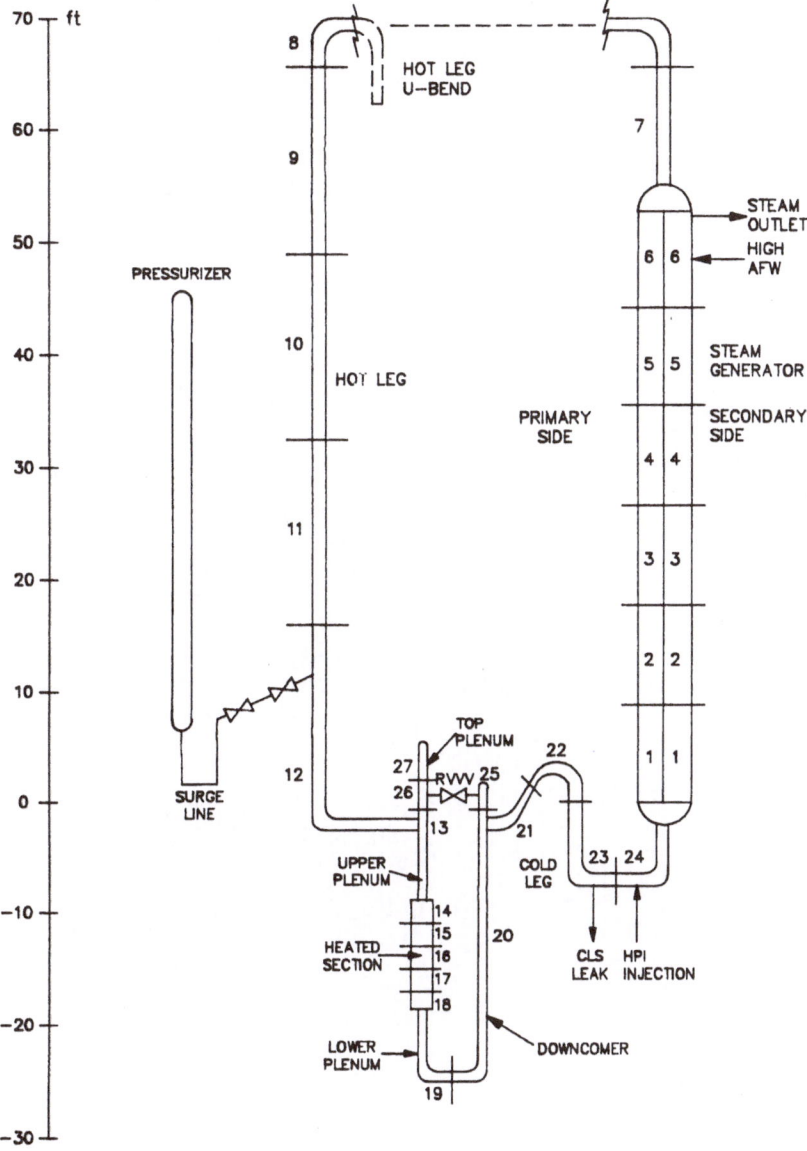

Figure 2: OTIS loop nodalization.

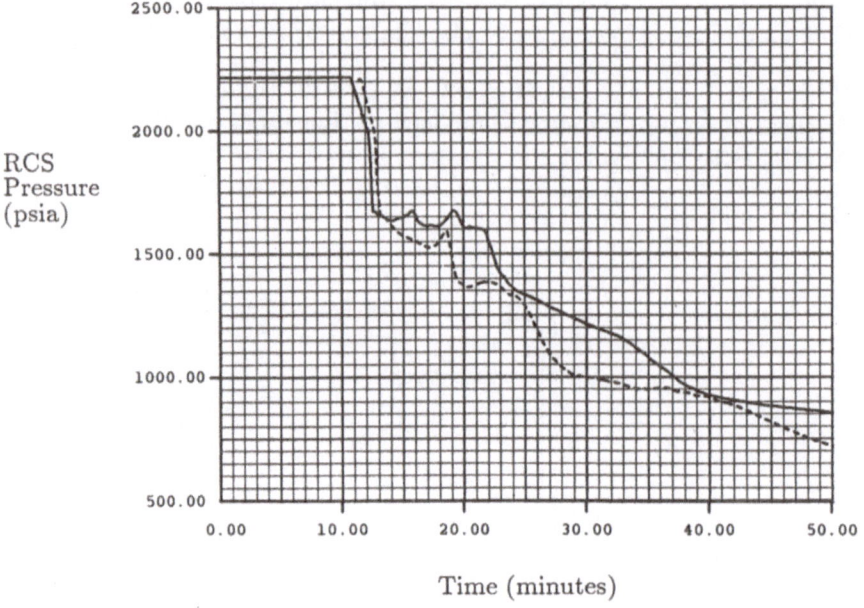

Figure 3: RCS pressure (psia)

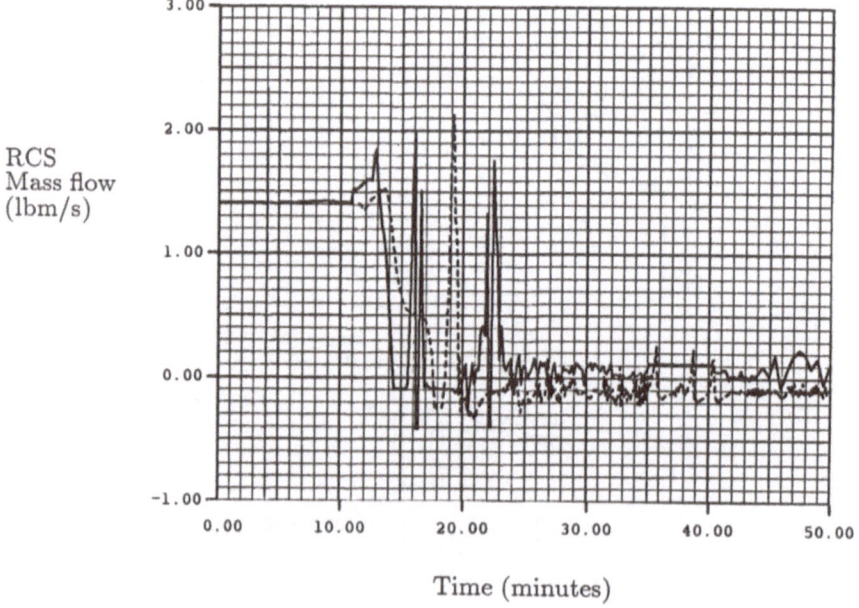

Figure 4: RCS cold leg mixture mass flow (lbm/s)

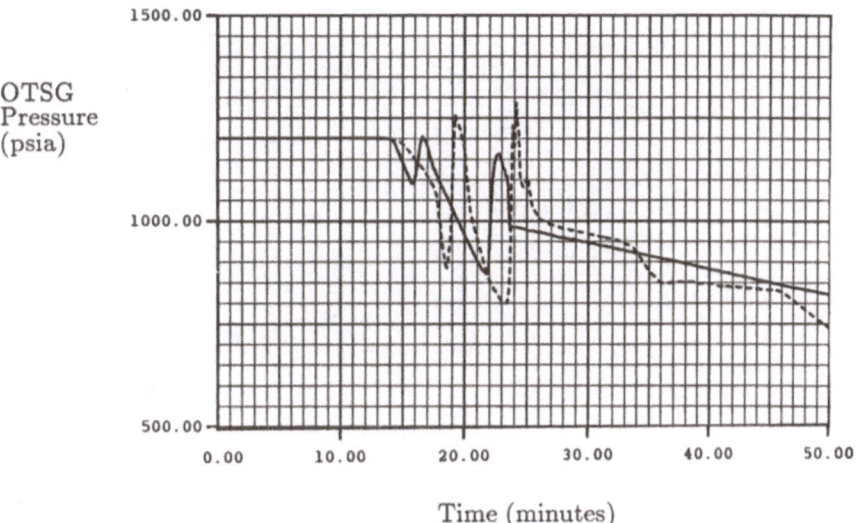

Figure 5: OTSG secondary pressure (psia)

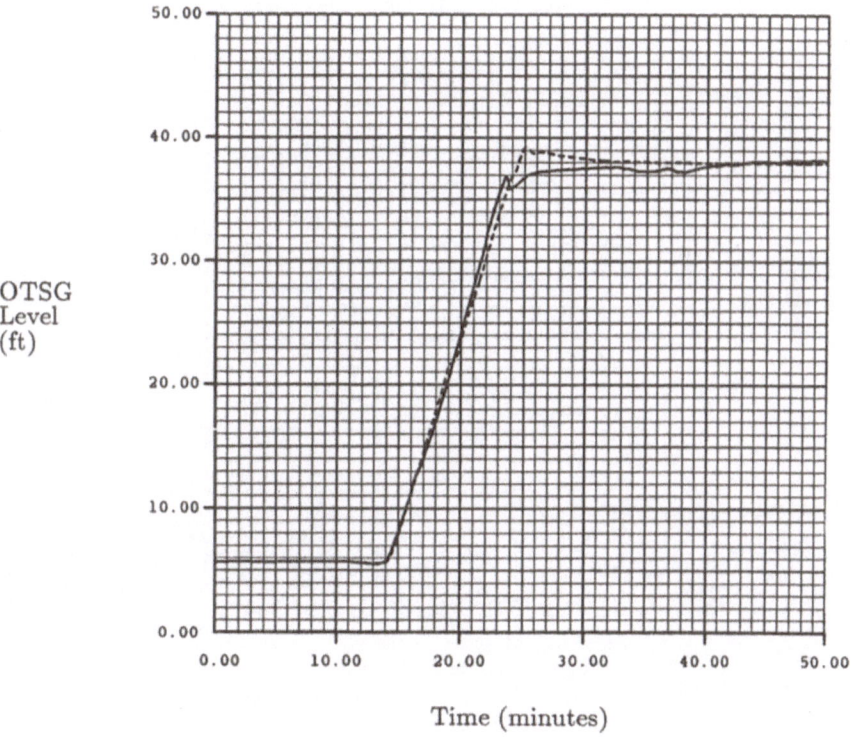

Figure 6: OTSG secondary side collapsed level (ft)

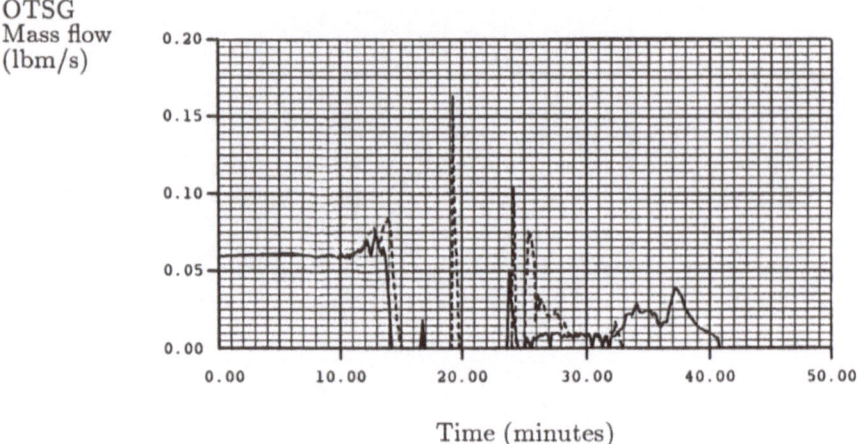

Figure 7: OTSG steam outflow (lbm/s)

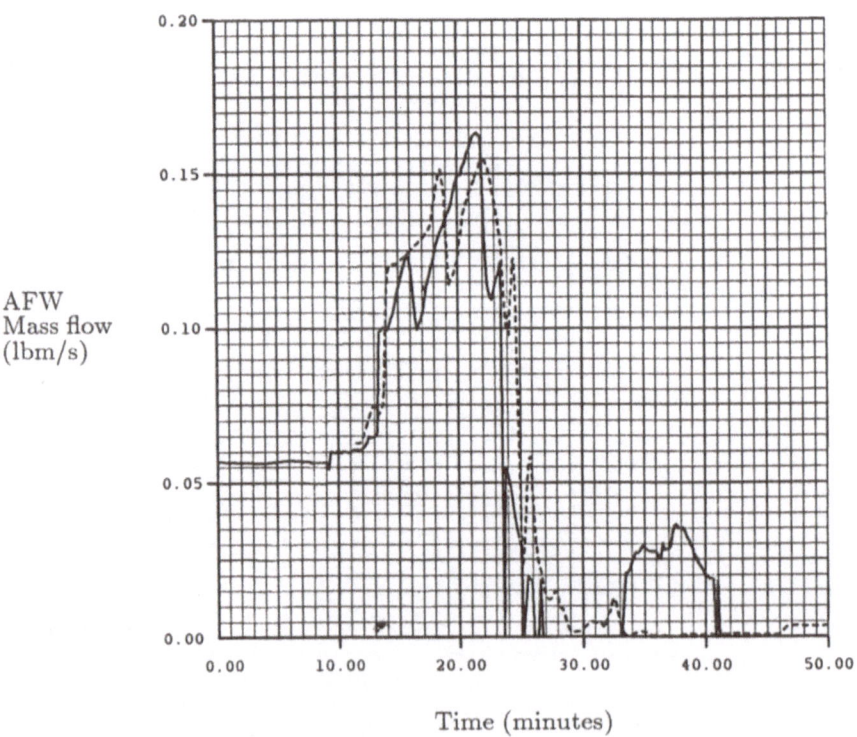

Figure 8: AFW mass flow (lbm/s)

Finite Element Analysis in Computational Fluid Mechanics

A. J. Baker

Department of Engineering Science and Mechanics

The University of Tennessee, Knoxville, TN USA

SUMMARY

This paper highlights and illustrates the fundamental aspects that the finite element discrete approximation method has in development of key requirements for CFD algorithms for fluid mechanics conservation law systems The fundamental decision of trial and test space leads in a natural way to asymptotic error estimates that determine accuracy improvement rate with the choice. The issue of stability, hence dissipation and dispersion error mechanisms, is developed using a Taylor series extension. The choice of time-integration procedure is fundamental and analyzed. Comparisons to alternative theories are made, and representative numerical solutions are cited to add substance to the theoretical developments.

INTRODUCTION

Aerodynamics, among the oldest branches in analytical fluid mechanics, utilized linear potential theory and incompressible boundary layer analysis to guide aeronautical evolution to midcentury. Lack of repeatability of supercritical (compressible transonic) aerodynamics experiments, in several wind tunnels in the 1960's, initiated detailed theoretical analyses in transonics. The scientific digital computer co-emerged from infancy, and Murmann and Cole, c.f., [1], published the first successful computational fluid dynamics (CFD) algorithm for prediction of shocked transonic potential flow. The CFD theory breakthrough was directional differencing in the supersonic region, which produced the required algorithm stability. Transonics algorithms and analyses mushroomed through the 1970's, ultimately leading to a full understanding of the mathematical requirements for entropy-consistent CFD procedures for the problem class. The key remained stabiity, generally achieved via numerical (artificial) viscosity, and the various independently-derived theories ultimately coalesced into the artificial-density transonic flow algorithm [2].

While widely used, the irrotational potential assumption excludes a wide range of practical flowfields. The next level of Navier-Stokes simplification neglects viscosity and heat conduction effects yielding the Euler equations. MacCormack [3] pioneered development of aerodynamics Euler algorithms, leading to the literal explosion in Euler CFD research in the 1970's. Explicit time integration dominated the early development, with key theoretical distinctions focusing on artificial dissipation mechanisms for algorithm stability. Capitalization on the hyperbolic conservation law form of the Euler equations eventually lead to development of implicit algorithms, with the founding development by Beam and Warming [4]. The implicit form relaxed the integration time-step size constraint, and approximate factorization of the delta-form algebra statement yielded an efficient numerical linear algebra procedure that capitalized on co-emergence of numerically-generated, body-fitted coordinate transformations, c.f., Thames, et.al. [5]. This CFD combination rapidly expanded to simplified viscous flow statements (thin-layer Navier-Stokes, parabolic Navier-Stokes, etc.) with detailed and imaginative development of "implicit" and "explicit" artificial dissipation mechanisms for algorithm stability, c.f., Chaussee, et.al. [6].

The Euler equations, as a homogeneous first degree hyperbolic conservation law system, also prompted detailed CFD analyses capitalizing on the associated characteristic eigenvalue-eigenvector structure. The goal was monotone, shock-capturing approximate solutions, which has yielded an extensive literature on flux-vector splittings [7-9], eigenvector split upwind methods [10], upwind schemes [11-13], upwind relaxation methods [14-15], total variation diminishing (TVD) schemes and entropy variables [16-17], and generalized and characteristic Galerkin methods [18-19]. The abiding character of these CFD theories is directional spatial derivative evaluation, again yielding algorithm stabilizing mechanisms.

For problem classes where viscous and heat conduction effects are important, and for all flowfields characterized as "turbulent," the sole realistic simplification to the Navier-Stokes equation system is some form of statistical manipulation, e.g., time-averaging, volume-averaging, etc. The resultant "Reynolds-averaged Navier-Stokes (RNS)" equations require a turbulence closure (constitutive) model, and may be specified for either incompressible or compressible fluids. Most compressible RNS algorithms are direct extensions on the Euler CFD methods, while incompressible algorithms must face the unique and fundamentally difficult constraint of producing a divergence-free velocity field.

The theoretical concept derived to computationally enforce continuity, i.e., $\nabla \cdot \vec{u} \simeq 0$, distinguishes the wide variety of incompressible RNS algorithms. The MAC algorithm c.f., Harlow [20], was the pioneer, wherein an unsteady differential equation was constructed for the variable "D," the measure of error in continuity, which was solved along with the momentum equations using explicit time integration. Concurrently, many CFD researchers developed streamfunction-vorticity algorithms, c.f., Roache [21], capitalizing on streamfunction as the vector potential function that analytically satisfies continuity. The formulation admits true time-accuracy, however vorticity is a rather ill-behaved variable, and the equation system suffers from intractible boundary conditions in three-dimensions. In the early 1970's, structural finite element penalty methods were transportd to incompressible CFD algorithms, and numerous groups published algorithms for laminar two-dimensional steady flows, c.f., Hughes [22]. Somewhat earlier, Spalding and his Imperial College students developed the "SIMPLE" algorithm, a semi-discrete pressure relaxation procedure, c.f., Patankar [23], used widely for steady flow prediction on Cartesian grids and including a k-ε turbulence model. The recent success of approximate factorization procedures for compressible Euler/NS algorithms, prompted rebirth of Chorin's [24], artificial compressibility concept, c.f., [25]. Analogous to MAC, an artificial time derivative is added to the continuity equation, and the resultant hyperbolic-behaving conservation law system is iterated to steady-state using Euler procedures with implicit/explicit artificial diffusion.

Thus, one recognizes the truly wide variety of theoretical constructions for CFD algorithms, e.g., finite difference, approximate factorization, finite volume, artificial density, flux vector splitting, eigen-vector split upwind, spectral, pseudo-spectral, finite element, spectral-element, etc. Each universally possess one feature in common, i.e., a computable approximation $q^h(\cdot)$ to the exact solution $q(\cdot)$ of the conservation law statement is sought, and is constructed by projection onto a finite-dimensional subspace S^h of the (Hilbert) space H^m of all admissible trial functions. That this occurs is more evident in some theories than others. For example, a spectral approximation is cast onto the first N eigenfunctions of a (global) Fourier expansion, a pseudo-spectral method uses Chebyshev or Legendre (global) polynomials, finite elements use local low degree polynomials (to "assemble" a globally piecewise continuous approximation), and spectral-element is a combination of the two methods. Conversely, the entire family of finite difference/finite volume/flux vector methods tends not to explicitly identify S^h, but does so intrinsically by assuming that certain (directional) differentiability requirements are met by the approximation.

The existence of the trial space, with or without explicit statement, is a necessary but not sufficient requirement for constructing a CFD approximate solution. A statement must also be made on how one elects to control (i.e., constrain) the error associated with the approximation. This leads to the concepts of accuracy and stability, and ultimately to definition of distinct ways to evaluate the spatial (and temporal) derivatives in the conservation law statement approximation. The breadth of finite difference/finite element/finite volume/flux vector methods can be interpreted as "weak statements," wherein the error is constrained in some integral sense over the union of elemental subdomains forming the computational mesh. This ultimately reduces to evaluation formulae for spatial derivatives on the mesh. For example, classical finite difference methods use nodal and mid-cell average values of $q^h(\cdot)$ directly, while finite volume algorithms integrate and use the divergence theorem to expose efflux integrals over cell boundaries. Flux and/or eigen-vector split methods redefine the dependent variables (for smooth transitions at critical points), and then employ one-sided forms (upwinding) of the divergence theorem. Finite element and spectral element methods use a Green-Gauss form of the divergence theorem, in a weak-statement, following specific definition of a set of test functions $v^h(\cdot)$ taken from an appropriate subspace in H^m, upon which one or more spatial derivatives can be projected. When the test and trial spaces are identical, which produces the "Galerkin" criteria, the (semi-discrete) approximation error is made orthogonal in S^h, hence is the "smallest" possible in some sense (norm). However, the mixed hyperbolic/incompletely elliptic character of Navier-Stokes equation systems can admit non-smooth solutions, which introduces the entire spectrum of upwinding and/or artificial diffusion CFD technology.

The goal of this paper is to illustrate the fundamental aspects that the finite element discrete approximation method has in development of key requirements for CFD algorithms for fluid mechanics conservation law systems. The fundamental decision of trial and test space leads in a natural way to asymptotic error estimates that determine accuracy improvement rate with the choice. The issue of stability, hence dissipation and dispersion error mechanisms, is developed using a Taylor series extension. The choice of time-integration procedure is fundamental and analyzed. Comparisons to alternative theories are made, and representative numerical solutions are cited to add substance to the theoretical developments.

PROBLEM STATEMENT

The conservation law system governing the kinematics, kinetics and thermodynamics of a viscous, heat-conducting fluid is termed the "Navier-Stokes equations." The Cartesian tensor indicial form for a compressible fluid is,

$$L(\rho) = \frac{\partial \rho}{\partial t} + \frac{\partial}{\partial x_i} (u_j \rho) = 0 \tag{1}$$

$$L(\rho u_i) = \frac{\partial(\rho u_i)}{\partial t} + \frac{\partial}{\partial x_j} (u_j \rho u_i + p\delta_{ij} - \sigma_{ij}) + \rho b_i = 0 \tag{2}$$

$$L(\rho e) = \frac{\partial(\rho e)}{\partial t} + \frac{\partial}{\partial x_j} (u_j \rho e + u_j p - \sigma_{ij} u_i + q_j) = 0 \tag{3}$$

$$p = p(e, \rho, \gamma) \tag{4}$$

where ρ is density, ρu_i is the momentum vector, p is pressure, δ_{ij} is the Kronecker delta, ρb_i is the body force and e is specific total energy. In (4), γ is the ratio of specific heats for a polytropic gas law fluid. Conversely, for an incompressible fluid, (4) is deleted, the time derivative term in (1) vanishes, and ρ becomes a reference constant except in ρb_i.

The expression of constitutive properties of the fluid is contained in the stress tensor σ_{ij} and heat flux vector q_j. For simple fluids and laminar flow, the accepted forms are

$$\sigma_{ij} = \mu\left[\frac{\partial u_i}{\partial x_j} + \frac{\partial u_j}{\partial x_i} - \frac{2}{3}\frac{\partial u_k}{\partial x_k}\delta_{ij}\right] \tag{5}$$

$$q_j = -k\frac{\partial T}{\partial x_j} \tag{6}$$

where the dynamic (molecular) viscosity μ and the thermal conductivity k are weak functions of temperature T. When modeling a turbulent flow, (1)-(4) are subject to a statistical rearrangement that introduces new variables. The well known example is mass-weighted time-averaging c.f., Cebeci and Smith [26], which employs the definitions,

$$
\begin{aligned}
\overline{\overline{u}}_i &\equiv \overline{\rho u_i}\,/\,\overline{\rho} \\
\overline{\rho u_i u_j} &= \overline{\rho}\,\overline{\overline{u}}_i\,\overline{\overline{u}}_j + \overline{\rho u'_i u'_j} \\
\overline{\rho H} &= \overline{\rho h} + \tfrac{1}{2}\,\overline{\rho}\,\overline{\overline{u}}_i\,\overline{\overline{u}}_i + \tfrac{1}{2}\,\overline{\rho u'_i u'_i} \\
\overline{\overline{H}} &= \overline{\overline{h}} + \tfrac{1}{2}\,\overline{\overline{u}}_i\overline{\overline{u}}_i + \tfrac{1}{2}\,\overline{\rho u'_i u'_i}\,/\,\overline{\rho} \\
\overline{\overline{h}} &= \overline{\rho h}/\overline{\rho} \equiv \overline{p}\,(\overline{\gamma}/\overline{\gamma}-1)/\,\overline{\rho}
\end{aligned}
\tag{7}
$$

Real gas effects are simulated in the last expression in (7) as $\overline{\gamma} = \overline{\overline{h}}/\overline{\overline{e}}$, the effective gamma, where the mass-weighted thermodynamic variables are total internal energy density ($\overline{\overline{e}}$), enthalpy ($\overline{\overline{h}}$) and stagnation enthalpy ($\overline{\overline{H}}$). For these variables, the constitutive law forms (4)-(6) become

$$
\begin{aligned}
\sigma_{ij} &= \overline{\sigma}_{ij} - \overline{\rho u'_i u'_j} \\
q_j &= \overline{q}_j + \overline{\rho H' u'_j} - \overline{u'_i \sigma_{ij}} - \overline{u}_i \overline{\sigma}_{ij}
\end{aligned}
\tag{8}
$$

Equation (8) thus introduces the Reynolds stress tensor $-\overline{\rho u'_i u'_j}$ and the turbulent heat flux vector $\overline{\rho H' u'_j}$ into the conservation law statement. This necessitates identification of a turbulence closure model; the most familiar is the two-equation system involving transport equations for turbulent kinetic energy k and isotropic dissipation function ε, and an algebraic stress constitutive relationship, e.g., Rodi [27]

$$\overline{u'_i u'_j} = k\left[\frac{2}{3}\delta_{ij} + \frac{(1-\alpha)\left(\dfrac{P_{ij}}{\varepsilon} - \dfrac{2}{3}\delta_{ij}\dfrac{P}{\varepsilon}\right) + (1-C_3)\left(\dfrac{G_{ij}}{\varepsilon} - \dfrac{2}{3}\delta_{ij}\dfrac{G}{\varepsilon}\right)}{C_1 + \dfrac{P+G}{\varepsilon} - 1}\right] \tag{9}$$

where α and C_i are model constants and,

$$P_{ij} = -\overline{\rho u'_i u'_k}\,\frac{\partial \overline{u}_j}{\partial x_k} - \overline{\rho u'_k u'_j}\,\frac{\partial \overline{u}_i}{\partial x_k}$$

$$P \equiv \tfrac{1}{2} P_{ij}\,\delta_{ij} = -\overline{\rho u'_i u'_j}\,\frac{\partial \overline{u}_i}{\partial x_j}$$

$$G_{ij} \equiv -\beta\left(g_i\,\overline{\rho u'_j \phi} + g_j\,\overline{\rho u'_i \phi}\right)$$

(10)

$$G \equiv \tfrac{1}{2} G_{ij}\,\delta_{ij}$$

In summary then, the conservation law statement applicable to a range of problem classes governed by the Navier-Stokes conservation law system can be written in the form

$$L(q) = \frac{\partial q}{\partial t} + \frac{\partial f_j}{\partial x_j} + s = 0 \qquad (11)$$

In (11), the array q contains the dependent variable set, f_j is the corresponding flux vector and s is a source/sink term. Equations (1)-(3) and (5)-(8), expressed in the form of (11), then yield the definitions,

$$q = \left\{\begin{array}{c} \rho \\ \rho u_i \\ \rho H - p/\gamma_0 \\ \rho k \\ \rho \varepsilon \end{array}\right\}, \quad f_j = \left\{\begin{array}{c} u_j \rho \\ u_j \rho u_i + p\delta_{ij}/\gamma_0 - \sigma_{ij} \\ u_j \rho H - q_j \\ u_j \rho k + k_j \\ u_j \rho \varepsilon + \varepsilon_j \end{array}\right\}, \quad s = \left\{\begin{array}{c} 0 \\ \rho b_i \\ 0 \\ -P + \rho\varepsilon \\ \dfrac{\varepsilon}{k}\left(-C_\varepsilon^1 P + C_\varepsilon^2 \rho\varepsilon\right) \end{array}\right\} \qquad (12)$$

following deletion of the various overbar and primed variable notations for clarity. The closure equations for p, σ_{ij} and q_j were given previously in (4)-(6) and (8)-(10); the additional definitions in (12) are

$$k_j = \left(C_k\,\overline{\rho u'_i u'_j}\,\frac{k}{\varepsilon} - \frac{\overline{\mu}}{Re}\delta_{ij}\right)\frac{\partial k}{\partial x_i}$$

$$\varepsilon_j = \left(C_\varepsilon\,\overline{\rho u'_i u'_j}\,\frac{k}{\varepsilon}\right)\frac{\partial \varepsilon}{\partial x_i} \qquad (13)$$

As written, (11)-(13) express a general form of the Reynolds-averaged Navier-Stokes (RNS) equations for a compressible, viscous and heat-conducting fluid and turbulent flow. The inviscid (Euler equation) simplification is recovered by deletion of the last two rows of terms in (12), and replacement of ρH by $\rho e + \gamma p$. Conversely, the incompressible RNS form is expressed in (12) by removal of the first entry (ρ) in q, the assumption elsewhere of its constancy (except in ρb_i), and deletion of (4).

THE TAYLOR WEAK STATEMENT CFD ALGORITHM

The convervation law system (11)-(12) is the general expression for problem statements in energetic fluid mechanics. A specific class, e.g., incompressible, compressible, inviscid, viscous-turbulent, is defined by specification of the constitutive relationships (4)-(6) and (8)-(10). Equation (11) guarantees existence of the time-derivative of q, provided the flux vector derivative and the source term are bounded. Hence, the Taylor series must exist,

$$q^{n+1} = q^n + \Delta t \, q_t^n + \frac{1}{2} \Delta t^2 q_{tt}^n + \frac{1}{6} \Delta t^3 q_{ttt}^n + \dots \tag{14}$$

where $\Delta t = t^{n+1} - t^n$ and subscript "t" denotes temporal derivative. The conservation law statement (11) permits exchange of the time and space derivatives. The first term is obviously expressible as,

$$q_t = -\frac{\partial f_j}{\partial x_j} - s = -\frac{\partial f_j}{\partial q} \frac{\partial q}{\partial x_j} - s \equiv -\overline{A}_j q_j - s \tag{15}$$

which serves to define the Jacobian $\overline{A}_j$ of the flux vector f_j. It is appropriate to resolve $\overline{A}_j$ into A_j, the Jacobian of the inviscid flux vector, and the remaining contributions resulting from the constitutive closure components. Thus, define,

$$\overline{A}_j \equiv \frac{\partial(E-E_v)}{\partial q} \hat{e}_1 + \frac{\partial(F-F_v)}{\partial q} \hat{e}_2 + \frac{\partial(G-G_v)}{\partial q} \hat{e}_3$$

$$\equiv A_j - \frac{\partial E_v}{\partial q} \hat{e}_1 + \frac{\partial F_v}{\partial q} \hat{e}_2 + \frac{\partial G_v}{\partial q} \hat{e}_3 \tag{16}$$

where $\{E, F, G\}$ are the corresponding scalar components in the $\hat{e}_i$ (generalized) coordinate system, and subscript "v" denotes the "viscous" component.

The inviscid flux vector Jacobian A_j plays a central role in defining n-dimensional functional expressions that manifest stabilizing mechanisms, as discussed in the Introduction. For example, for the transonic potential equation, $f_j = \rho u_j$, $q = \rho$, and the corresponding flux vector Jacobian A_j is

$$A_t^{TP} = \frac{\partial f_j}{\partial x_j} = \frac{\partial(\rho u_j)}{\partial \rho} = u_j + \rho \frac{\partial u_j}{\partial \rho} = u_j(1 - M^{-2}) \tag{17}$$

where M is the Mach number. Conversely, for the Euler equations with $q = \{\rho, \rho u_i, \rho e\}$, $f_j = \{u_j, \rho u_i u_j + p\delta_{ij}, u_j(\rho e + p)\}$, and the polytropic gas law $p = (\gamma-1)(\rho e - m^2_i/2\rho)$, one can directly derive,

$$A_j^E \simeq \begin{bmatrix} 0 & , & \delta_{ij} & , & 0 \\ -u_i u_j + \tfrac{1}{2}(\gamma-1)v_k^2 \delta_{ij} & , & (3-\gamma)u_j & , & (\gamma-1)\delta_{ij} \\ -\gamma e u_j + (\gamma-1)u_j v_k^2 & , & \gamma e \delta_{ij} - \tfrac{1}{2}(\gamma-1)(v_k^2 \delta_{ij} + 2u_i u_j) & , & \gamma u_i \end{bmatrix} \qquad (18)$$

where the free index i denotes the corresponding scalar component of the convection velocity $u_i \equiv \rho u_i/\rho$. The inviscid flux vector Jacobians for other RNS categories are equally easy to determine.

Returning to (14), using (11), the second term in the Taylor series is,

$$q_{tt} = -\left(\frac{\partial f_j}{\partial x_j} + s\right)_t = -\frac{\partial}{\partial x_j}\left(\frac{\partial f_j}{\partial t}\right) - s_t \qquad (19)$$

and since $f_j = f_j(q)$, then

$$\frac{\partial f_j}{\partial t}_t = \frac{\partial f_j}{\partial t} q_t = \bar{A}_j q_t$$

$$= \frac{\partial f_j}{\partial x_j}\left(\frac{\partial f_k}{\partial x_k} + s\right) = -\bar{A}_j\left(\bar{A}_k \frac{\partial q}{\partial x_k} + s\right) \qquad (20)$$

The principal interest in the Taylor series expansion is not a high order accurate integration formula, but rather to identify functional expressions corresponding to the inviscid limit of (11). Thus, for the RNS system, we assume that s can be neglected and that A_j can be approximated by the inviscid contribution. The sum of the two expressions given in (20) then yields [28],

$$q_{tt} \simeq \frac{\partial}{\partial x_j}\left[\bar{\alpha} A_j q_t + \bar{\beta} A_j A_k \frac{\partial q}{\partial x_k}\right] \qquad (21)$$

where $\bar{\alpha}$ and $\bar{\beta}$ are arbitrary to within a convex constraint. Similar arguments applied to the third derivative term in (14) produces,

$$q_{ttt} \simeq \cdots \equiv \frac{\partial}{\partial x_j}\left[\bar{\gamma}\left(A_j A_k \frac{\partial}{\partial x_k} + \frac{\partial}{\partial x_k}(A_j A_k)\right)q_t \right.$$

$$\left. + \bar{\mu}\left(A_j A_k \frac{\partial}{\partial x_k} + \frac{\partial}{\partial x_k}(A_j A_k)\right)A_\ell q_\ell\right] \qquad (22)$$

where $\bar{\gamma}$ and $\bar{\mu}$ are arbitrary but form a convex sum. Inserting (15), (21) and (22) into (14) and collecting terms then yields

$$
\frac{q^{n+1}-q^n}{\Delta t} = -\frac{\partial f_j^n}{\partial x_j} - s^n + \frac{\Delta t}{2}\frac{\partial}{\partial x_j}\left(\bar{\alpha}\, A_j\, q_t + \bar{\beta}\, A_j\, \frac{\partial f_k}{\partial x_k}\right)^n
$$

$$
+\frac{\Delta t^2}{6}\frac{\partial}{\partial x_j}\left(\bar{\gamma}\left(A_j A_k\frac{\partial}{\partial x_k}+\frac{\partial}{\partial x_k}(A_j A_k)\right)q_t + \bar{\mu}\left(A_j A_k\frac{\partial}{\partial x_k}+\frac{\partial}{\partial x_k}(A_j A_k)\right)\frac{\partial f_\ell}{\partial x_\ell}\right)^n
\tag{23}
$$

The left side of (23) expresses the discrete approximation to q_t on the interval Δt. Taking the limit as Δt goes to zero, but retaining the two higher order terms, yields a non-discretized conservation law statement modification to (11) of the form

$$
L_w(q) \equiv q_t + \frac{\partial f_j}{\partial x_j} + s - \Delta t\frac{\partial}{\partial x_j}\left[\alpha\, A_j\, q_t\right] - \Delta t^2\frac{\partial}{\partial x_j}\left[\gamma\, A_j\frac{\partial}{\partial x_k}(A_k\, q_t)\right]
$$

$$
- \Delta t\frac{\partial}{\partial x_j}\left[\beta\, A_j A_k\frac{\partial q}{\partial x_k}\right](A_k q_t) - \Delta t^2\frac{\partial}{\partial x_j}\left[\mu\, A_j\frac{\partial}{\partial x_k}\left(A_k A_\ell\frac{\partial q}{\partial x_\ell}\right)\right] + \ldots = 0
\tag{24}
$$

for the definitions

$$
\alpha \equiv \frac{\bar{\alpha}}{2} + \frac{\bar{\gamma}\,\Delta t}{3}\frac{\partial A_k}{\partial x_k}\ ,\ \beta \equiv \frac{\bar{\beta}}{2} + \frac{\bar{\mu}\,\Delta t}{3}\frac{\partial A_\ell}{\partial x_\ell}\ ,\ \gamma \equiv \frac{\bar{\gamma}}{2}\ ,\ \mu \equiv \frac{\bar{\mu}}{2}
\tag{25}
$$

Equations (25) and (11) are identical in the limit $\Delta t \to 0$. However, in the CFD sequel, Δt is never zero, and evaluations and/or simplifications (through $\bar{\alpha}$, $\bar{\beta}$, $\bar{\gamma}$ and $\bar{\mu}$) to the additional terms forms the structure of many independently-derived algorithm stability mechanisms. For example [28], the α term stabilizes a VLSOR transonic artificial density algorithm [2] and a flux vector splitting procedure [8]. The β term contains the artificial viscosity functional for essentially all dissipative algorithms. The γ term appears in the original Taylor-Galerkin formulation [29], and as "implicit artificial viscosity" in the RNS approximate factorization algorithm [6], while the μ term contains the upwind differencing of flux vector splittings as well as characteristic methods [18].

As discussed in the Introduction, a weak statement construction for a CFD algorithm identifies the key theoretical issues of approximation and error minimization. The preamble is that one seeks an approximte solution $q^h(\mathbf{x},t)$ to the dependent variable set $q(\cdot)$, the solution to (11). The space of functions from which one can select this approximation is extremely large, spanning everything from local truncated polynomials in $\mathbf{x}$ to a finite Fourier series. Thus, the fundamental statement of approximation is,

$$q(x,t) \quad \approx \quad q^h(x,t) \quad \equiv \quad \sum_j Q_j(t)\,\psi_j(x) \tag{26}$$

where $Q_j(t)$ are the unknown (time-dependent) expansion coefficients of the series, and the trial function set $\psi_j(x)$, $1 \leq j \leq J$, supports the approximation on a finite dimensional subspace S^h contained in H^m, the (Hilbert) space of all functions whose mth derivatives are square-integrable.

With (26) the statement of approximation, the practical requirement is to extremize (i.e., minimize) the error $e^h = q^h - q$ in the approximation. Since the semi-discrete approximation error $e^h(\mathbf{x},t)$ is a function, minimization corresponds to requiring it to be orthogonal to another function set $v(\mathbf{x})$, termed the "test function," which in turn is approximated by a series expansion, i.e.,

$$\mathbf{v}(\mathbf{x}) \quad \approx \quad v^h(\mathbf{x}) \quad \equiv \quad \sum_j V_j\,\psi_j(x) \tag{27}$$

There are some technicalities that boundary conditions place on the series (27), but selection of $\psi_j(\mathbf{x})$ for the test function expansion guarantees that e^h will be orthogonal to the approximation trial space, hence minimum in the classical sense.

With (26)-(27), forming the weak statement for minimization of the approximation error is direct. Substituting (26) into the conservation law statement (11) yields $L(q^h)$, which upon subtraction from $L(q)$ yields $L(e^h) = L(q^h)$, i.e., the conservation law statement satisfied by the approximation error. This is minimized by requiring it to be orthogonal to v^h on the domain Ω of (11), hence,

$$\int_\Omega v^h(x)\,\mathrm{L}(e^h)\,dx = \int_\Omega v^h(x)\,\mathrm{L}(q^h)\,dx \equiv 0 , \quad \textit{for all } v^h(x) \in V^h \subset H^m \tag{28}$$

Equation (28) must hold in the most general sense, i.e., the specific choice for $v^h(\mathbf{x})$ must be arbitrary, which translates into the trial space expansion coefficient set V_j, $1 \leq j \leq J$, i.e., the column matrix $\{V\}$ being arbitrary. Similarly denoting the trial space set ψ_j as $\{\psi\}$, and differentiating (28) with respect to $\{V\}$, which accounts fully for the arbitrariness, produces the CFD algorithm weak statement (WS) constraint on the unknown expansion coefficient set $\{Q(t)\}$ in (26) as,

$$WS^h(\{Q\}) \equiv \int_\Omega \{\psi\} \left(q_t^h + \frac{\partial f_j^h}{\partial x_j} + s^h \right) dx \equiv \{0\} \tag{29}$$

where (11) has been substituted directly for clarity. Following the same procedure, the weak statement on the Taylor series extension (24) of (11) produces the semi-discrete "Taylor Weak Statement (TWS)" in the form,

$$
\begin{aligned}
TWS(\{Q\}) \equiv \quad & \int_\Omega \{\psi\} \left(q_t^h + \overline{A}_j^h \frac{\partial q^h}{\partial x_j} + s^h \right) dx \\
& + \Delta t \int_\Omega \frac{\partial \psi}{\partial x_j} A_j^h \left(\alpha\, q_t^h + \beta A_k^h \frac{\partial q^h}{\partial x_k} \right) dx \\
& + \Delta t^2 \int_\Omega \frac{\partial \psi}{\partial x_j} A_j^h \frac{\partial}{\partial x_k} \left(\gamma A_k^h q_t^h + \mu A_k^h A_\ell^h \frac{\partial q^h}{\partial x_\ell} \right) dx \\
& - \Delta t \oint_{\partial\Omega} \{\psi\} A_j^h \left(\alpha\, q_t^h + \beta A_k^h \frac{\partial q^h}{\partial x_k} \right) \cdot \hat{n}_j\, d\sigma \\
& - \Delta t^2 \oint_{\partial\Omega} \{\psi\} A_j^h \frac{\partial}{\partial x_k} \left(\gamma A_k^h q_t^h + \mu A_k^h A_\ell^h \frac{\partial q^h}{\partial x_\ell} \right) \cdot \hat{n}_j\, d\sigma \equiv \{0\}
\end{aligned}
\tag{30}
$$

The first term in (30) is identical to (29), while the last two terms denote integrals created over the solution domain surface $\partial\Omega$ resulting from use of a Green-Gauss form of the divergene theorem. A surface integral will also result from the first term upon resolution of the flux vector into its Euler and constitutive closure components. Recalling (16), the first TWS^h term is reexpressed using the Green-Gauss divergence theorem to yield

$$
\begin{aligned}
\int_\Omega \{\psi\} \left(q_t^h + \frac{\partial f_j^h}{\partial x_j} + s^h \right) dx &= \int_\Omega \{\psi\} \left(q_t^h + \frac{\partial(E_j^h - E_{vj}^h)}{\partial \eta_j} + s^h \right) dx \\
&= \int_\Omega \{\psi\} \left(q_t^h + \frac{\partial E_j^h}{\partial \eta_j} + s^h \right) dx + \oint_{\partial\Omega} \{\psi\} E_{vj}^h \cdot \hat{n}_j\, d\sigma - \int_\Omega \frac{\partial \{\psi\}}{\partial \eta_j} E_{vj}^h\, dx
\end{aligned}
\tag{31}
$$

where $E_j^h = \{E,F,G\}^h$ is the inviscid flux vector approximation, $E_{vj}^h = (E_v, F_v, G_v)^h$ is the constitutive closure flux vector, and $\partial/\partial\eta_j$ is the divergence operator in the $\hat{e}_j$ coordinate system. The closed surface integral term on $\partial\Omega$ is then available to enforce all Neumann-type boundary conditions as natural constraints, c.f., [30].

In actual practice, the global trial function set $\psi_j(\mathbf{x})$ is usually replaced by the union of components spanning a generic computational cell Ω_e of the discretization Ω^h of Ω. Figure 1 illustrates representative discretization geometries; the functional modification to (26) is,

$$q(x_i, t) \approx q^h(x_i, t) \equiv \sum_j Q_j(t)\,\psi_j(x_i) \equiv \bigcup_e \left\{ N_k(\eta_i) \right\}^T \{Q(t)\}_e \qquad (32)$$

a)   b)

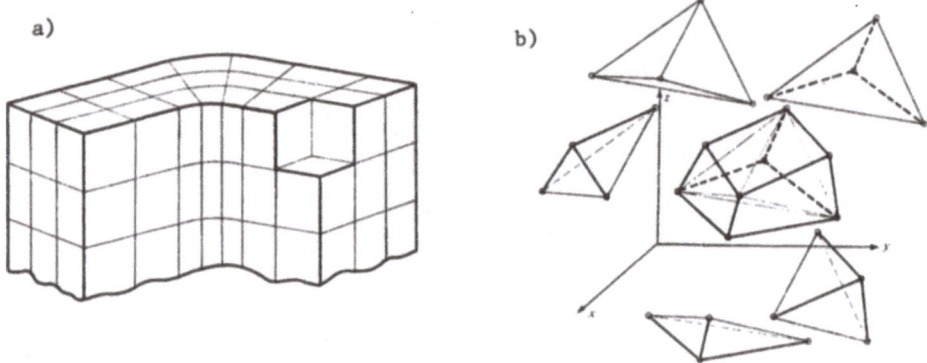

Figure 1. Solution Domain Discretizations, a) a hexahedra discretization of $\Omega \subset R^3$, b) composite hexahedra with eight nodes and its subdivision into five tetrahedra

where the trial space basis $\{N_k(\eta_i)\}$ contains elements which are polynomials complete to degree k and expressed in the intrinsic coordinate system. A basis $\{N_k\}$ can be constructed for any computational cell geometry, and the elements in $\{Q(t)\}_e$ are the values taken by $Q_j(t)$, recall (26), at the "nodes" of the computational cell which defines its vertex geometry. For the definition (32), the TWS CFD algorithm then takes the specific computational form,

$$\int_\Omega v^h\, L_w(q^h)\, dx \equiv \bigcup_e \int_{\Omega_e} \{N_k(\eta)\}\, L_w(q^h)\, dx = \{0\} \qquad (33)$$

where all integrals are now defined on the generic discretization domain Ω_e. These are easy to form, since the trial space basis $\{N_k\}$ is readily differentiated k times, and products of $\{N_k\}$ and its derivatives are at worst rational polynomials.

The Taylor weak statement CFD algorithm (33), or (30), yields a matrix ordinary differential equation on time, since only a (spatial) semi-discretization has been defined. The functional form of (33) is,

$$M\left(\{N_k\}, \alpha, \gamma\right) \frac{d\{Q\}}{dt} + \{R(\{N_k\}, \beta, \mu, \{S\}, \mathbf{A}, \Delta t)\} = \{0\} \tag{34}$$

which denotes the functional dependence of the TWS algorithm arguments on trial space basis $\{N_k\}$ and conservation law augmentation ($\alpha, \beta, \gamma, \mu, \Delta t, A_i$). The fully discrete linear algebra statement then results from employing (34) in the evaluation of the discrete Taylor series,

$$\{Q\}^{n+1} = \{Q\}^n + \Delta t \left.\frac{d\{Q\}}{dt}\right|^{n+\theta} \tag{35}$$

wherein $0 \le \theta \le 1$ defines the derivative evaluation implicitness. Substituting (34) and rearranging yields the CFD algorithm linear algebra statement,

$$\{F\} = M\{\Delta Q\} + \Delta t \left(\theta\{R\}^{n+1} + (1-\theta)\{R\}^n\right) = \{0\} \tag{36}$$

where $\{\Delta Q\} = \{Q\}^{n+1} - \{Q\}^n$ and superscripts $n+1$, n denote evaluation at the corresponding time level. The generic procedure for solving (36) is the Newton algorithm,

$$J_p^{n+1}\{\delta Q\}_{p+1}^{n+1} = -\{F\}_p^{n+1} \tag{37}$$

where J_p is the Jacobian of $\{F\}^{n+1}$ evaluated with the pth solution estimate $\{Q\}^{n+1}$. The non-iterative, single step delta form of (37) is obtained by linearization about t_n yielding

$$\left[M + \theta\Delta t \frac{\partial\{R\}}{\partial\{Q\}}\right]\{\Delta Q\} = -\Delta t \{R\}^n \tag{38}$$

Equations (37)-(38) present the matrix algebra statement for the fully discrete TWS algorithm, and the integration implicitness parameter θ constitutes another option requiring a decision. The linear algebra statements (37)-(38) are amenable to solution using any consistent numerical approximate procedure, e.g., Gauss-Seidel, line relaxation, etc.

ACCURACY, CONVERGENCE and STABILITY

The truly fundamental decision is choice of the approximation trial space $\{\psi_j(x)\}$, hence the associated trial space basis $\{N_k(\cdot)\}$, recall (26) and (32). This decision directly affects algorithm accuracy, hence the asymptotic rate of convergence to the exact solution. Rigorous asymptotic error estimates for TWS algorithms for RNS systems are not published to date. However, Galerkin estimates are available for model equations of the appropriate form, c.f., Oden and Reddy [31]. For a regular elliptic boundary value problem, and recalling $e^h \equiv q - q^h$, the asymptotic error estimate is [31, eqn. 8.80]

$$\|e^h\|_{H^1(\Omega)} \leq C\left(h_m^{\gamma_1}\|f\|_{H^r(\Omega)} + h_m^{\gamma_2}\|g\|_{H^s(\partial\Omega)}\right) \tag{39}$$

where $\|\cdot\|_{H^p}$ is the p-Sobolev norm, defined on the solution domain Ω and its boundary $\partial\Omega$, f and g are the corresponding data, $r \geq 0$ is the measure of differentiability of the exact solution and $s \geq 1/2$. For sufficiently small mesh measure h, C is a constant independent of h_m, the extremum mesh measure, $\gamma_1 = min(k, r+1)$ while $\gamma_2 = min(k, s+1/2)$, where k is the complete degree of the trial space basis $\{N_k\}$.

For a scalar linear parabolic initial-value problem, the error estimate at time $t = n\,\Delta t$ is [31, eqn. 9.66],

$$\|e^h(\cdot\ ,n\Delta t)\|_{H^1(\Omega)} \leq C_1 h_m^k \|q(n\Delta t)\|_{H^{k+1}(\Omega)} + C_2\Delta t\|Q_0\|_{H^1(\Omega)} \tag{40}$$

where Δt is the time step, Q_0 is the nodal interpolation of $q(\mathbf{x},t=0)$, and C_1 and C_2 are constants independent of h for sufficiently small h. For a smooth solution to a linear hyperbolic equation, the form of (40) is augmented to [31, eqn. 9.108]

$$\|e^h(\cdot\ ,n\Delta t)\|_{H^1(\Omega)} \leq C_1 h_m^{k+1}\|q\|_{H^{k+1}(\Omega)} + C_2\Delta t\|Q_0\|_{H^1(\Omega)}$$
$$+ C_3 h^k \int_0^{n\Delta t} \|q\|_{H^{k+1}(\Omega)}\, dt \tag{41}$$

where, for sufficiently small h and Δt, convergence is dominated by the last term involving evolution of the true solution in H^{k+1} over the entire time interval. For non-smooth solutions to a hyperbolic equation, $r = 0$, hence the estimate must be in H^0 or L^1. While none are published, even for a model problem, viewing (40)-(41) one might anticipate the form to be,

$$\|e^h(\cdot\ ,n\Delta t)\|_{H^0(\Omega)} \leq C\,h_m^p\,\|q\|_{H^0(\Omega)} + \ \cdots \ ,\qquad p \geq 0 \tag{42}$$

The analysis for TWS algorithm stability is conducted on the terminal linear algebra statement, i.e., (37) or (38), and it can be completed for arbitrary parameters α, β, γ, μ, θ and Δt upon definition of a specific $\{N_k\}$. The procedure uses Fourier analysis; for the model linear one-dimensional equation $q_t + aq_x - \varepsilon q_{xx} = 0$, and definition of a uniform discretization and the linear basis $\{N_1\}$, Baker and Kim [28] show that (38) takes the recursion relation form

$$\left[1 + a_A \Delta_0 + \left(\frac{1}{6} - a_B\right)\delta^2 + a_C \delta^2 \Delta_{\mp}\right]\Delta Q_j^{n+1}$$

$$= -\left[c\Delta_0 - ca_D\delta^2 + ca_E\delta^2\Delta_{\mp}\right]Q_j^n \qquad (43)$$

where Δ_0, δ^2 and $\delta^2\Delta_{\mp}$ are the second-order finite difference forms for the first, second and third upwind spatial derivatives and $\Delta Q^{n+1} = Q^{n+1} - Q^n$. The coefficients in (43), in terms of the original TWS parameters are,

$$a_A \equiv c\left(\theta - \bar{a}/2 - \bar{\gamma}\Delta t\,a_x/3\right)$$

$$a_B \equiv c^2\left(\bar{\gamma}/6 + \theta(\bar{\beta}/2 + \bar{\mu}\,\Delta t\,a_x/3)\right)$$

$$a_C \equiv \theta c^3 \bar{\mu}/6$$

$$a_D \equiv c\left(\bar{\beta}/2 + \bar{\mu}\,\Delta t\,a_x/3\right)$$

$$a_E \equiv c^2\,\bar{\mu}/6\,, \qquad c \equiv a\Delta t/\Delta x \qquad (44)$$

where $c = a\Delta t/\Delta x$ is the Courant number. The typical term in the Fourier expansion of the fully discrete solution is

$$q_\Delta(j\Delta x, t + \Delta t) = g\,q^h(j\Delta x, t) \qquad (45)$$

where g is the amplification factor Substituting (45) into (43) yields

$$g = 1 - \frac{c\left[\Delta_0 - a_D\delta^2 + a_E\delta^2\Delta_{\mp}\right]q^h}{\left[1 + a_A\Delta_0 + (1/6 - a_B)\delta^2 + a_C\delta^2\Delta_{\mp}\right]q^h} \qquad (46)$$

Using well known identifies for the operators in (46), the expansions for the fully discrete solution dissipation error $\omega(\Delta t)\bar{\bar{\delta}}$ and phase dispersion error $\omega\Delta t(a-\bar{\bar{\beta}})$ are [28]

$$\omega(\Delta t)\,\bar{\bar{\delta}} = c\left[m^2\left(\frac{c}{2} - a_1\right) + m^4\left(-\frac{c^3}{8} + \frac{a_1 c^2}{2} - (\beta_2 - \beta_0)c\right.\right.$$

$$\left.\left. + (\beta_1 - \beta_0 a_1)\right) + m^6(\cdot\ ) + \dots\right] \qquad (47)$$

$$(\omega \Delta t)(a - \overline{\overline{\beta}}) \;=\; c\left[m^3\left(\frac{c^2}{3} - a_1 c + (\beta_2 - \beta_0) \right) \right.$$

$$+ m^5\left(-\frac{c^4}{30} + \frac{a_1 c^3}{6} - \frac{(\beta_2 - \beta_0)c^2}{2} + (\beta_1 - \beta_0 a_1)c \right.$$

$$\left. \left. - \left((\gamma_2 - \gamma_0) - \beta_0(\beta_2 - \beta_0) \right) \right) + m^7(\cdot \;\;) + \dots \right] \qquad (48)$$

where $m \equiv \omega \Delta x$ and the various Greek-letter parameters $a_i, \beta_i, \gamma_i, i \geq 0$ are algebraic functions of the TWS parameter set $a_A - a_E$, hence $\overline{a} - \overline{\gamma}$ in (24)-(25). For example, the coefficient $(c/2 - a_1)$ on the m^2 term in (47) governs the lowest order, i.e., dominant, artificial dissipation mechanism. Substituting the definition for a_1 yields,

$$\frac{c}{2} - a_1 \;=\; \frac{c}{2} - \left(a_A + a_D \right)$$

$$=\; c\left(\frac{1}{2} - \theta + \frac{\overline{a}}{2} + \frac{\overline{\gamma}\,\Delta t\,a_x}{3} \right) - \frac{\overline{\beta}}{2} - \frac{\overline{\mu}\,\Delta t\,a_x}{3} - \frac{\varepsilon}{ha} \qquad (49)$$

which verifies how algorithm stability is intrinsically controlled by the TWS parameters, the flux vector Jacobian (a and a_x), integration implicitness (θ), and (finally) physical diffusion (ε). The dominant dispersion error term is controlled by the coefficient of m^3 in (48); substituting the definitions yields,

$$\left(\frac{c^3}{3} - a_1 c + \beta_2 - \beta_0 \right) = c^2\left(\frac{c}{3} - c(a_A + a_D) + a_B + a_A^2 - a_A a_D \right) \qquad (50)$$

which involves every coefficient except a_E, as defined in (44). Baker and Kim [28] document sixteen published CFD algorithms that are determined to date to fit the TWS theory. **Table 1** summarizes the key order-of-accuracy and stability aspects of representative decisions for the TWS parameter set.

DISCUSSION AND RESULTS

The TWS algorithm is being applied to a wide range of problem classes in CFD, to reduce the parametric arbitrariness for a given implementation. This section briefly highlights current results and developing procedures for a range of RNS definitions.

Model Problems:

Baker and Kim [28] summarize the performance of a range of TWS parameter specifications for the one-dimensional model problem defined by

Table 1. Dissipation and Dispersion Error Orders for TWS Algorithms

METHOD	TWS PARAMETERS					DISSIPATION ORDER		DISPERSION ORDER		STABILITY CONSTRAINT
	$\bar{a}$	$\bar{\beta}$	$\bar{\gamma}$	$\bar{\mu}$	θ	Δx^2	Δx^4	Δx^3	Δx^5	
Galerkin					$\frac{1}{2}$		0	$c^2/12$		None
					1	$-c/2$	$3c^3/8$	$c^2/3$		None
Donor Cell Upwind		$(\text{sgn } a)/c$			0	$(c-1)/2$		$(1-3c+2c^2)/12$		$\|c\| \leq 1$
Lax-Wendroff		1			0		$c(c^2-1)/8$	$(1-c^2)/6$		$\|c\| \leq 1$
Euler-Taylor Galerkin		1	1		0		$c(c^2-1)/24$	0	$f(c)/180$	$\|c\| \leq 1$
Euler-Characteristic Galerkin		1		1	0		$-c(c-1)^2/24$	0	$f(c)/180$	None
Swansea-Taylor Galerkin		1			0		$c(c^2-1/3)/8$	$-c^2/6$		$c < \sqrt{1/3}$
Petrov-Galerkin	$v(\text{sgn } a)/c$	$(1-v)/c^2$			0	$(c-v)/2$		$f(c,v)/6$		$c \leq v < 1$
	$(\text{sgn } a)$	$(1-v)/v^2$			0		$c(3c^2+2c-3)/24$	$(1-c-c^2)/6$		$c = v < 2/3$
Raymond-Garder	$v(\text{sgn } a)/c$	$v(\text{sgn } a)/c$			$\frac{1}{2}$		$-v/12$	$c^2/12$		None
	$v(\text{sgn } a)/c$	$v(\text{sgn } a)/c$			1	$-c/2$	$3c^3/8 - v/12$	$c^2/3$		None

89

$$L(q) = q_t + a(q)q_x - \varepsilon q_{xx} = 0 \qquad (51)$$

Figure 2 summarizes linear basis steady-state nodal solutions achieved with TWS algorithms applied to the linear form (a=constant) of (51) as a function of ε and discretization. The analytical solution to (51) is exponential in the argument $Pe(x/L-1)$, where the Peclet number definition is $Pe = aL/\varepsilon$. For $Pe \approx 1$, the physical diffusion is ample to control the dispersion error of the basic Galerkin algorithm, see the first row in Table 1. Conversely, for $Pe \geq 10^2$, the uniform grid Galerkin algorithm solution is dominated by phase dispersion error, Fig. 2b). This is functionally corrected by the TWS specification corresponding to the Raymond-Garder [32] formulation (the last entry in Table 1), see Fig. 2c). This solution identifies where the sharp solution gradient lies; mesh adaption to a non-uniform discretization, and return to the Galerkin form produces an excellent quality solution, Fig. 2d).

A non-linear model problem is established from (51) by the definition $a(q) \equiv q$ and $\varepsilon \equiv 0$, the so-called "inviscid Burgers equation." Figure 3 presents the steady-state nodal solutions obtained by the TWS algorithm for the Raymond-Garder specification (Table 1) with $v = 2/3$. The test is a squarewave propagating inwards from each end of the solution domain, with a steady-state achieved when they collide in the center. The steady shocks are monotone and spread over two-three cells for the range of shock strengths tested. A demanding unsteady linear square wave problem is a model of species intrusion, for example in an oil recovery process. The front propagates with constant velocity a and $\varepsilon = 0$ is assumed. Figure 4 summarizes Galerkin TWS algorithm solutions at a specific time point, for both linear $(k=1)$ and quadratic $(k=2)$ bases, recall (32) and the error estimate dependence on k, (40). Using the $\theta=1/2$ integration formula produces unacceptable dispersion error, Fig. 4a)-b). Recalling from (19) that $\theta > 1/2$ introduces a low order artificial diffusion mechanism, Figures 4c)-d) confirm a significant improvement in the $k=2$ solution while the $k=1$ results are still unacceptable.

These model problems can be extended to $n > 1$ dimensions to produce demanding test cases. The n-dimensional generalization of (51) is

$$L(q) = q_t + \underline{a} \cdot \nabla q - \varepsilon \nabla^2 q + s = 0 \qquad (52)$$

For example, forced convection of an axi-symmetric $(n=2)$ or point-symmetric $(n=3)$ scalar field is particularly demanding regarding artificial dissipation and phase dispersion error mechanisms. For the non-viscous specification, $\varepsilon = 0$ in (52), the analytical solution is exact preservation of the initial condition distribution. For a Gaussian, Figure 5 summarizes Galerkin TWS algorithm solutions for $k=1$ and $k=2$ basis implementations and $\theta=1/2$, following one

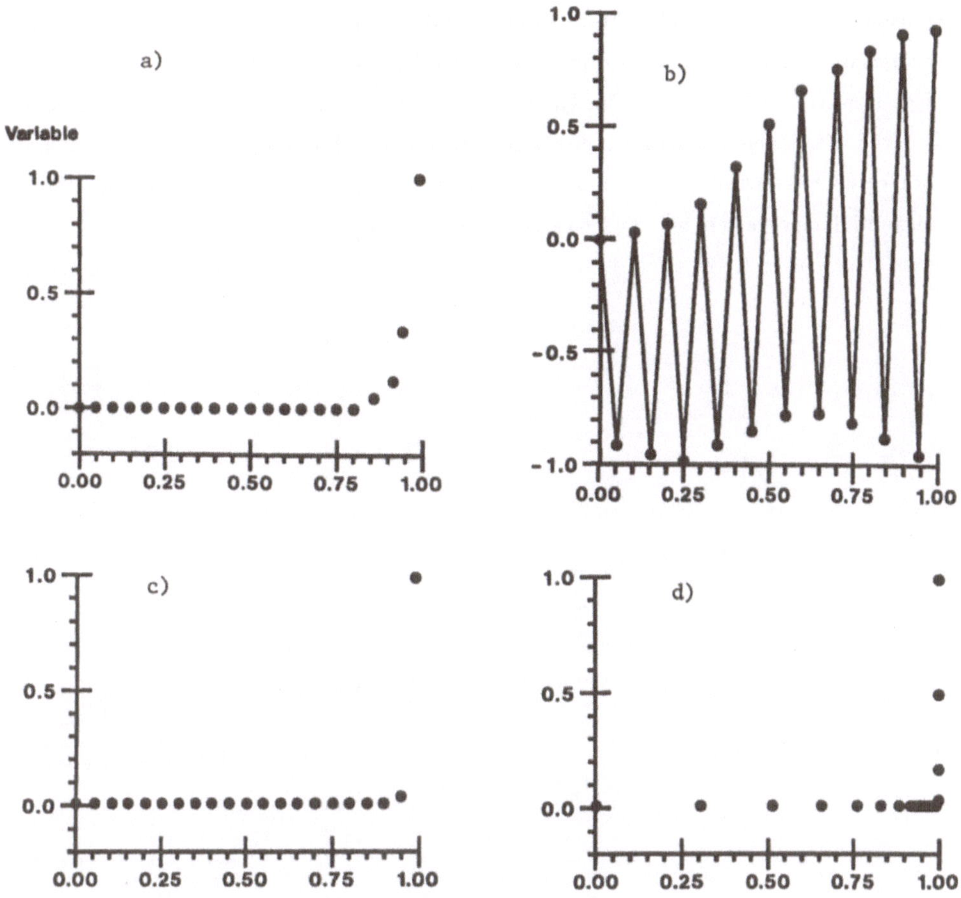

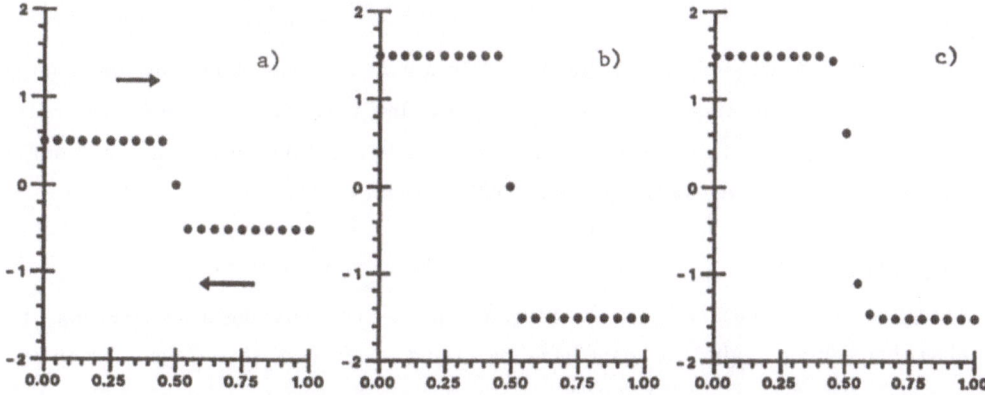

Figure 2. TWS Algorithm Nodal Steady-State Solutions, Advection-Diffusion Model Problem, M = 21 Grid, a) Pe = 1, b) Pe = 100, c) Pe = 100, v = 1/2, d) Pe = 100, v = 0, non-uniform grid.

Figure 3. TWS Algorithm Nodal Steady-State Solutions, Burgers Equation, M = 21 Grid, v = 2/3, a) $\Delta U = 1.0$, b) $\Delta U = 3.0$, c) $\Delta U = 3.0$ without stagnation point node.

complete revolution around the x-y plane. The middle graphs give the perspective view of both solutions, and the $k=1$ algorithm suffers relatively from a significant dispersion error coalescence manifested as the trailing wake. The X-X and Y-Y distributions confirm that the dispersion error distortion is confined to the convection direction. The cross-stream solutions are free from dispersion and artificial dissipation; the indicated loss in peak concentration for the $k=1$ result has cascaded into the dispersion error wake. The $k=2$ algorithm solution is a substantial improvement over $k=1$ on all comparison bases.

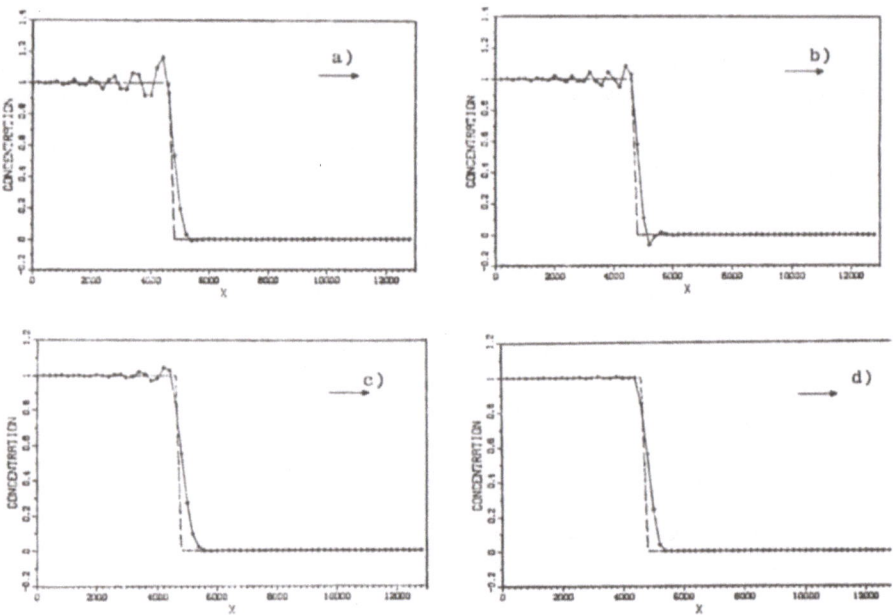

Figure 4. TWS Galerkin Algorithm Unsteady Nodal Solution, Linear Advection of a Square Pulse, $c = 0.24$, $\varepsilon = 0$, a) $k=1$, $\theta = 0.5$, b) $k=2$, $\theta = 0.5$, c) $k=1$, $\theta = 0.6$, d) $k=2$, $\theta = 0.6$.

These conclusions are confirmed for linear $n=3$ test cases as well. The $k=1$ steady-state solution for the $n=2$ generalization of the nonlinear Burgers test is shown in Figure 6. The initial conditions are given in Fig. 6a), and the monotone coalescence of the steady-state shock in Fig. 6b) was obtained using the Raymond-Garder TWS form with $v = 2/3$.

Inviscid Problems

The Euler equation form of the RNS system provides a rich environment for assessing CFD algorithm solution quality for shock (non-linear) and rarefaction wave (linear) prediction. Nozzle flows provide an excellent test venue in a quasi-one-dimensional formulation. The 1986 GAMM Workshop double-throat nozzle profile distribution is shown in Figure 7a). As a function

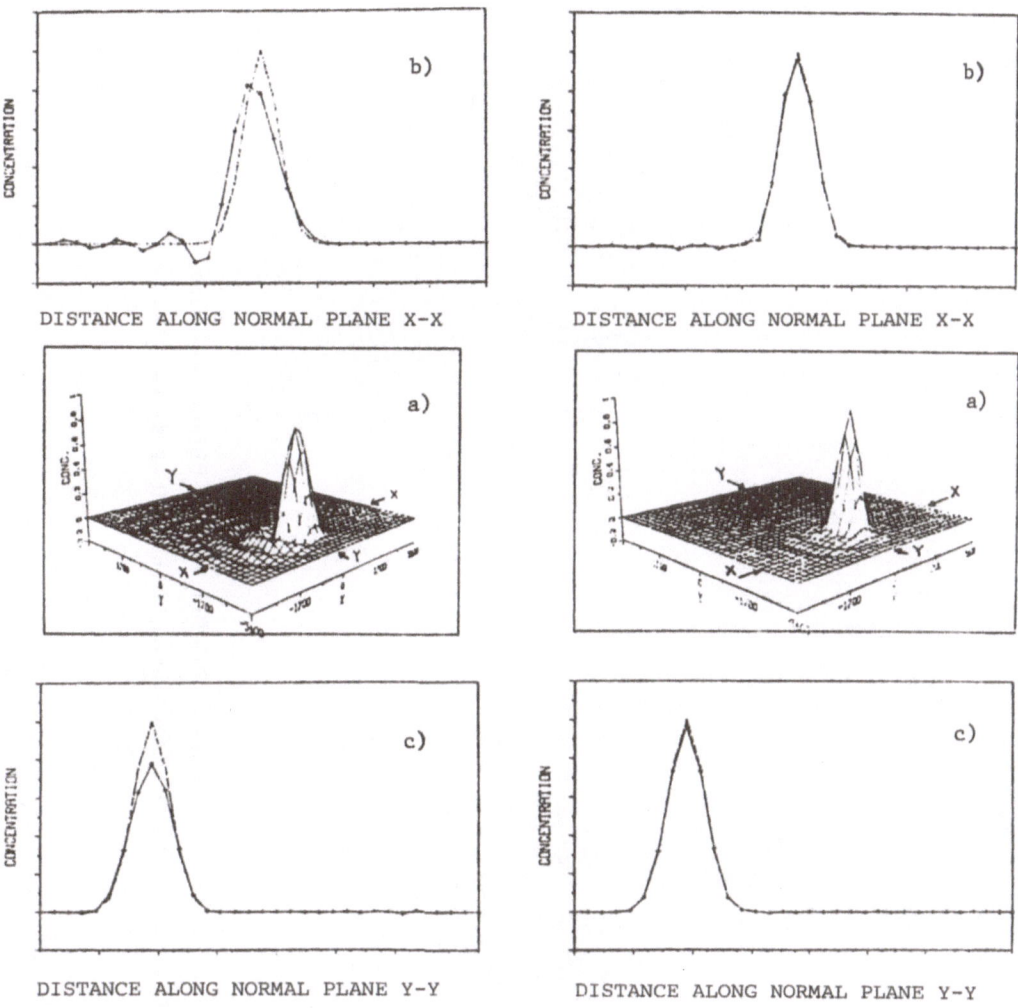

Figure 5.   TWS Galerkin Algorithm Unsteady Solution, Solid Body Rotation of Gaussian Hill Following One Revolution, $M = 33^2$ Uniform Grid, $c_m = 0.35$, $\theta = 0.5$, a)-c) $k = 1$ basis perspective view, profile through plane X-X, profile through plane Y-Y, d)-f) $k = 2$ basis perspective view, profile through plane X-X, profile throughplane Y-Y.

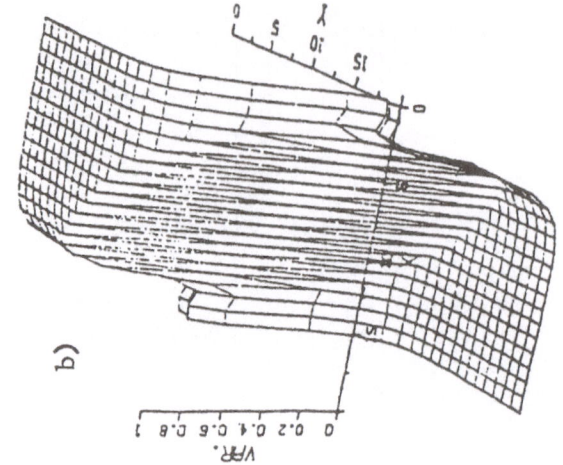

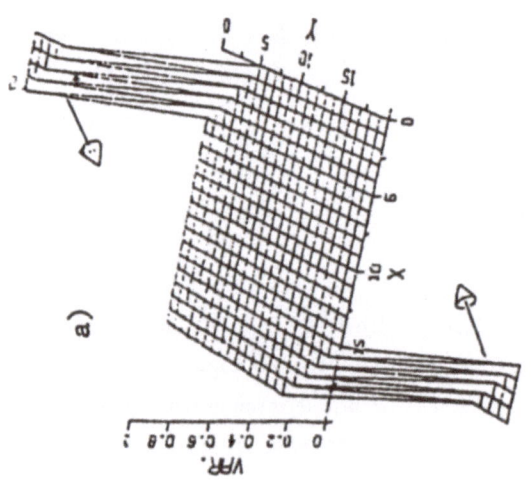

Figure 6.   TWS Algorithm Unsteady Solution. Two-Dimensional Burgers Equation, $M = 21^2$ Grid, $k = 1$, $\nu = 2/3$, a) initial condition, b) centroidal steady shock disk.

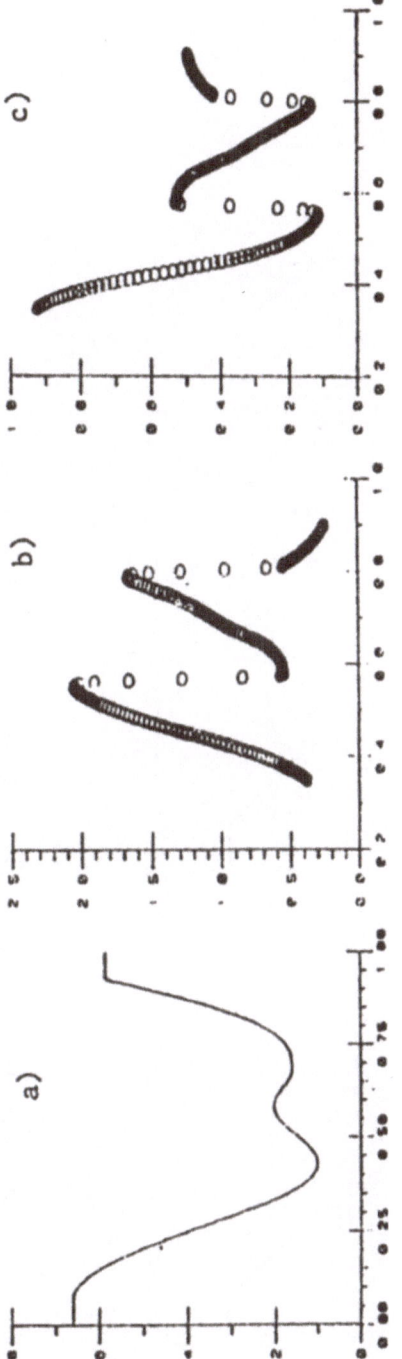

Figure 7.   TWS Algorithm Steady-State Nodal Solutions, Quasi-one-dimensional Euler Equations, GAMM Double Throat Nozzle [33], a) cross-section geometry, b) Mach Number, c) pressure.

96

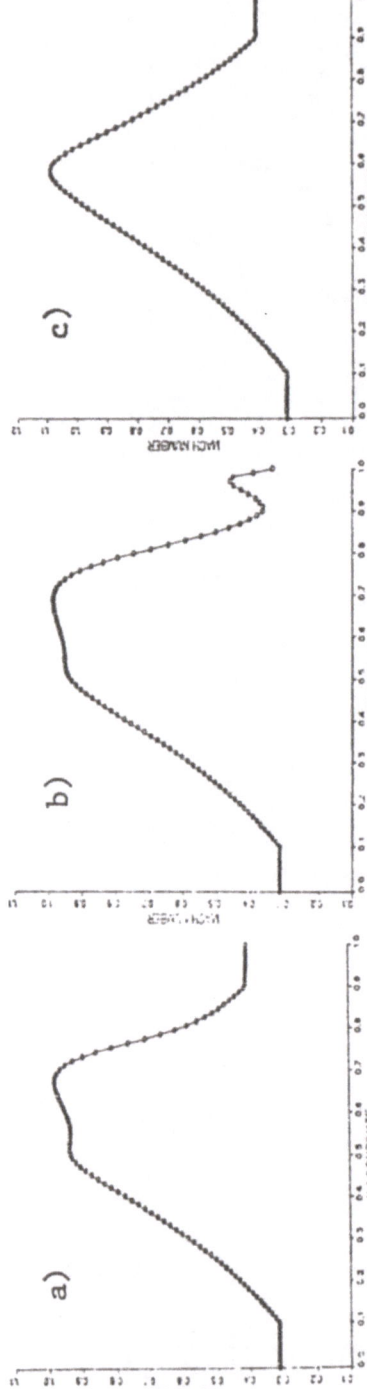

Figure 8.   TWS Algorithm Mach Number Profiles, Unsteady Euler Equations, de Laval
Nozzle, $\bar{\alpha}$, $\bar{\beta}$ from Baker [33], a) implicit Runge-Kutta algorithm, b) trapezoidal
rule, $\theta = 1/2$, c) backwards Euler, $\theta = 1.0$.

of exit pressure, the existence of two throats admits prediction of two steady-state shocks of different strengths. Figures 7b)-c) summarize the steady-state TWS algorithm nodal solutions for Mach number and pressure, as obtained using the Raymond-Garder form (Table 1) generalized to a system of equations, c.f., Baker [33,Ch.8]. The shock Mach numbers are $M_S = 2.2$ and $M_S = 1.7$, and the shocks are monotone and interpolated across 3-4 element domains.

An unsteady deLaval nozzle flow permits comparison of TWS algorithm solution quality as a function of time integration algorithm (θ). Figure 8 summarizes nodal Mach number solutions, as obtained after instantaneously depressing the exit pressure from a subsonic steady-state flow state. The accurate $k = 1$ solution, Fig. 8a), was obtained using a new implicit Runge-Kutta generalization of the θ algorithm family and clearly shows the rarefaction wave at the throat and the impending shock. The $\theta = 1/2$ trapezoidal rule results are comparable, Fig. 8b), except for the incorrect downstream oscillations. The first order accurate $\theta = 1$ backwards Euler results are stable but totally inaccurate, Fig. 8c).

The one-dimensional Riemann shock tube has served as an Euler algorithm test case for many years, c.f. [34]. A two-dimensional test can be created, wherein the shock waves generated by puncturing an L-shaped diaphragm intersect in the central region yielding an enhancement. Figure 9 summarizes the TWS Raymond-Garder solution distribution at a time point after the diaphragm rupture. All distributions are monotone, the shock waves are interpolated across 3 cells, all variable gradients are crisp, and the temperature extremum following the shock intersection is twice the simple shock level. The one-dimensional Riemann shock test has been evaluated to validate the (assumed) asymptotic error estimate form given in (42). Figure 10 summarizes the data generated using the $k = 1$ TWS algorithm, as used to estimate the convergence rate in both H^0 and L_1. For all variables evaluated, a line of unit slope ($p = 1$) interpolates the data.

Incompressible Viscous Problems

The TWS algorithm construction is being evaluated for a range of incompressible Navier-Stokes formulations, as distinguished by the theory for enforcing continuity, recall the discussion in the Introduction. To date, it has been completed for the streamfunction-vorticity and velocity-penalty formulations in two-dimensions. One standard benchmark problem is the "driven cavity," constituted of a unit square domain Ω bounded by three stationary walls with the fourth wall moving parallel to itself with unit constant velocity. As a consequence, the problem is dominated by corner singularities, a particular challenge for the vorticity formulation. Figure 11 summarizes measured asymptotic convergence for streamfunction (ψ^h) for $k = 1$ quadrilateral basis for $Re \leq 10^2$, Fig. 11a), and triangles for $Re \leq 10^3$, Fig. 11b) In both instances, on sufficiently refined grids, the convergence rate is in excellent agreement with the asymptotic

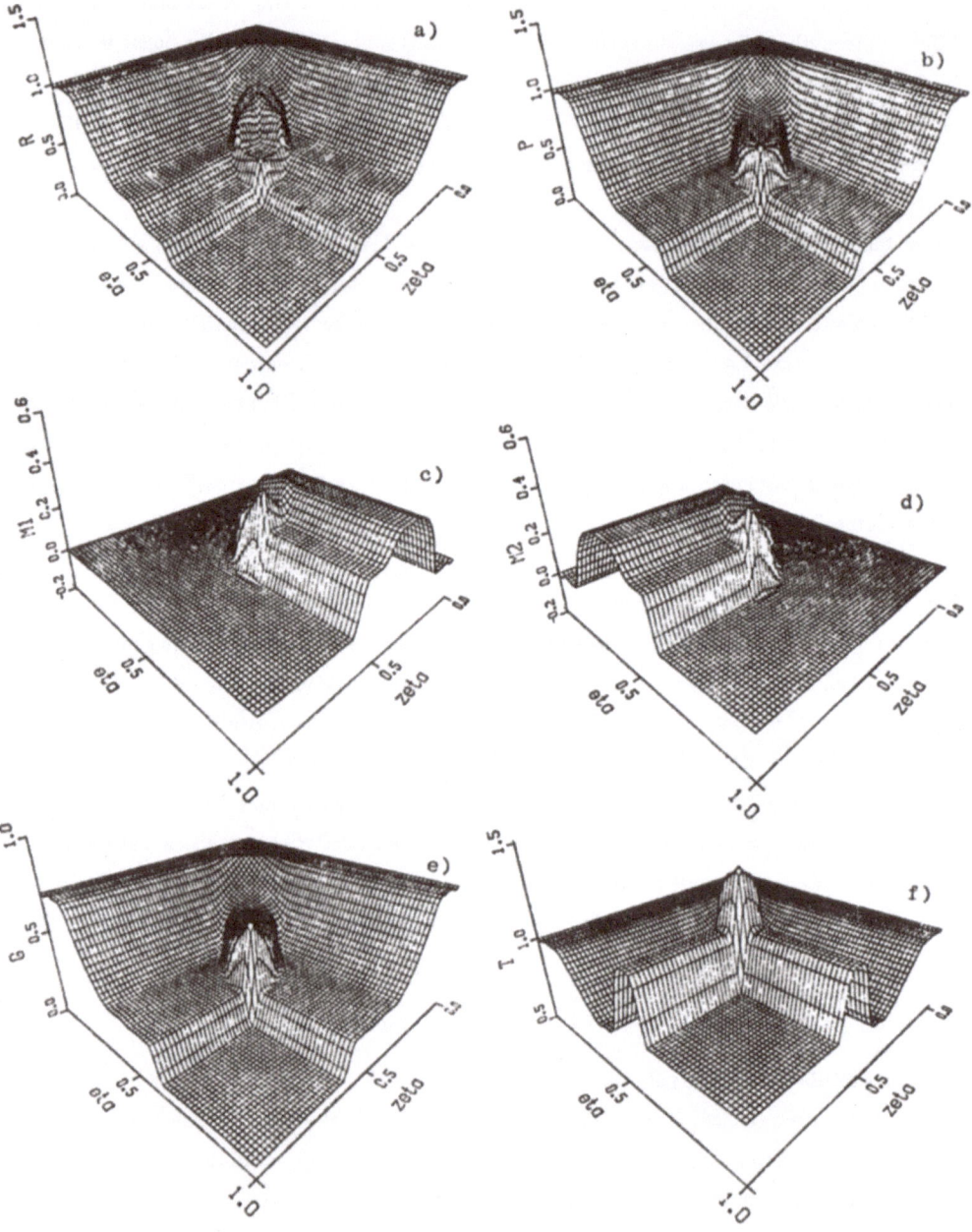

Figure 9.   TWS Euler Algorithm Prediction of Unsteady Intersecting Riemann Shocks, $k = 1$ Basis, $M = 65 \times 65$ Uniform Grid, $\bar{\alpha}$, $\bar{\beta}$ from Baker [33], a) density, b) pressure, c) momentum ρu_1, d) momentum ρu_2, e) total energy, f) temperature.

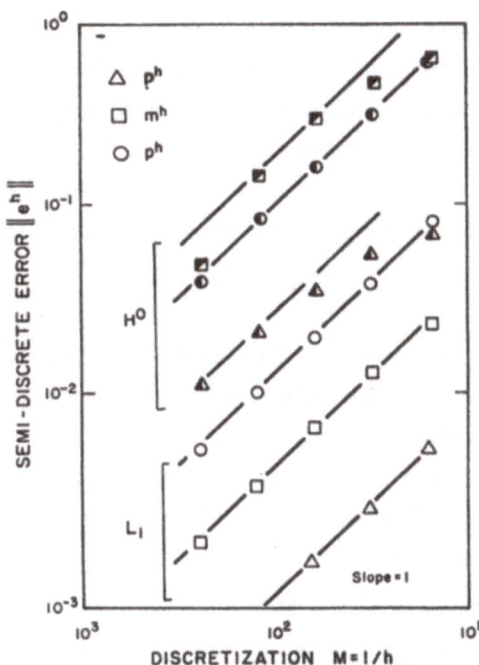

Figure 10. Asymptotic Convergence for TWS Algorithm Unsteady Solutions to Euler Equations, Riemann Shock Tube of Sod [34], TWS Parameter Specification of Baker [33].

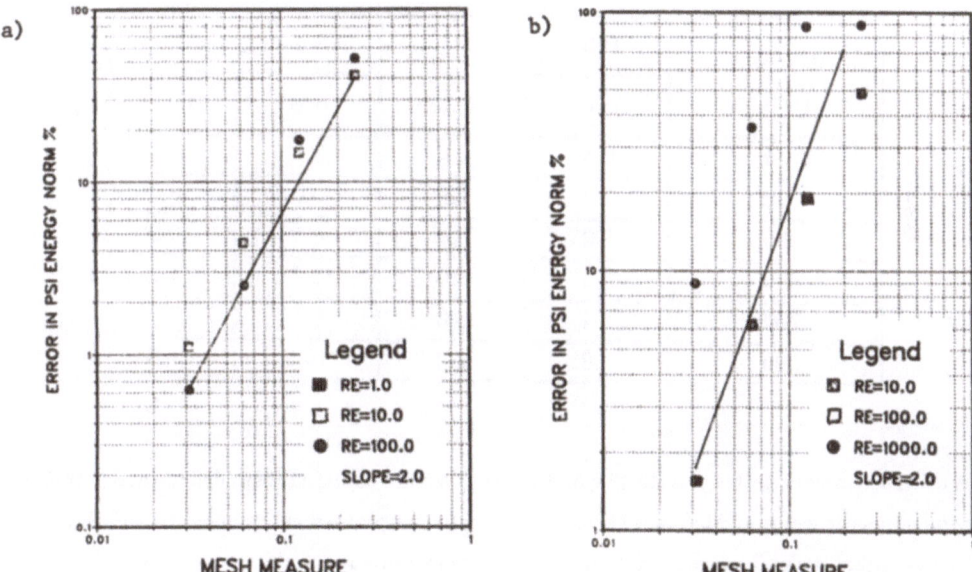

Figure 11. Asymptotic Convergence for TWS Algorithm Steady-State Solutions to Stream function Equation, Driven Cavity Problem, $1 \leq Re \leq 10^3$, a) bilinear quadrilateral basis, b) linear triangular basis.

error estiamte (39). At large Reynolds number (Re), the dispersion error mechanism grows to dominate the vorticity discrete solution ω^h. Figure 12a) shows perspectively the steady-state solution grid oscillation characteristic of this error mode. The Raymond-Garder TWS formulation generates a highly phase selective dissipation mechanism (recall Fig. 2b), which is effective in annihilating this error while not diffusing other solution detail, Fig. 12b). This again yields determination of the high gradient regions; a non-uniform regridding and return to a Galerkin TWS formulation produces the desired monotone solution, Fig. 12c), and the resultant detailed velocity field, Fig. 12d).

A second benchmark problem is flow through a duct with a sudden enlargement in cross-section, the so-called "step-wall diffuser." For the ratio of step height to full duct width of about one-half, experiment [35] verifies creation of multiple zones of separated recirculating flow on both walls as a function of Reynolds number. Both the vorticity and penalty TWS algorithms predict this character, with the former predicting somewhat larger region extents for all Reynolds numbers $200 \leq Re \leq 1000$, see Table 2. Figure 13 shows the steady-state velocity distribution computed from ω^h, as obtained using the Galerkin TWS formulation at $Re = 600$.

Table 2:   **Primary and Secondary Recirculation**
Region Spans, Step Wall Diffuser, h/H = 0.5

Reynolds Number	Primary Cell $\Delta x/h$		Secondary Cell $\Delta x/h$	
	(ω, ψ)	(u, v, λ)	(ω, ψ)	(u, v, λ)
1	0.4	0.4	0.0	0.0
50	2.2	2.0	0.0	0.0
150	6.0	4.3	4.3	0.0
400	6.8	6.0	8.7	4.5
700	7.4	6.8	9.1	6.5
1000	7.8	7.2	10.5	7.8

A third benchmark problem is the thermal cavity, a square domain Ω bounded top and bottom by solid adiabatic walls, and left and right by solid walls maintained at a temperature differential. The resultant natural convection-induced flow field exhibits recirculation cell multiplicity as a function of Rayleigh number (Ra). Figure 14 shows steady-state solutions, obtained on two uniform griddings for $Ra = 10^5$, using the TWS Galerkin algorithm form for vorticity-streamfunction. The coarser grid solutions, Figs. 14a)-c), are non-monotone confirming an inadequate discretization to support the solution gradients. While artificial dissipation ($\bar{\beta}$) could

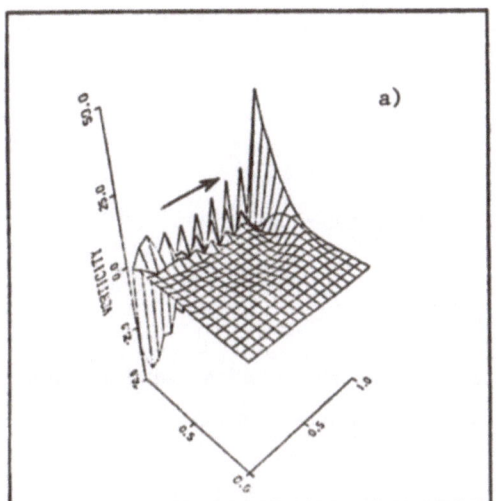
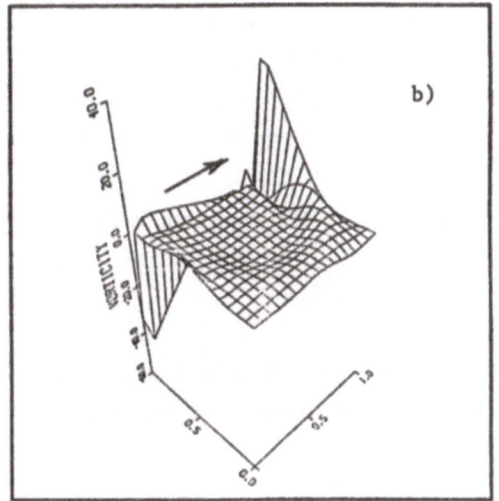
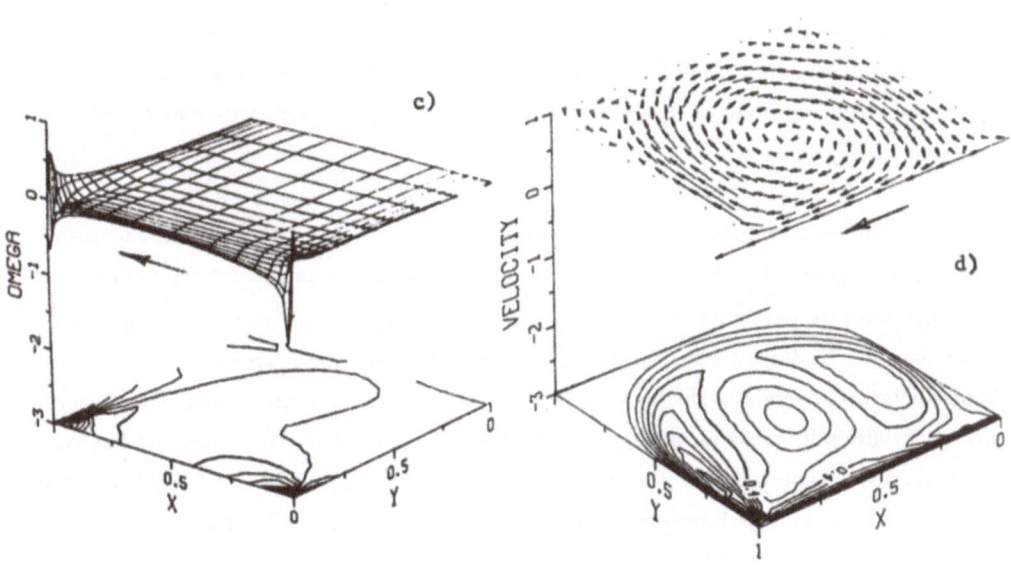

Figure 12.   TWS Algorithm Steady-State Solutions, Driven Cavity, ω^h-ψ^h Formulation,
Re = 1000, a) M = 17^2 uniform grid ω^h, v = 0., b) M = 17^2 uniform grid ω^h, v = 0.2,
c) M = 17^2 non-uniform grid, v = 0., d) velocity field computed from ω^h, arrow
denotes moving wall location in each plot.

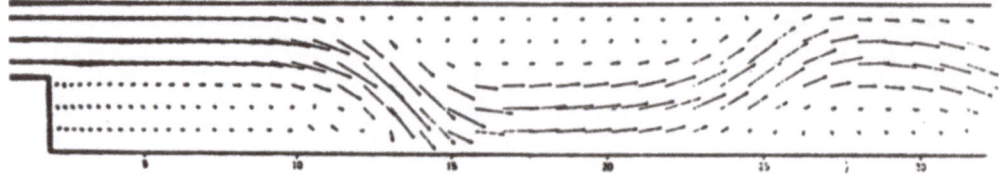

Figure 13.   TWS Algorithm Steady-State Velocity Vector Distribution, Step Wall Diffuser, ω^h-ψ^h Formulation, Re = 600.

smooth these errors, the proper correction is a refined grid. The solution in Fig. 14d) then shows the proper double recirculation cell, as manifested by the two interior extrema in ψ^h.

CONCLUSIONS

The intent of this paper was to introduce and document the theoretical robustness present in a finite element weak statement formulation for various problem classes in computational fluid mechanics. The theoretical structure clearly identifies the essential ingredients in forming a discrete approximate solution, and in defining and addressing the key issues of accuracy, convergence and stability of the final formulation. Numerical examples have been discussed to verify and expand on the basic theoretical considerations for a wide range of problem classes in energetic fluid mechanics.

Lengths constraints exclude discussion on many allied topics, e.g., dynamic solution-adaptive meshing, grid generation, numerical linear algebra techniques and computer programming. The last topic is particularly critical, since the emergence of multi-tasking vector supercomputers, and/or attached parallel processors, has opened a wide spectrum for algorithm implementation strategies. The highly organized structure of the finite element CFD theory, with its applicability to a wide spectrum of problem classes, i.e. Reynolds, Mach and Rayleigh numbers, etc., prompts consideration for design of a "general purpose CFD code" highly tailored to one or more specific architectures. Parallel processing is an intrinsic architecture feature across the board, and the TWS algorithm structure exhibits a hierarchic parallelism directly amenable to systolic processing, cf. [37]. The prospect for significant development of this topic in the near term is highly stimulating and exciting.

ACKNOWLEDGMENTS

This work is the culmination of many years research efforts by many colleagues and graduate students. I am particularly indebted to Mr. P.D. Manhardt and Mr. J. A. Orzechowski, who have been invaluable in converting theory into code practice. Most of the benchmark test results were

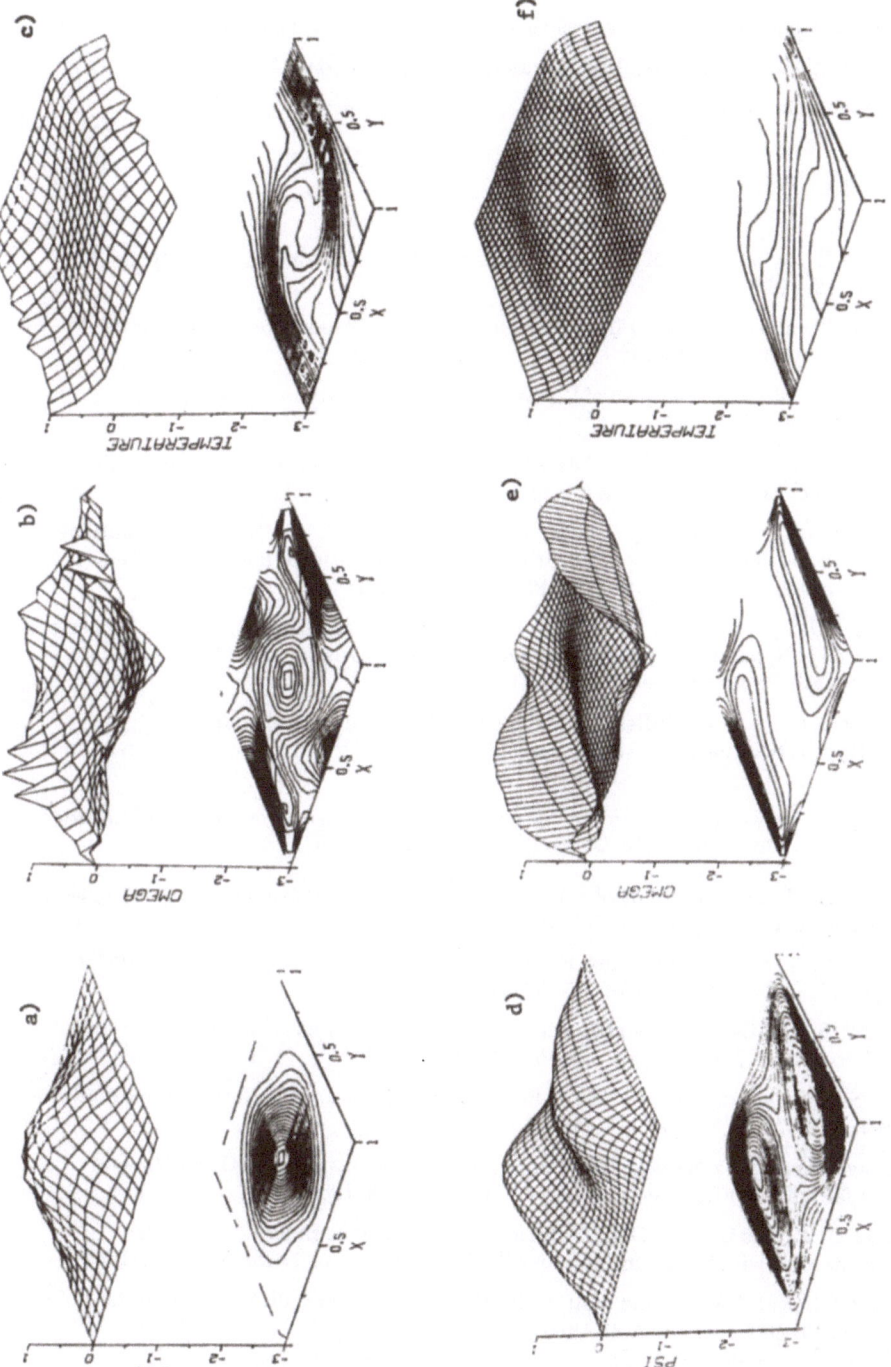

Figure 14. TWS Algorithm Steady-State Solutions, Thermal Cavity Problem, ω^h-ψ^h Formulation, Ra = 10^5, a)-c) M = 17^2 uniform grid, ψ^h, ω^h, T^h, d)-f) M = 33^2 uniform grid, ψ^h, ω^h, T^h.

produced by my graduate students, pursuing their academic programs, and I wish to acknowledge in particular Messrs. Wilbert Noronha, Jin Kim, Jim Freels, Joe Iannelli and Paul Williams. The computer resources have been graciously provided by the University of Tennessee and the Computational Fluid Dynamics Laboratory. I am indebted to Control Data Corporation for their support of this paper.

REFERENCES

1. Murman, E. M., "Analysis of Embedded Shocks Calculated by Relaxation Methods, " AIAA J., 12, 626-633 (1974).

2. Habashi, W., and Hafez, M., "Finite Element Solutions of Transonic Flow Problems, " AIAA J., 20, 1368-1377 (1982).

3. MacCormack, R. W., "The Effect of Viscosity in Hyper-velocity Impact Cratering, " Tech. Paper AIAA-69-354 (1969).

4. Beam, R. M., and Warming, R. F., "An Implicit Factored Scheme for the Compressible Navier-Stokes Equations, " AIAA J., 16, 393-402 (1978).

5. Thames, F. C., Thompson, J. F., Martin, C. W., and Walker, R. L., "Numerical Solutions for Viscous and Potential Flow about Arbitrary Two-Dimensional Bodies using Body-Fitted Coordinate Systems, " J. Comp. Phys, 24, 245 (1977).

6. Chaussee, D. S., Patterson, J. L., Kutler, P., Pulliam, T. H. and Steger, J. L., "A Numerical Simulation of Hypersonic Viscous Flows Over Arbitrary Geometries at High Angle of Attack, " Tech. Paper AIAA-81-0050 (1981).

7. Godunov, S. K., "A Finite Difference Method for the Numerical Calculation of Discontinuous Solutions of the Equations of Fluid Dynamics," Mat. Sb., 47, 271-290 (1959).

8. van Leer, B., "Towards the Ultimate Conservation Difference Scheme, V. A Second Order Sequel to Godunov's Method," J. Comp. Phys., 32, 101-136 (1979).

9. van Leer, B., "Flux Vector Splitting for the Euler Equations," in Lecture Notes in Physics, V. 170, Springer-Verlag, NY, pp. 507-512 (1982).

10. Lombard, C. K., and Venkatapathy, E., "Universal Single Level Implicit Algorithm for Gasdynamics, " Tech. Rpt. NASA CR-166531 (1984).

11. Engquist, B., and Osher, S., "Stable and Entropy Satisfying Approximations for Transonic Flow Calculations," Math. Comp., 34, 45-75 (1980).

12. Osher, S., and Chakravarthy, S., "High Resolution Schemes and the Entropy Condition," SIAM J. Num. Anal., 21, 955-984 (1984).

13. Chakravarthy, S. R., and Osher, S., "Numerical Experiments with the Osher Upwind Scheme for the Euler Equations," AIAA J., 21, 1241-1248 (1983).

14. Walters, R. W., and Dwoyer, D. L., "Efficient Solutions to the Euler Equations for Supersonic Flow with Embedded Subsonic Regions," Tech. Rpt. NASA TP 2523, (1987).

15. Thomas, J. L., van Leer, B., and Walters, R. W., "Implicit Flux-Split Schemes for the Euler Equations," Tech. Paper AIAA-85-1680 (1985).

16. Harten, A., "On the Symmetric Form of Systems of Conservation Laws with Entropy," Tech. Report, NASA ICASE 81-34 (1981).

17. Harten, A., "High Resolution Schemes for Hyperbolic Conservation Laws," J. Comp. Phys., 49, 357-393 (1983).

18. Morton, K. W., and Parrott, A. K., "Generalized Galerkin Methods for First Order Hyperbolic Equations," J. Comp. Phys., 36, 249-270 (1980).

19. Morton, K. W., "Generalized Galerkin Methods for Hyperbolic Problems," Comp. Mtd. Appl. Mech. Engr., 52, 847-872 (1985).

20. Harlow, F. H., "Numerical Methods in Fluid Dynamics, an Annotated Bibliography," Tech. Rpt. LA-1281, Los Alamos Nat. Lab. (1969).

21. Roache, P. J., **Computational Fluid Dynamics**, Hermosa Publishers, Albuquerque, NM (1971).

22. Hughes, T. R. J., Liu, W. K. and Brooks, A., "Review of Finite Element Analysis of Incompressible Viscous Flows by the Penalty Function Formulation, " J. Comp. Phys., 30, 1-60 (1979).

23. Patankar, S. V., **Numerical Heat Transfer and Fluid Flow**, Hemisphere Pub. Corp., N.Y. (1980).

24. Chorin, A. J., "A Numerical Method for Solving Incompressible Viscous Flow Problems," J. Comp. Phys., 2, 12-26 (1967).

25. Chang, J. L. C. and Kwak, D., "On the Method of Pseudo-Compressibility for Numerically Solving Incompressible Flows", Tech. Paper AIAA 84-0252(1984).

26. Cebeci, T. and Smith, A. M. O., **Analysis of Turbulent Boundary Layers**, Academic Press, New York (1974).

27. Rodi, W., **Turbulence Models and their Application in Hydraulics - A State-of-the-Art Review**, IAHR Publication, Delft, Holland(1980).

28. Baker, A. J. and Kim, J. W., "A Taylor Weak Statement for Hyperbolic Conservation Laws," Int. J. Num. Mtd. Fluids, 7, 489-520 (1987).

29. Donea, J., "A Taylor Galerkin Method for Convective Transport Problems," Int. J. Num. Mtd. Engr., 20, 101-119 (1984).

30. Dutt, P., "Stable Boundary Conditions and Difference Schemes for Navier-Stokes Equations," Tech. Report NASA ICASE 85-37(1985).

31. Oden, J. T. and Reddy, J. N., **An Introduction to the Mathematical Theory of Finite Elements**, Wiley, New York (1976).

32. Raymond, W. H., and Garder, A., "Selective Damping in a Galerkin Method for Solving Wave Problems with Variable Grids," Mon. Weather Rev., 104, 1583-1590 (1976).

33. Baker, A. J., **Finite Element Computational Fluid Mechanics**, Hemispher/Harper & Row, New York (1983).

34. Sod, G. A., "A Survey of Several Finite Difference Methods for Systems of Nonlinear Hyperbolic Conservation Laws," J. Comp. Phys., 27, 1-31(1978).

35. Armaly, B. F., Durst, F., Pereira, J. C. F., and Schonung, B., "Experimental and Theoretical Investigation of Backward Facing Step Flow," J. Flu. Mech., 127, 473-496 (1983).

36. Zienkiewicz, O. C., **The Finite Element Method**, McGraw-Hill, London (1977).

37. Baker, A. J., Manhardt, P. D., and Soliman, M. O., "On Recent Advances and Future Research Directions for Computational Fluid Dynamics," in A. K. Noor (ed.), **Proc. Sym. Future Directions of Comp. Mech**, ASME (1986).

Prediction of Fluid Behaviour during Reactor Transient Analysis using Coupled 1D and 3D Models

D KIRKCALDY, P J PHELPS AND N RHODES

Concentration Heat and Momentum Ltd
Bakery House
40 High Street
Wimbledon SW19 5AU

Abstract

The detailed understanding of the behaviour of a pressurised water reactor during transient operation is crucial to the design of the reactor.

Hitherto mathematical simulation effort has largely concentrated on one-dimensional loop analysis of the reactor system; this gives a reasonable understanding of the consequences of a transient, such as the blowdown, but does not yield adequate information on features such as pressure effects in the reactor vessel. For greater understanding of such effects, a three-dimensional reactor model is required

Two- and three-dimensional transient simulations of reactor vessels and of individual components of the circuit have, in the past, been performed in order to provide pressure and temperature fields for input to structural analysis programs. Such decoupled calculations have however tended to underestimate the real situation, since boundary conditions have been specified from stand-alone loop calculations; furthermore, there has been no simple way of incorporating fluid-structure interaction effects .

This paper describes the application of the PHOENICS program to predict the fluid behaviour in a PWR during a hypothetical blowdown. The analysis features two technical novelties. The first is the use of a three-dimensional model for the reactor vessel, directly linked with one-dimensional loop models for the primary water circuits; the second is the dynamic coupling of a fluids code (PHOENICS) with a stress code (ABAQUS) to provide a first step towards the modelling of fluid-structure interaction effects .

INTRODUCTION

Previous Work and Present Contribution

Reactor safety considerations demand detailed knowledge of the behaviour of the reactor internals during normal operation, and in particular during transient operation. Because of the cost and difficulty in extracting test data from a full-size reactor or from scaled down tests [1,2], computer simulations have become increasingly important. Considerable development has gone into the creation of computer programs to simulate the transient behaviour of a reactor system under the severe conditions that occur during a system blowdown. One such program, RELAP4/MOD6 [3], features a one-dimensional representation of the reactor circuit, with the non-uniform flow in the reactor represented as parallel but unconnected flow paths. Another model, TRAC/P1 [4], represents the reactor vessel three-dimensionally, and the circuit one-dimensionally.

PHOENICS has been used previously to simulate blowdown transients, similar to those modelled by RELAP and TRAC, [5,6]. However these simulations have been limited to one-dimensional loop analyses, with the reactor modelled as one-dimensional, parallel, non-interacting flow paths. Two- and three-dimensional analyses of single components in isolation have also been performed. Generally these have been transient simulations [7,8], designed to provide data for input to stress-analysis codes.

As such, they have suffered from two shortcomings: first, the boundary conditions specified for the analyses have been obtained from decoupled one-dimensional system simulations, so that reactor-to-loop interactions are not properly accounted for . Secondly, account has not been taken of fluid-structure interaction effects , arising from deformation of the vessel structure and its subsequent influence on the fluid flow . This latter omission has been due to the complications inherent in interfacing fluid and structural codes, there being no single code with dual capability .

This paper represents a significant advance over the previously reported work by presenting the outcome of applying PHOENICS to the three-dimensional analysis of a reactor vessel coupled with simultaneous one-dimensional analyses of the reactor loops. In addition, the simulation represents a further step towards the representation of fluid-structure interaction effects , arising from the interactive coupling with the stress code ABAQUS [9].

Nature of the Process Simulated

The reactor system considered was a three loop 800 MWe PWR system. Two of the loops were geometrically identical with the third loop differing only with the addition of a pressuriser. The transient was to be initiated by considering one of the non–pressuriser loops to be 'broken', with the break occurring near the reactor cold–leg inlet. For the purposes of the analysis, each of the three loops was divided into a number of regions, as depicted in Figure 1. These regions could then be further sub–divided into a number of finite–difference control cells for integration purposes.

The ends of the loops were attached to the three–dimensional reactor, with the cold legs connected to the reactor downcomer and the hot legs connected to the reactor upper–core region. A lack of symmetry in the arrangement of these inlets and outlets meant that the full 360° of the reactor vessel would have to be modelled (see Figure 2).

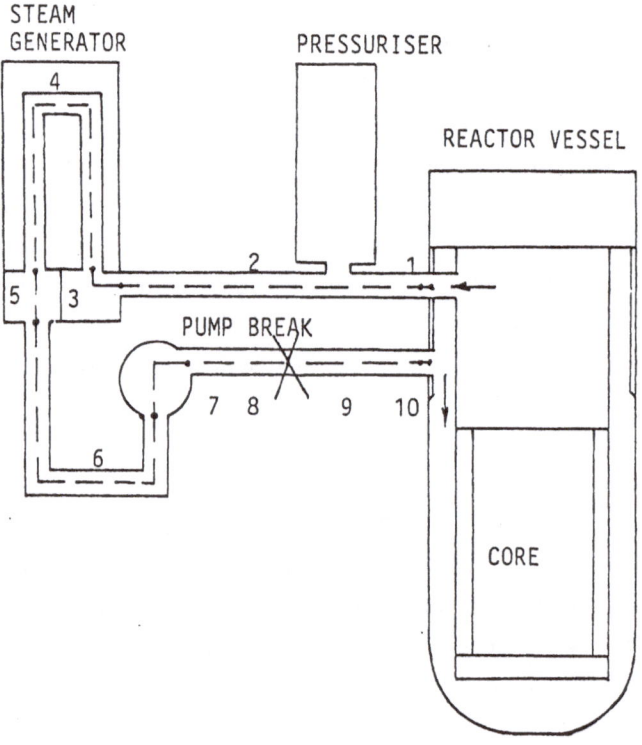

Figure 1. Configuration of the circuit showing region distribution

Region	Component	flow area (m^2)	Volume (m^3)
1	Reactor vessel outlet	0.442	0.049
2	Hot-leg pipe	0.442	3.525
3	Steam generator inlet plenum	2.681	4.289
4	Steam generator tubes	1.247	21.940
5	Steam generator outlet plenum	2.681	4.289
6	Steam generator to pump piping	0.442	4.938
7	Pump	0.442	1.774
8	Cold-leg pump side of break	0.442	1.593
9	Cold-leg reactor side break	0.442	1.593
10	Reactor vessel inlet	0.442	0.049

Table 1 Regions of circuit Model.

Objectives of the Study

The objectives of the study were as follows:

a. Develop a model, using the PHOENICS code, to predict the transient flow (and in particular the pressures), in the circuit of a PWR. This model would include one-dimensional analysis of the reactor loops coupled to a three-dimensional analysis of the reactor vessel.

110

b.   Link up the PHOENICS calculation to run in parallel with a structural-analysis code, in this case ABAQUS.

c.   Compute the pressure-wave behaviour during the first 50 msecs of a hypothetical blowdown.

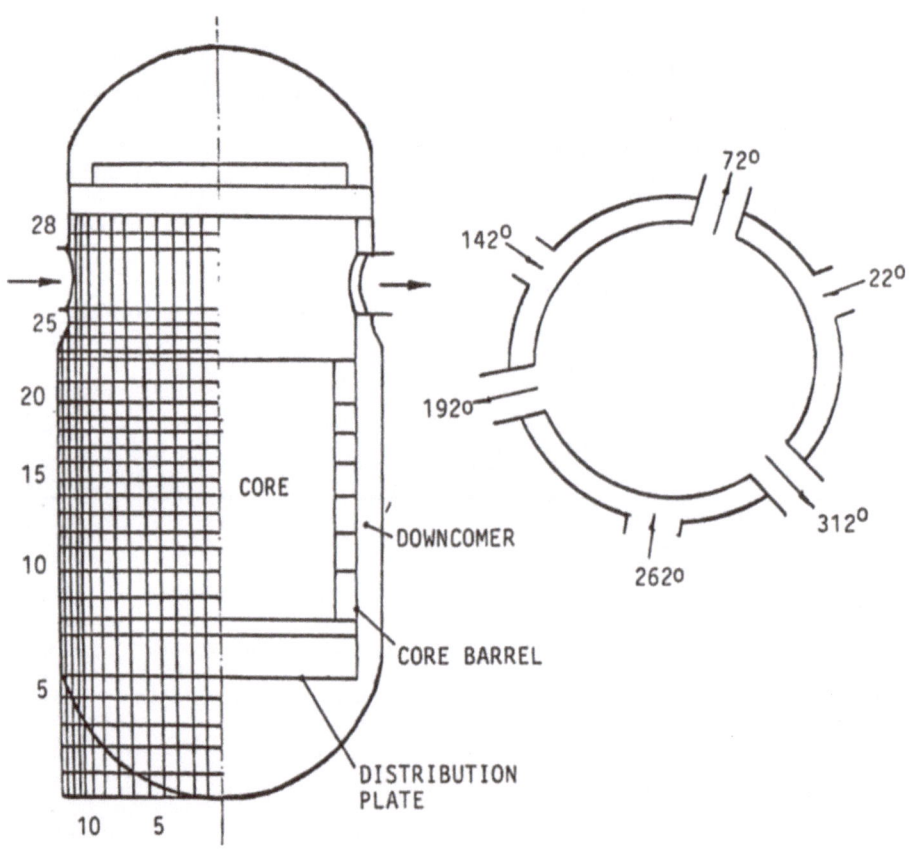

Figure 2. Grid distribution and main features of reactor-vessel model

THE MATHEMATICAL MODEL

The Differential Equations

The independent variables around the loops were the distance (x) and the time (t). In the reactor these variables were the three spatial degrees of freedom (x, y and z) and the time (t).

The dependent variables were: the steam and water velocities U1 and U2 in the loops, and U1, U2, V1, V2, W1 and W2 in the reactor; steam and water volume-fractions R1 and R2; the pressure, P; and steam and water enthalpies, h_1 and h_2. The governing equations for all the above dependent variables can be expressed in the following standard form:

$$\frac{\partial(\rho_i R_i \phi)}{\partial t} + \text{div} (\rho_i R_i \vec{U_i} \phi + \Gamma_\phi \, \text{grad}\phi) = S_\phi \qquad (1);$$

where ϕ and S_ϕ stand for the dependent variable and the source term, respectively; and subscript i refers to the phase in question.

For the dependent variables R_1 and R_2, $\phi=1$ and equations (1) represent the phase continuity equations. The source terms S_ϕ include the pressure-gradient terms and the interphase-friction terms for the momentum equations; the rate of interphase heat-transfer for the enthalpy equations; and the rate of interphase mass transfer for the continuity equations.

The Reactor Vessel Model

The reactor geometry

Figure 2 depicts the principal features of the reactor vessel geometry. Some minor simplifications were made to assist in the modelling of the vessel. The major geometric features can be summarised as follows:

*　　　The full 360° of the vessel were simulated, requiring the incorporation of a cyclic boundary condition;

* Features such as the shape of the vessel, the core barrel and the hot leg outlet pipes, were modelled by use of cell volume and area blockages;

* Internal features such as the reactor core, the core-support structures, the flow distribution plate and the above-core structure were modelled using partial porosities.

The reactor boundary conditions

The boundary condition specification for the reactor vessel model was as follows:

Cold leg inlets: The inlets from the cold legs of the loops to the reactor downcomer were considered as 'prescribed flux' boundaries, with appropriate values taken as the outlet values from the corresponding circuit analysis. The updating and interchange of these inlet boundary values is described elsewhere in this paper.

Hot leg outlets: The outlets from the reactor to the loops were situated in the vessel wall, and connected to the upper core region through pipes spanning between the vessel wall and the core-barrel wall. For the initial steady-state analysis, a 'prescribed external pressure' type of boundary condition was imposed, with the external value obtained from the appropriate loop analysis. For the transient simulation, prescribed flux conditions were imposed.

Special source terms

The following special sources were used in the modelling of the reactor vessel.

Heat-transfer from core to fluid: The heat-transfer from the core to the fluid was represented as a uniform heat source per unit volume over the region occupied by the fuel rods.

Pressure drops for internal baffles: The friction force, F_i, acting on each phase, due to the presence of an internal baffle, was given by:

$$F_i = 0.5 \, \rho_i \, R_i \, f \, U^2_i \, A \qquad\qquad (2);$$

where: ρ_i is density of phase i; R_i is volume fraction of that phase; U_i is velocity of that phase; A is flow area through the baffle; and f the friction factor calculated as in [10]. The following devices were included in the simulation:

* The flow distribution plate;

* Top and bottom formers;

* Internal formers;

* Fuel element supports; and,

* The grid plate.

Core bundle friction: Sink terms were inserted into the momentum equations in the region occupied by the core bundle, to account for the irrecoverable momentum losses arising from effects at the tube surfaces. For simplicity, standard pressure-drop correlations e.g. those cited in [11] were employed for approximating axial-flow and crossflow pressure drops; these were scaled appropriately where the flow was locally oblique to the bundle axes.

For crossflow, the pressure-drop in the in-line (square-pitch) bundle was estimated from the correlation of Gunter and Shaw, namely:

$$-\Delta p = 0.5 \, f_{gs} \, R_i \, \rho_i \, V_{mi}{}^2 \, (\frac{d_v}{S_t})^{0.4} \, \frac{\Delta s}{d_v} \qquad\qquad (3);$$

where Δs is the cell-length in the direction of velocity V_{mi}; V_{mi} is the velocity based on the minimum free area in the bundle; d_v the volumetric hydraulic diameter; and S_t the transverse pitch of the bundle. The friction factor f_{gs} was a function of Reynold's Number.

Axial friction was calculated from the formula:

$$F_i = 0.5 \ f_z \ R_i \ \frac{L}{d} \ \rho_i \ V^2 \ A \qquad\qquad (4);$$

where L is the length in axial direction; d the hydraulic diameter of the tubes, A the free cross-sectional area. f_z was estimated in this instance from the Reynold's analogy formula, ie. $f_z = 0.184 \ Re^{-0.2}$

The Circuit Model

The circuit geometry

All three loops of the primary circuit were modelled; the first loop containing the pressurizer, the second identical, except for the absence of a pressurizer, and the third representing the 'broken' loop. To model the three loops, a two-dimensional finite-difference grid was set up. Two rows of cells were used to represent the unbroken loops, and two rows of cells were used to represent the broken loop and the containment. All links between the loops (except a small area at the break between the broken loop and the containment) were disconnected by blocking the intervening cell-face areas.

Each of the loops was divided into a number of regions (as given in Table 1) which were connected by flow paths. These regions were further subdivided into a number of finite-domain control volumes for integration purposes.

The circuit boundary conditions

The following boundary conditions were specified for the modelling of the reactor circuits.

Cold leg outlets: The cold leg outlets were connected to the reactor vessel downcomer and provided mass flow and property values for the reactor vessel.

Hot leg inlets: The inlet flow rates and properties were updated from the reactor outlet values. The updating of these inlet conditions is explained later in the paper.

Special source terms

The following special sources were incorporated into the equations at appropriate locations in the loop models.

Expansion and contraction pressure losses: Pressure losses due to expansion and contraction losses were calculated as in Equation (2), with friction factors calculated as follows:

for expansion:

$$f = (1 - \text{flowarea/area downstream})^2 \qquad (5);$$

and for contraction:

$$f = 0.45 \, (1 - \text{flowarea/area upstream}) \qquad (6).$$

Primary pumps: The primary pumps were modelled as simple momentum sources, which ensured the correct steady-state pressure-drop with the specified steady-state mass flows.

Steam generator heat-transfer: Heat removal from the water by the steam-generators was simulated by considering these components as simple heat sinks.

Pressuriser: The pressuriser was modelled numerically as a fixed pressure boundary (during the steady-state simulation only).

Modelling of break

The break was initiated between the Reactor-Pressure-Vessel and the Steam-Generator, in loop 3. The break was simulated by opening the area between loop 3 and the containment over two cells. The break opening time was taken as 15 msecs.

Coupling of Reactor Vessel and Loops

Coupling of the reactor vessel and the loops required careful attention, to ensure no numerical discontinuties occurred. The method adopted was to retain a cell in each link as a cell common to both the vessel and the loop. Because the vessel model used a polar coordinate grid system, it was necessary to adjust the dimensions of the common cells in the vessel using volume and area porosities, to ensure correct matching of the links. The advantage of this method of transferring only mass flows, momentum and properties, was that the pressures in the common cells were calculated independently, thereby giving a measure of the convergence by monitoring the degree of matching between pressures.

Because of the implicit nature of the solution scheme in PHOENICS, it was necessary to iterate between the reactor model and the circuit model a number of times during a time-step. The time-step length was dictated by the structural program, ABAQUS; consequently the number of sweeps required for convergence of each model at a given step varied.

Auxiliary Relations

Physical property data

Steam and water property data required as inputs to the program were evaluated from a property package [12]. The property values required were the vapour and liquid densities and saturation enthalpies. Saturation properties were functions of pressure only; phase densities were functions of pressure and enthalpy.

Interphase friction

The interphase friction force per unit slip-velocity, F_{ip}, was evaluated from the relation:

$$F_{lp} = C_{f,ip} \ R_g R_\ell \rho_g \times \text{cell volume} \qquad (7);$$

The parameter $C_{f,ip}$ was specified as a constant, whose value was obtained from comparisons between model predictions and experimental data for similar process conditions.

Interphase heat and mass transfer

Interphase heat-transfer rates were evaluated from the relations:

$$\dot{q}_{\ell i} = C_{\ell i} \ R_\ell \ (1-R_\ell)(h_\ell - h_{\ell,sat}) \times \text{cell volume} \qquad (8);$$

$$\dot{q}_{g i} = C_{g i} \ R_\ell \ (1-R_\ell)(h_g - h_{g,sat}) \times \text{cell volume} \qquad (9);$$

where $q_{\ell i}$ and $q_{g i}$ are the total rates of heat transfer from the liquid and from the gas to the interface between the phases.

For the blowdown transient, the coefficients $C_{g i}$ and $C_{\ell i}$ are assigned constant values. These were calculated from comparisons with experimental results for similar flow situations.

The rate of interphase mass transfer was evaluated from a heat balance at the vapour-liquid interface. This gave the following expression for the mass transfer from liquid to vapour:

$$m_{\ell g} = \frac{a_g(h_g - h_{s,g}) + a_\ell (h_\ell - h_{s,\ell})}{(h_{s,g} - h_{s,\ell})} \qquad (10);$$

where $a_g = C_{g i} \ R_\ell \ R_g \times \text{cell volume};$

$a_\ell = c_{\ell i} \ R_\ell \ R_g \times \text{cell volume};$

and the interphase-heat-transfer coefficients $C_{g i}$, $C_{\ell i}$ were as defined above.

Structural Analysis Coupling

As mentioned above, the principal aim of the study was to determine the affects of pressure on the reactor vessel structure, eg. bending of fuel pins, damage to core barrel, fracture of bolts etc. In order to analyse the structural effects of the transient, it was necessary to compute in parallel the stresses and strains using a finite-element structural code, in this case ABAQUS. Two reasons made it necessary to run in parallel: firstly, the amount of pressure data that would be required to model the transient would create extreme difficulties with data handling; and, secondly, the structural analysis would be non-linear and it would therefore be necessary to use a variable time-step procedure. A description of the structural modelling of the vessel and its internals is however beyond the scope of this paper.

When running in fully-coupled mode, the timestep prescription was automatically controlled by ABAQUS. For a given time interval, PHOENICS would perform a thermohydraulic analysis to furnish the pressure field values, which, converted to a suitable form, were passed to ABAQUS. ABAQUS would perform the stress analysis, and in turn provide PHOENICS with data on the deformation of the structure. This information could be incorporated into the subsequent PHOENICS calculations, and so on. In the present study this last step was omitted, but could be activated for future studies of interation effects.

PRESENTATION AND DISCUSSION OF THE RESULTS

Computations Performed

The computations were divided into two distinct stages. The first stage was to perform a steady-state calculation to provide the initial conditions for the transient. The second stage was to simulate 50 msecs of the blowdown transient.

The steady-state field values were obtained by means of temporary alterations to the boundary links between the one-dimensional circuit calculations and the three-dimensional vessel calculations. The boundary-condition modifications involved prescribing the mass flows at the inlet to all three circuits and the reactor. These restrictions were removed for the transient calculations.

The transient calculation was initiated by opening up the areas in the break cells, between the circuit and the containment, over a period of 15 msecs. The transient simulation covered a period of 50 msecs, during which it was expected that the peak pressure-differences in the reactor core would have occurred.

Results of Transient

A limited set of results is presented here to depict the pressure effects in the reactor during the blowdown transient. Briefly, the sequence of events predicted is as follows: Pressure waves are created by the rapid depressurisation from the initial pressure to the saturation pressure during an initial period of 5 msecs. The waves travel in both directions from the break, through the circuit. As the pressure wave reaches the reactor vessel, large pressure differences are produced inside the vessel, particularly across the core barrel. It was these pressure-differences inside the vessel that were transferred to the structural program, ABAQUS, for analysis. Figure 3 shows the pressures in the broken loop on either side of the break. Here the rapid pressure decay, resulting from the break, can be clearly seen.

In Figure 4 the pressures at various locations in the broken loop are presented. These show depressurisation effects from both the inlet and outlet of the reactor vessel. The pressure at the entrance to the pump shows an increase towards the end of the transient. This is a consequence of the interaction of the two pressure waves, combined with the boiling of the water at this point.

Pressures in the reactor downcomer at various angles around the core-barrel circumference, are given in Figure 5. As would be expected the pressure next to the inlet, Plot 7, shows a close resemblance to the pressure at the exit of the circuit (as seen in Figure 4). Also, as expected, the pressures around the barrel show a phase-shift away from the inlet.

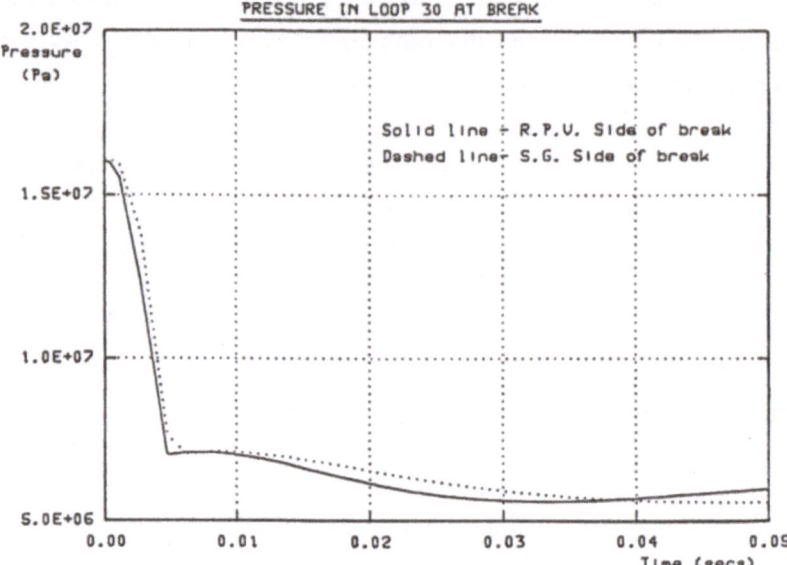

Figure 3. Pressures in loop on either side of break.

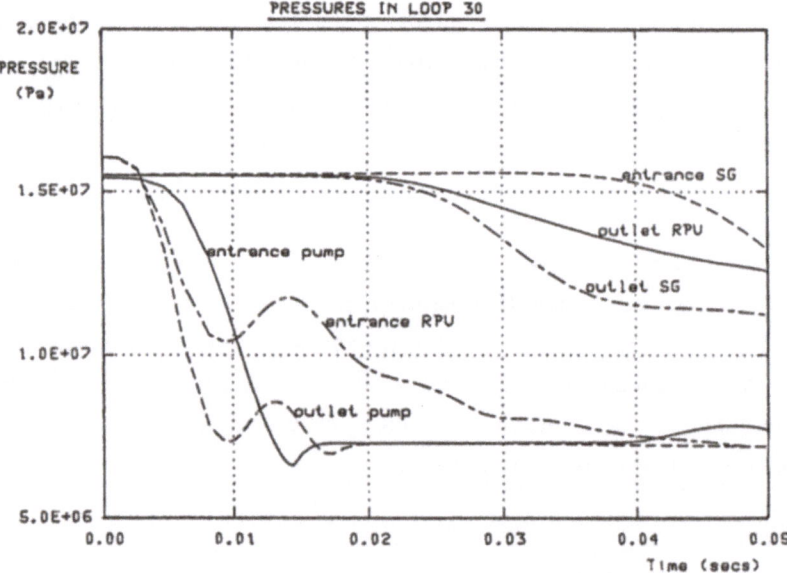

Figure 4. Pressures in broken loop at various locations.

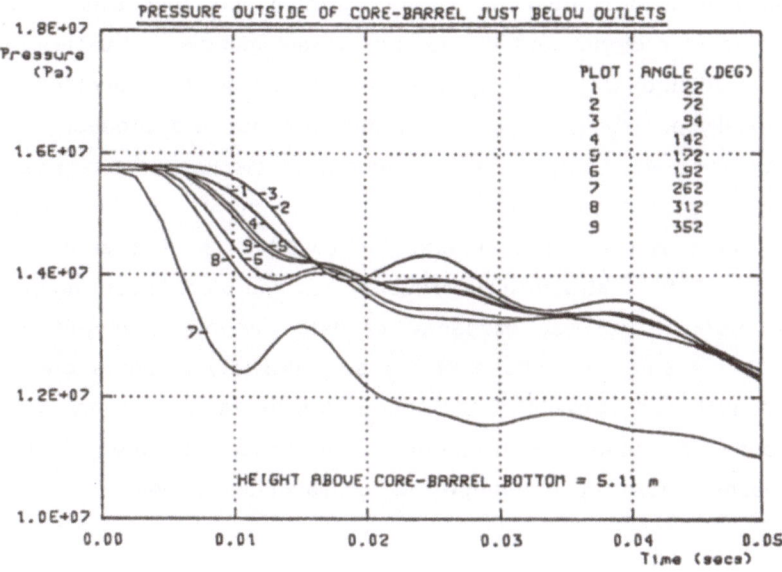

Figure 5. Pressures in reactor-vessel at different circumferential locations.

CONCLUSIONS

This paper has described the development of a theoretical model to predict transient effects in a pressurized water reactor, in particular the dynamic effects of a violent transient, such as a reactor blowdown, on the internals of the reactor vessel. This required two major features not normally available with heat transfer and fluid dynamics codes, such as PHOENICS. These are the simultaneous heat transfer and fluid dynamic modelling of the three one-dimensional reactor circuits coupled with the three-dimensional modelling of the reactor vessel, in combination with the transfer of pressure information to a structural-analysis program, in this case ABAQUS.

The limited results presented have demonstrated the successful coupling between the reactor-circuit and the reactor-vessel models. Pressure waves have been predicted to travel through the circuit to the reactor-vessel, where potentially-damaging large pressure-differences are produced. The extent of the damage, if any, to the reactor internals has been analysed by passing these reactor pressure data to the structural program ABAQUS, which was being run in parallel with PHOENICS. A first step towards representation of fluid structure interations has therefore been made with the present work. Two-way exchange of data was not employed in this case, due to the small magnitude of the structural deformations predicted, but the potential and capability have been demonstrated. There remains however scope for future improvement in the linkage between fluids and structural codes before the technique can be routinely applied.

ACKNOWLEDGEMENTS

The authors gratefully acknowledge the technical assistance of: Mr. H. Moldenhauer, who constructed and tested the structural model; the staff of Control Data Gmbh; and of TUEV Stuttgart e. V., who also sponsored the project.

REFERENCES

1. REEDER D. L. 'LOFT System and test Description (5.5 ft. Nuclear Core 1 LOCEs)' NUREG/CR-0247, TREE-1208, July 1978.

2. PATTON M. L. 'Semiscale MOD-3 Test Program and System Descriptions', TREE-NUREG-1212, July 19778.

3. RELAP4/MOD6. 'A Computer Program for Transient Thermal-Hydraulic Analysis of Nuclear Reactors and Related Systems'. EG. and G., Idaho, Inc., January 1978.

4. 'TRAC-P1: An Advanced Best-Estimate Computer Program for PWR LOCA Analysis. Vol I: Methods, Models, User Information, and Programming Details, Los Alamos Scientific Laboratory Report LA-7279-MS. Vol I (NUREG/CR0063), June 1978.

5. MARKATOS N. C., RAWNSLEY S. M., and TATCHELL D. G. 'Analysis of a Small-Break Loss-of-Coolant Accident in a Pressurized Water Reactor', I Mech E Conference Publications 1983-4, Heat and Fluid Flow in Nuclear and Process Plant Safety, Paper C103/83, pp. 121-134 (1983).

6. MARKATOS N. C., RAWNSLEY S. M., and SPALDING D. B. 'Heat Transfer During a Small-Break Loss-of-Coolant Accident in a Pressurized Water Reactor-A Parametric Study for a 4 in. Lower-Plenum Break.' Int. J. Heat Mass Transfer Vol. 27, No. 8, pp. 1379-1394, (1984).

7. KIRKCALDY D., PHELPS P. J., and VAN ESSEN D. 'PHOENICS Code Thermal Hydraulic Analysis of the SNR-300 IHX' ASME Winter annual meeting 1985, HTD-Vol. 51 pp. 9-16.

8. MES H., VAN ESSEN D., KIRKCALDY D., and PHELPS P. J. 'PHOENICS Code Thermal Hydraulic Analysis of a Prototype LMFBR Straight Tube Steam Generator'. Presented at the ASME Winter Annual Meeting in Anaheim California. Dec. 1986.

9. HIBBITT H. D. 'A General Purpose Finite Element Code With Emphasis on Nonlinear Applications', Nuclear Engineering Design 77 pp. 271-297 (1984), North-Holland, Amsterdam.

10. IDEL'CHIK I. E. 'Handbook of Hydraulic Resistances.' AEC Translations, 6630, USAEC 1960.

11. KAYS W M and LONDON A L 'Compact Heat Exhangers', Mc Graw Hill, New York 1958 (2nd. ed. 1964).

12. AGEE L. 'Functional Fits of Steam Table Data'. Private Communication 1976.

ENEA Nuclear Power Plant Engineering Simulator: Mathematical Models, Perfomances, Opportunities of Usage

G. M. Mancini, A. Mattucci, ENEA, Italy

1. Introduction

Enea, the Italian Commission for Nuclear and Alternative Energy Sources, has acquired an Engineering Tool.for Design and Analysis of Pwr Nuclear Power Plants: an Engineering Simulator, which reproduces in real time the behaviour of the whole plant, presenting all informations available to the operator in Control Room in a compact way.
The Simulator reproduces a specific Pwr plant, but will be slightly modified to become the Pun Simulator, where Pun means italian unified project for pressurized nuclear power plants.
The need of Simulators for design and verification purposes has increased in the last few years, accordingly to the new impressive capacity and speed of digital computers, in order to enhance all possible preventive safety defenses in the Nuclear field. The main subject remains evidently operator training, since human factor is considered of vital importance in plant operation. There are, though, some aspects that have had, in the past, no chance to be treated adequately on Simulators, like the possibility of testing new hardware for Protection and Control Systems, or the evaluation of operator errors during recovery from severe accidents.

In order to be able to accomplish those tasks, it is necessary to build a Simulator which has, like a full scope, real time and completness, bud adds precision and reliability in results and an efficient way of presenting and recording data for further and detailed analysis.
With this target an Engineering Simulator has been designed, realized and tested by Enea and Westinghouse.
Here we show a general view of adopted mathematical models, a synthesis of most significant verification done and a prospect of activities already active and on schedule.

2. Mathematical Modles

Basically, mathematical models are a crucial point for application that
involve nuclear safety; in the past it was not possible to reproduce real
time dynamic behauiour of a Nuclear Power Plant with enough accuracy, so
that it was not possible to evaluate complex phenomena, like loss of
primary or secondary coolant.

The purpose of the advanced models of the Enea Simulator, is to be able
to take in account such events.

Here we will present major hypotheses done and the consequent structure
of the most critical models:

Reactor Coolant System and Steam Generator.

In particular are presented mass, energy, volume and mementum equations;
an accent will be posed on aspects like mass correction and flow equation
solution.

Fundamental hypotheses and solution methodology are also reported.

2.1 RCS Model

2.1.1 Definitions

Reactor Coolant System includes vessel, loops, Steam Generator U tubes,
Pumps and Pressurizer.

Elementary components of the model are:

- internal node, fixed volume in which mass and energy balances are performed for both liquid water and steam;
- external or boundary node, which contains only informations on pressure
  and enthalpy and is connected to internal nodes by critical links;
- non critical links, where a mementum equation is associated, that
  connects only internal nodes;
- critical links, in which flow is a function of difference of pressure
  between internal and boundary node.

A nodal representation of RCS is reported in fig. 1. It includes
16 internal nodes, connected ty 19 non critical links.

2.1.2 Major hypotheses

Internal nodes allow distiction between liquid and vapor, which are not
forced to be in equilibrium. Energy and mass balances are executed
separately for each phase.

Link flows are divided into liquid water and dry steam, by using a drift flux model; futhermore, all liquid goes into liquid zone in the next node, all steam is divided proportionally to the existing masses in the arrival node.

One of the fundamental hypotheses is that defined as global compressibility, which is meaning: Density variations, during a transient, due to the differences between global pressure and local pressure, are small enough in order to evaluate the termodynamic condition of coolant.

It could be shown that this is equivalent to assume sound speed infinite. Volume balance is used to evaluate global pressure.

Another approximation done is that regarding the use of "hydraulic" specific volumes other than "termodynamic", for flow computing.

Termodynamic specific volumes are those obtained from steam tables, Hydraulic are the hypotetical specific volumes that the coolant should have in order to respect volume nodal balance, which allows volume conservation for each model.

2.1.3 Mathematical methods

The need to operate all the simulation in real time forced to use a fixed time step of .25 seconds in the RCS model, which means a frequency of four executions per second, the highest in the periodical programs. Another important decision to make was the choice of the execution order for the subroutines which realize RCS model.

First of all masses are computed, later enthalpies, then global pressure. At this point a mass correction is performed, due to the fact that at a given pressure and enthalpy the already computed masses may be do not respect nodal voulume constraint.

This is done to avoid iteractions, which are time consuming. The mass correction is used in the flow computation, which is solved on the basis of an overall network equation. Local pressures are then derived as a function of global pressure and head losses. Mass balances are explicitely integrated. The order of solution of entahlpy equations is due to a Gauss method applied to the matrix of the system, solving first steam enthalpies. The complete computing flows is as follows:

- internal nodes mass solution;
- energy balance and enthalpy evaluation;
- global pressure determination;
- nodal interface processes computation (evap./condes.);

- nodal levels;
- non critical links flows and local pressures;
- boric acid concentration;
- nitrogen 16 concentration.

2.2 Steam Generator Model

2.2.1 Definitions

In this case there are only internal nodes, in which are executed mass
and energy balances.

The adopted simulation scheme is shown in fig. 2.

There is to point out that nodes 1 and 8 have, as a constraint, the sum
constant, but they are variable nodes. This is done because they are
monophase node, respectively liquid and steam.

Despite of the appearance, there are no links, because the flow is computing using an overall momentan equation and it is adjusted between
nodes taking in account mass variations.

2.2.2 Major hypotheses

All nodes are considered to be in termodynamic equilibrium, for water
property evaluations and heat exchange factors.

In the six two-phase nodes there is no phase separation. Heat exchange
versus primary side is considered in five nodes, corresponding to U-Tubes.
Since primary side has only two nodal temperatures available, to the
extent of heat exchange the temperature behaviour has been linearized
along R-tubes. Metal temperature is taken in account, having three radial
temperatures in the tube.

For the fluid-metal heat exchange coefficient are made other hypotheses:

- It is taken as a constant at very low flow;
- in monophase conditions depends only on the flow using a Dittus-Boelter
  correlation;

Under nucleate boiling, Thom's correlation is used:

- when film boiling is present, heat exchange coefficient is taken linear
  in function of void fraction;
- it is taken as a constant when in monophase steam.

To identify the status, Thom correlation is adopted.

Pressure in the Steam Generator is considered to be global.

Contribution to pressure changes are taken from:

- volume balance;
- energy balance.

Flows are computed from an overall momentum balance.

2.2.3 Mathematical methods.

Since time step of 0.25 sec. is large for rapid transients, some limitations have been made to avoid numerical instability:

- evaporation rate is limited to 10% of nodal mass in each time step;
- an independent path is used to solve only liquid situations;
- a nodal mass correction is performed;
- pressure variations are limited by a factor less than 1, that is
  equivalent to a lag, and cannot be more than 10% of previous pressure
  value;

The routine organization is as follows:

- primary-secondary heat exchange;
- evaporation/condensation rates;
- nodal masses update;
- node enthalpy
- overall pressure;
- steam flows to turbine, PORV, Safety valves, breaks (if existing);
- nodal mass correction;
- level calculation;
- internal flows.

3. Simulator Performances

3.1 General aspects

High Quality of an Engineering Simulator, can be represented by:

- capability of the hardware to perform real time operation of the
  simulated plant;
- fidelity of simulated plant configuration;
- component operability;
- well known range of reliability on results in every possible configuration;
- repeatability;
- good documentation.

All these aspects have been taken into account in the project, and have
required verification and validation tests.

Complete procedure for those test have been written, including expected
results, by experts in plant operation and safety analysis. Hardware was
tested preliminarly and it is not relevant to the purpose of this
article.

3.2 Verification

Mainly verification tests regard fidelity of configuration and component operability, observing en passant real time capability. Every display representing plant system or subsystem was checked versus Process & Instrumentation Diagrams, every operable component was tested on the display, documenting by a full report any discrepancy from expected or known condition. This precedure ended to a corrective action and a new test, iterationg until a satisfactory condition was obtained. In this phase about four hundred of such report were made and then successfully cleared.

Furthermore, any possible configuration was tesed.

This, for evident reason, resulted a preliminary job, in order to be more confident to be able of executing validation tests.

3.3 Validation

There is a significant difference of importance that ditinguisher this step from the previous. In fact the main purposes of validation are:
- excitation of all possible states of the models;
- comparison of Simulator with design data, experimental data, plant data, results coming from other codes.

It becomes evident that is impossible to have available reference data for every reachable condition. As a consequence, engineer judgement supplies in many cases the lack of information.

In the case of ENEA Engineering Simulator we have performed a set of tests, covering in particular:

- numerical drift tests, consisting in a one hour run in two different plant conditions, recording all principal parameters and verifying they do not drift by more than one percent;
- accuracy tests, for five different levels of power, where principal parameter have been compared with plant design data, allowing deviations of no more than oen percent within oen hour test;
- plant operation tests, including:
  . Hot Functional tests, or tests executed prior core loading, like pressurizer relief tank and CVCS tests;
  . Startup tests, which are relative to control systhem stability and plant manoevring verification;
  . Normal Operations which explore all the possible normal plant status, going from nominal condition to refueling condition and viceversa;
- analysis of malfuctions consequences, for a total ot 120 failures in various plant systems;
- Special Transients tests, analysis of plant conditions beyond DBA:
  . Prz steam space leak (3");

. Steam line break, upstream isolation valves;
. Steam Generator tube rupture;
. Steam Generator tube rupture with overfilling;
. Small LOCA in Cold Leg;
. Loss of all Feedwater;
. Steam line break inside Containment;
. Station Blackout;
. Steam Generator tube rupture;
. ATWS following feedwater pump trip.

During all these transients, operator actions followed operating and or emergency plant procedures, to ensure that all the required action could be performed on the Simulator.

The execution required the same configuration of the mathematical models, to avoid that any modification could affect already tested systems.

The execution and analysis of those tests required three man-year work. Some of the Special Transients by using RELAP code were performed, too, results of comparison showed good agreement (see fig. 3).

A complete report with full analysis of results has been made, whose conclusions were very positive, leading to release of the Simulator for all the planned activities.

4. Opportunities of usage

Many are the activities the Simulator can be involved with.

Actually it has be planned to be employed for:

- plant procedures design and verification, for normal and emergency conditions;
- supporting safety analysis, particulary to take in account operator mistakes during normal and abnormal plant conditions,
- test of new hardware for the plant, like Protection and Control related system;
- man-machine interface study.

The contribution of an Engineering Simulator to plant procedure evaluation has been made possible by the completness of representation and by the advanced capability of models. In fact it will be remarkable to study the operator actions during extremely serious accident, like large loss of primary or secondary coolant, still having good responses in terms of fidelity and accuracy of results.

The function of supporting safety analysis is to be intended as a screening of many different condition to find out which on is more limited for a particular plant parameter. Then the analysis has to be refined on a restricted number of transients, using more precise tools and adopting typical conservative hypotheses.

Very interesting and useful can be the use of the Simulator for testing
new hardware, particularly for Control and Protection System C & P, so
the feedback can be tested completely and setpoints, time constants,
gains can be adjusted.

In the case of Enea Simulator, three steps have been planned:

- build a model of C & P system, with a structure similar to the real
  system, particularly with the same input/output;
- test the model alone and integrated with all other models in the
  Simulator;
- design an interface and connect the prototipe to the Simulator,
  replacing the model.

The Engineering Simulator shows all the plant status in
a compact way (figg. 4,5), and operator has to know where to find the
needed information, so it appears more useful to study how a single
information can be made more meaningful, other than for global considerations. Anyway it can be used to enhance the quality of information,
adding special displays on the plant status, based on event trees, or
giving informations on variables not measureable, like enthalpies, void
fractions, local levels etc.

Conclusions

In this paper has been shown an Engineering Simulator for nuclear power
plants of the kind Pwr (Pressurized Water Reactors), giving informations
on the principal mathematical models, on the qualification tests that
the Simulator has passed and on the activities actually performed and
planned for the future.

This system has been developed by Westinghouse and Enea and is now of
Enea property; the entire process required more than 20 man year work.

References

A. Mattucci, O. Modonesi
Progetto del Simulatore avanzato di Ingegneria di impianto Pwr
Westinghouse
NWEE 2TP4B 86062   ENEA-NIRA

G. M. Mancini
Prove die accettazione e validazione
NWEE 2TPAG 85026   ENEA-NIRA

D´Apice, Lombardi, Mancini
Prove di accettazione relative ai malfunzionamenti ed eventi
incidentali speciali
NWEE 2TP4B 86054   ENEA-NIRA

D´Apice, Mancini, Mattucci   et al.
Installazione e prove in sito del Simulatore die Ingegneria
NWEE 2TP4B 87015   ENEA-NIRA

Mattucci, Negrenti
Studies of severe transients in Nuclear power plants by using
the ENEA Engineering Simulator.

G. M. Mancini et al.
I modelli avanzati del Simulatore d´ingegneria d´impianto Pwr.

A. Mattucci
Plano di Qualificazione per la personalizzazione del Simulatore die
ingegneria all´impianto Pun.
NWEE 2TP4H 86046   ENEA-NIRA

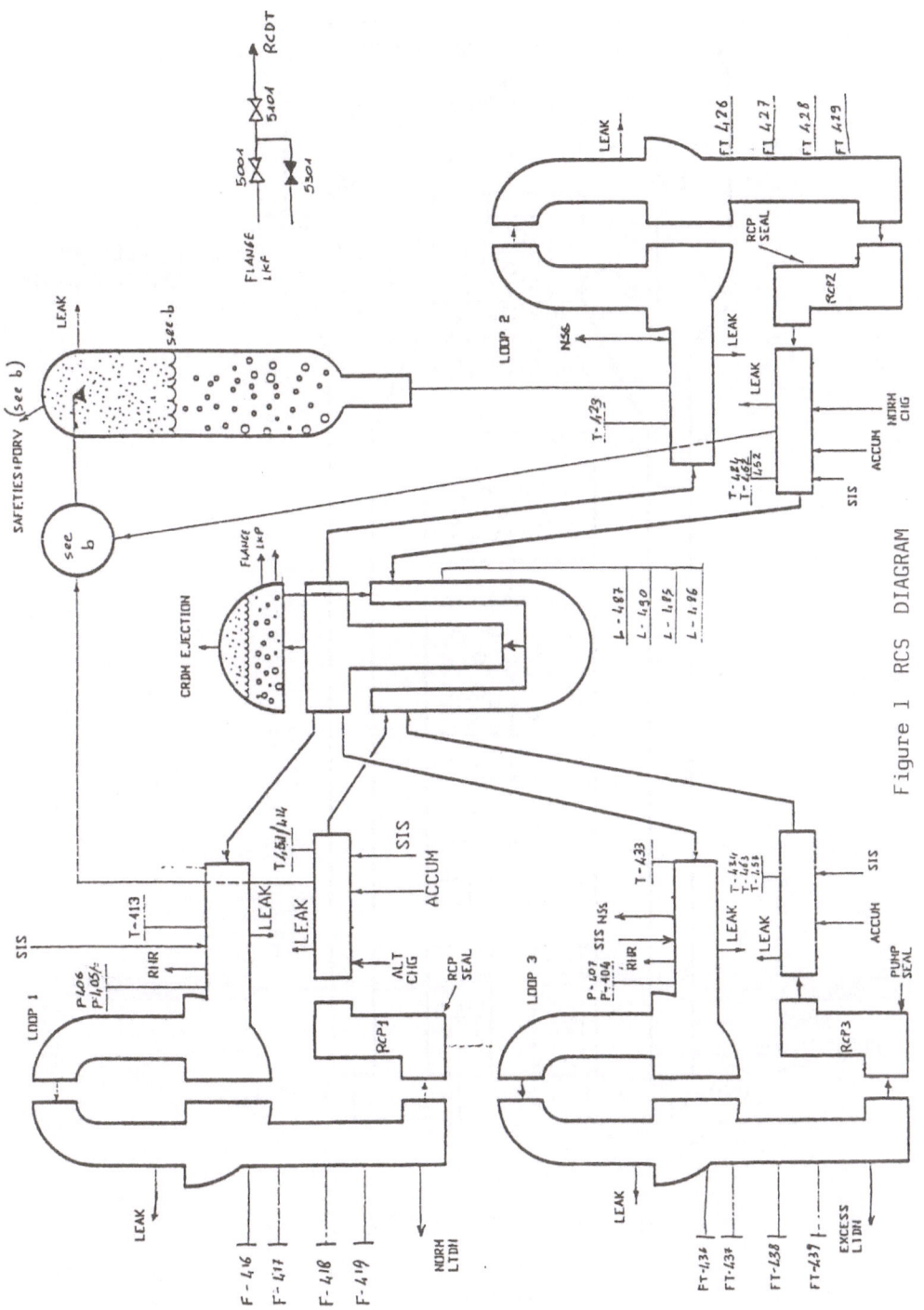

Figure 1 RCS DIAGRAM

134

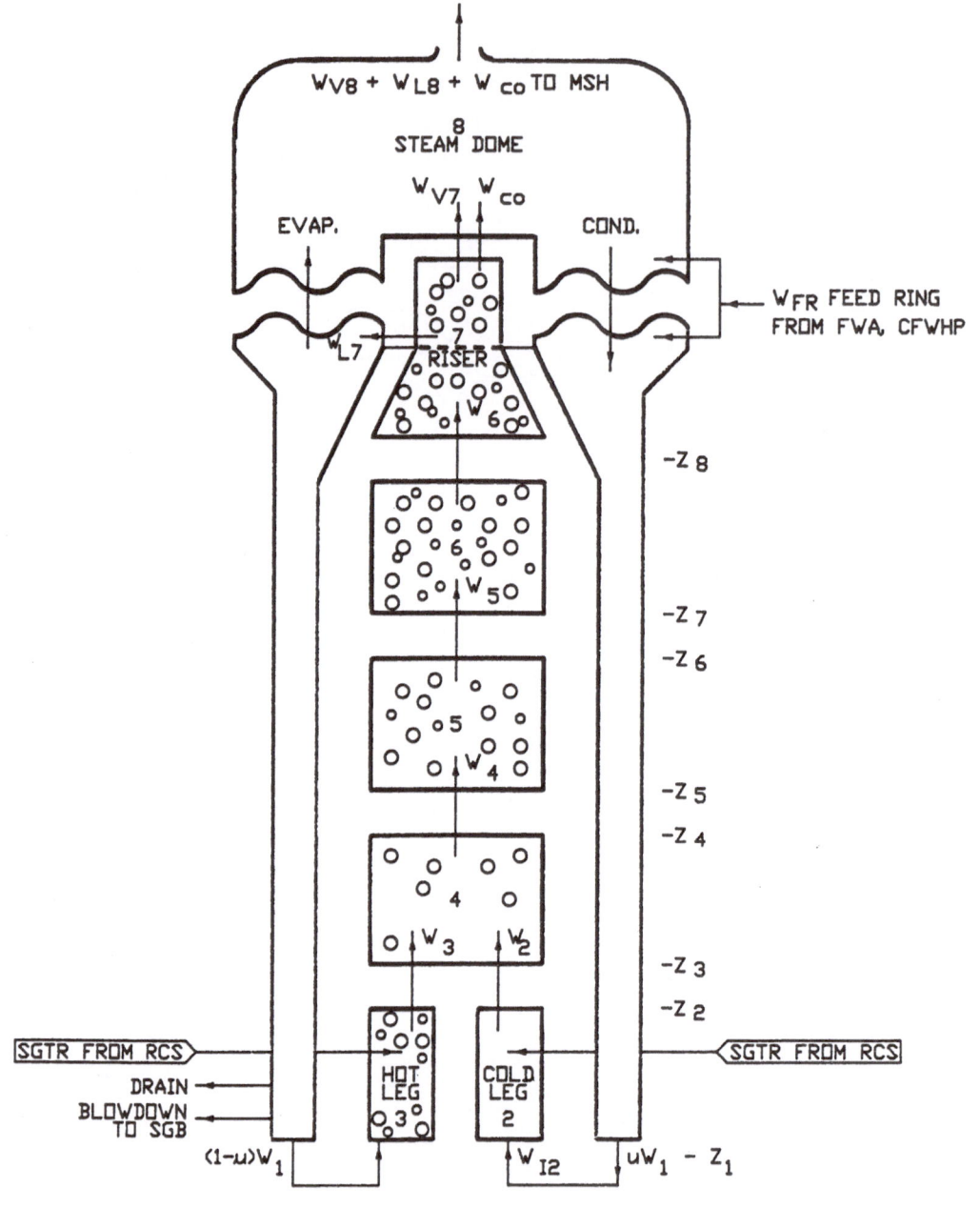

STEAM GENERATOR SCHEMATIC

Figure 2

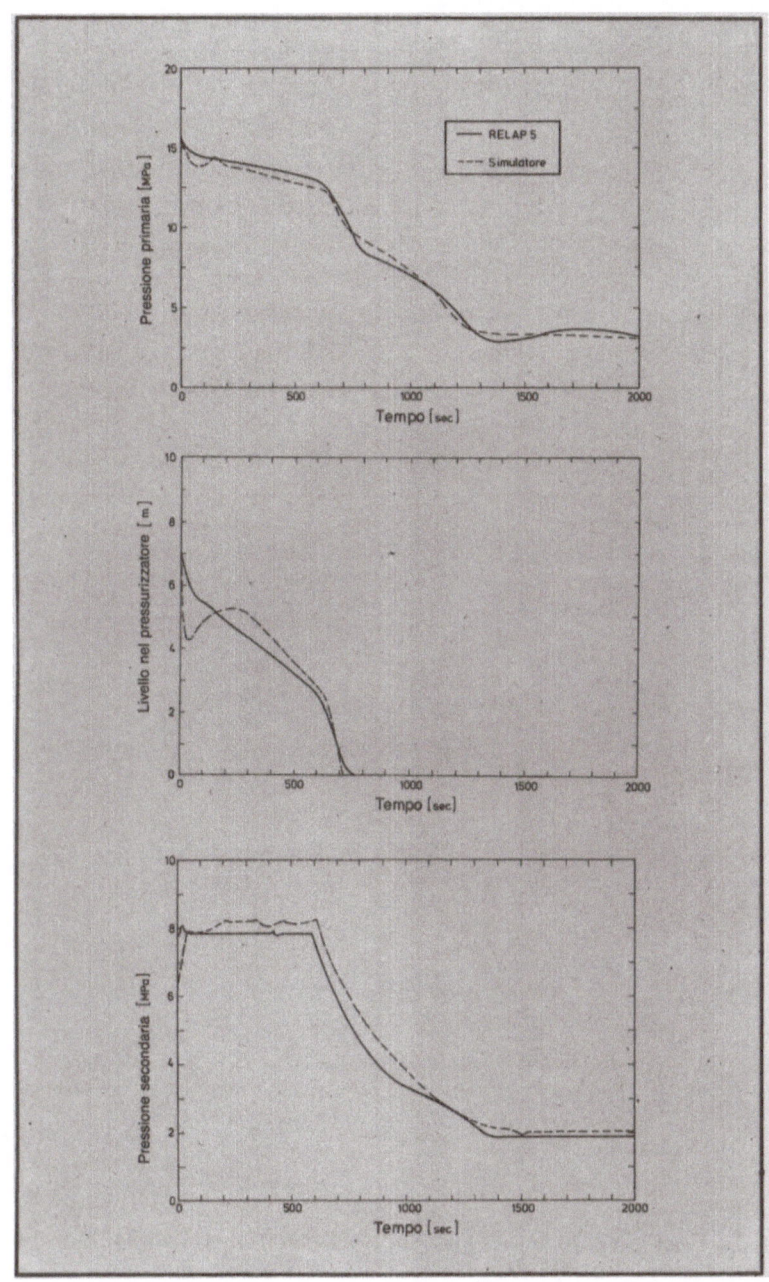

Figure 3   RELAP VS SIMULATOR

Figure 4 SIMULATOR CONSOLE

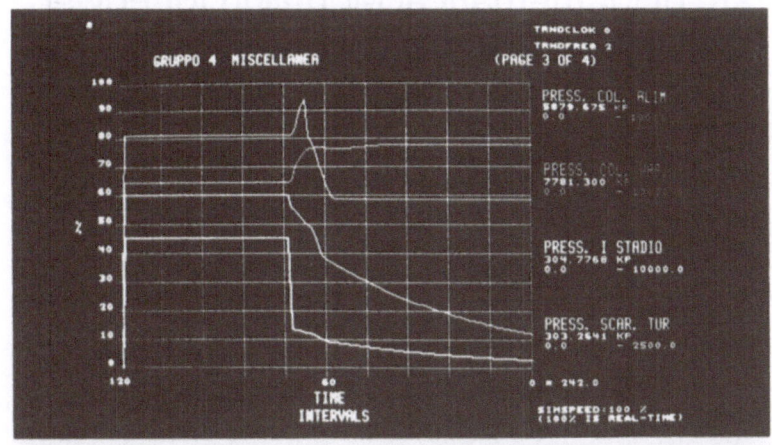

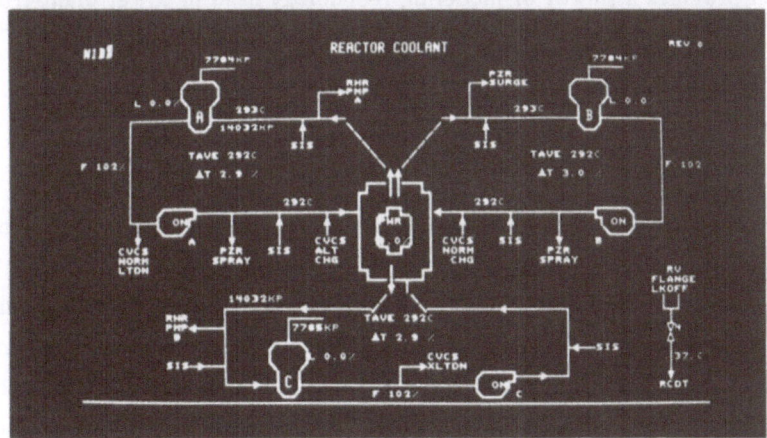

Figure 5   SIMULATOR DISPLAYS

The State of Mathematical Modeling for Power Plant Training Simulators

L. R. FOULKE

Nuclear Services Integration Division
Westinghouse Electric Corporation
Pittsburgh, Pennsylvania

Summary

This paper provides a survey of modeling technology used in power plant training simulators, identifying the capability of existing models, and commenting on the areas where modeling advances are expected to occur.

INTRODUCTION

One may consider two classes of simulators: those which are used for training and those which are used for engineering analysis. The distinction is somewhat blurred, however, as many of the mathematical models used in training simulators are sufficiently accurate for use in engineering analysis. In fact, some nuclear training simulator models have been driven and adapted from the analysis codes such as RELAP (28), RETRAN (26), and the Nuclear Plant Analyzer of Wulff (36,37).

The question that motivates this paper is "How good do training simulators need to be?" In a few more years, computer capabilities may outrun the needs for training simulator fidelity. In the future, will we be running engineering codes in real time for training purposes, or will today's training simulator models be sufficient?

Training simulators being delivered today are very good. The modeling and mathematics which have been developed for simulators are currently able to deal with training scenarios that could not be handled only a few years ago. Methods which have been developed and validated over the past six years provide stable, high fidelity representations using a fixed and relatively long integration time step.

However, training simulators still cannot be used to represent abnormal situations with confidence prior to validation and benchmarking. Microscopic phenomena associated with exchanges of mass, energy, and momentum in multiphase, multicomponent systems with complex flow patterns, which are hard to address from first principles, are neglected as long as the trends of observable, macroscopic variables are plausible. In many cases, training simulator requirements still assume that if the operator cannot observe a variable or phenomenon on the instrumentation of the main control board, it need not be simulated with high fidelity. Unfortunately, there is no guarantee that the macroscopic phenomena will be correct in all circumstances if the microscopic phenomena are neglected.

To study the issue of how good training simulators need to be, it is necessary to understand that the important phenomena for operator training are dynamic in nature:

- How fast do things change?

- How big are the maxima or minima, and when do they occur?

- Is the first derivative positive or negative, and how much control can the operator exert to change the first derivative?

The standard simulator acceptance criteria, ANSI/ANS-3.5 (1) provides a basis for scope and accuracy requirements of training simulators. However, the transient qualification criteria of the Standard are subjective, and the steady-state qualification criteria are not closely related to training objectives. The EPRI report on simulator qualification (2) presents a methodology for simulator development needs based on a training perspective.

More testing of simulator results against benchmark results, in line with the EPRI report, is required in order to assess the dynamic phenomena listed above. Until such testing is forthcoming, we must continue to emphasize the following considerations which are important for high fidelity transient behavior:

- Conservation equations which correctly account for all terms that govern rate processes (e.g., energy balances which account for the condensation of vapor on subcooled surfaces)

- Correct values of rate process parameters (e.g., cross sections and radioactive decay constants)

- Stable numerical integration schemes which maximize simultaneity of solution

- High resolution of variable definition along all dimensions which show sensitivity to training cues or stimuli (e.g., high resolution of axial phenomena in reactor cores) or sensitivity to perturbations (e.g., two-phase fluid system connections to the reactor coolant system)

SIMULATION MODELS

A typical simulator for a commercial nuclear power plant with either a pressurized
water reactor (PWR) or boiling water reactor (BWR) consists of approximately 60 to
70 separately modeled systems. These systems may be grouped into categories of
systems which are similar in terms of modeling techniques or in terms of technology.
These categories of systems are as follows:

- Core Models
- Fluid Systems Models
  -- Multiphase, multicomponent, nonequilibrium systems
  -- Multiphase, equilibrium systems
  -- Single-phase systems
- Electrical Systems Models
- Instrumentation and Control Systems Models
- Material Transport Systems Models

The mathematical models of a simulator must be capable of reproducing all operational aspects of the reference plant in a plant-specific manner. The models must
generate all data and variables required by other simulation models as well as that
needed to drive control board instrumentation. The responses of all variables being
simulated must be predicted accurately by the models, regardless of the initial operating point or the magnitude and severity of the disturbances caused by operator or
instructor action.

Each plant system model interfaces with other program models and the control board
input/output systems through the main-memory-resident datapool. The datapool
variables are continually updated as each plant system model completes a time step.
Typically, the "iteration rate" of each plant system model is fixed to vary between
10 iterations per second (10PS)--a integration time step size of 0.1 second--to one
iteration per 5 seconds (0.2PS)--an integration time step size of 5 seconds.

The selection of iteration rate for each model is based on the characteristic time
constants of the model, the frequency at which the information is needed for driving
other models, and the requirement to run the simulation in real time. Real time
means that the trainee must not be able to distinguish any speedup or slowdown of
events perceived on the simulator as compared to the real plant. Real-time behavior
is assessed in terms of duty cycle. A simulator duty cycle of 110 percent would
imply that the simulator is "slipping" real time; that is, 10 seconds of plant time
would take 11 seconds to be played out on the simulator. Therefore, simulator duty

cycle must always be less than 100 percent even for the most computationally demanding transient. Simulator specifications normally require sufficient computation power to permit 30 to 50 percent spare duty cycle. This guards against slippage of real time and provides margin for future upgrading of the scope or fidelity of the mathematical models. This emphasis on real-time behavior discourages use of variable integration time steps or iterative solution techniques which may cause duty cycle to increase during certain evolutions.

Training exercises start from initialization points referred to as "initial conditions." An initial condition is a "snapshot" of all simulator variables (temperatures, flows, pressures, control rod positions, power level, xenon concentration, time in core life, and the like) for a particular condition of the plant. Many different snapshots are stored in disk-resident memory for repetitive use to establish desired starting points for training exercises.

Simulated malfunction effects are designed to realistically represent the result of defined equipment failures or other specifically identified causes. Selected remote plant equipment controls, referred to as local operator actions (LOAs), are included to permit the performance of specified remote plant operations such as valve isolations and surveillance tests. Control of these remote functions, along with malfunction insertion, initial condition selection, and other instructor-selectable parameters, is accomplished from a stand-alone instructor's station. After initiation of a training scenario by the instructor, the dynamic behavior of the plant is determined by the operators' actions and the mathematical models of the simulated systems. No precalculated or tabulated transients are used in modern power plant simulators.

The design of a full-scope, high fidelity simulator will normally include the equipment, instrumentation, and controls that enable an operator to control the reference plant. The simulator is a realistic reproduction of the actual control room equipment; there will seldom be discernible differences between the simulated control room and the reference plant control room. It should be possible to experience all modes of plant operation in the simulated control room--commencing at the completion of refueling, including core physics testing, prestartup checks, plant heatup or cooldown, shutdown, maneuvering through the power range, surveillance testing, and normal, abnormal, and emergency operating conditions. All remote activities needed to support the simulation are accomplished by the instructor using the instructor's station. The categories of system models listed above are discussed in the following paragraphs, in terms of performance expectations, fundamental issues, methods used, validation status, and anticipated future developments.

CORE MODELS

The fidelity of the reactor core model in training simulators is directly related to the computation power available. Core models of a decade ago used either point kinetics or a very coarse collection of nodes coupled together with equations of the form of the point kinetics equations. These core models were also limited in their treatment of the buildup and release of fission products. Reactor core modeling today is very good; it will continue to improve as more computation power becomes available. Today's methods for the space- and time-dependent behavior of the neutron flux can be classified as either coupled kinetics methods or expansion-based methods, as explained below. The former have often been referred to as nodal methods; the term nodal is used here to express the subdivision of a spatial solution domain for either method.

Expectations on Performance

Current expectations for performance of a training simulator core model include the representation of all core-related phenomena that an operator trainee can monitor through instrumentation, and all responses that an operator trainee can produce through control room manipulations. Recent specifications for simulator performance in the core modeling area include the following:

- Core reactivity as a function of rod positions for various core and plant states, and rod bank overlap conditions

- Effects of changes of fuel characteristics representing various times in the fuel cycle

- Effects of xenon, boron, samarium, pressure, voids, and temperatures (fuel and moderator) on net core reactivity and control rod worths throughout the core cycle

- Effects of detector locations on detector indications

- Effects of control rod motion, moderator temperature, and fuel temperature on xenon spatial behavior

- Neutron flux space-time effects generated by the neutron sources, including the effects of source strength and location and subcritical multiplication during startup and shutdown

- Effects of asymmetric flow and temperature on reactivity and ex-core nuclear instrument indications

- Effects of fuel-cladding gap on fuel temperature and fuel temperature reactivity feedback

- Thermal conductivity of fuel and cladding as a function of temperature

- Hydrogen and fission product release as a function of cladding temperature, power history, length of time the fuel is uncovered, and pressure difference between cladding gap and coolant

- Effects of core and downcomer voids on ex-core nuclear instrument indications

- Release of N-16 from the core as a function of neutron flux

Fundamental Issue

The fundamental issue in reactor core modeling for real-time simulation is the incorporation of as much detail as possible in the temporal and spatial domains while maintaining real-time behavior. Even with today's computer capability, compromises must be made in the selection of time step and spatial mesh size. The core models currently being installed on training simulators do a very good job in general, but correct representation of some phenomena requires empirical assistance.

Review of Methods

In the coupled kinetics methods (33, 14, 19, 20), the three-dimensional extent of the core is subdivided into smaller three-dimensional regions which may be referred to as nodes. The neutron balance equation is solved in each node considering all neutron production and destruction processes. Neutron transfer between nodes is treated with a neutron coupling approximation which ranges from the formulation of Avery (4) to more modern formulations of Popa, et al. (30). There are simulators still in operation which use only one node for the entire core and are, therefore, simply using point kinetics.

A typical coupled kinetics model for a reactor core ranges from 16 to a few hundred solution points. More recent models have a substantial number of solution points [Deaton, et al. (15): 6 axial by 25 radial; Gregory et al. (18): 10 axial by 24 radial; Shesler, et al. (33): 4 axial by 133 radial; Popa et al. (30): 12 axial by 24 radial]. Relative nodal powers (the three-dimensional power distribution) are recomputed anywhere from twice per second (30) to once every 5 seconds (33). Recomputation of power distribution is expensive because the number of calculations required increases faster than the number of solution points.

The amplitude function scales the power distribution. It is calculated with an implicit integration technique from an equation which resembles the point kinetics equation at a rate of at least four times per second (4PS). Detector readings and more detailed axial power distributions may need to be constructed from a weighting of nodal fluxes (in the case of an in-core or ex-core detector reading) or a fitting of fine mesh fixed distributions (as in the case of a detailed axial power shape for an in-core flux map traverse). It is this lack of spatial detail which has been the biggest weakness of coupled kinetics models of the past.

In the expansion-based methods, the neutron flux is normally expanded into a summation of products of unknown amplitudes of reduced dimension and time, and known distributions (modes) in the remaining dimensions.

The general form of the expansion methods may be expressed as

$$\phi\,(x, y, z, t) = \sum_{k=1}^{K} \sum_{n=1}^{N} Z_{kn}\,(t)\,P_{n}\,(z)\,H_{nk}\,(x,y,z,t)$$

where n indexes a spatial nodalization, if any, and k indexes the expansion over a set of basis functions (modes).

Ward, et al. (35) describe a modal expansion method based on the use of unknown amplitudes, $Z_k(t)$, and known harmonic distributions, $H_k(x,y,z)$. The harmonic distributions are precomputed on a 28 by 28 by 14 nodalization. In this method, $N = 1$, and K may go up to 8 in order to include the azimuthal and axial overtone lambda modes of order two.

The general form of another expansion method [Chan (8,10)] has been used successfully in more than 15 training simulators. The axial nodalization of the core is indexed by n and K equals 1. It is based on expanding the three-dimensional flux in terms of unknown amplitudes, $Z_n(t)$, and spatial distributions, $P_n(z)\,H_n(x,y,t)$, where $P_n(z) = 1$ for z in axial node n and is zero otherwise. The radial flux distributions, $H_n(x,y,t)$, are updated during simulation to represent radial and azimuthal phenomena.

The method uses up to 24 nodes in the axial direction. The flux amplitude, $Z_n(t)$, in each axial node is determined at a rate of 4PS. The time-dependent power at any point in the core is constructed by scaling $H_n(x,y,t)$, the radial flux distribution appropriate for the axial elevation by the $Z_n(t)$ at that axial node. The radial mesh is fine enough that each of the 193 fuel assemblies in a four-loop Westinghouse PWR is represented by at least one radial node. More nodes may be represented in the radial direction with the expansion methods because the radial distribution is not recomputed at each time step but, normally, only when the radial distribution of macroscopic cross sections has changed significantly, as when a control rod has moved.

Important training phenomena for operator trainees are primarily axial phenomena because of the way cores are operated. For this reason, the expansion method described by Chan has provided satisfactory behavior at relatively low duty cycle cost. It provides a direct and detailed treatment of axial phenomena and a high degree of spatial detail in the radial direction. However, three-dimensional phenomena may be treated more directly using the coupled kinetics methods. The vector processing capability of the next generation of super minicomputers, coupled with improved methods (Chao et al., 11,12,13), is expected to permit coupled kinetics models to be run with nodalizations approaching that used with present expansion methods.

With either an expansion approach or the coupled kinetics approach, the success of the core model appears to be a function of the care with which the basic cross section information is developed and used, and the treatment of the thermal-hydraulic phenomena which alter the cross sections. Unique cross section sets which consider the effects of operating history for individual nodes, and a detailed calculation of the effects of void distribution on the cross sections, have been shown to be necessary for representing such refinements as the reverse power effect in boiling water reactors (30).

Methodologies for core simulation which are based on fuel design methods should be best. If the simulator model can be developed from core design information, this core model will be very good irrespective of whether it is a coupled kinetics model or an expansion-based model.

Validation of Models

Validation work with the core models has been fairly extensive. Validation is done by comparison with higher-order calculations (18,7,8) or by comparison of results with design data and startup physics test data (10).

Future Developments

Future developments in core models will be (a) further refinement of mechanistic fuel failure models to represent fuel failure, (b) further refinement of the effect of core voids on in-core and ex-core detector re-sponses, and (c) more detailed flux shape behavior resulting from faster and more powerful digital computers (vector machines), or more advanced core modeling and computation methods (11,12,13), with the attendant better representation of reactivity phenomena and spatial detail.

FLUID SYSTEMS MODELS

The modeling and mathematics developed for the modeling of fluid systems in a training simulator have taken two directions of interest: achievement of high fidelity in multiphase, multicomponent systems models, and the use of expert systems to model single-phase fluid systems. In the following paragraphs, the advances in the high fidelity modeling areas are described first; then the recent work in the application of expert systems is described.

MULTIPHASE, MULTICOMPONENT FLUID SYSTEMS MODELS

Complex multiphase, multicomponent systems such as the primary coolant system in either a BWR or PWR, are being modeled to a very high level of fidelity (3,16,17,22, 23,25,27,41,42). These models are currently able to handle training scenarios that could not be treated only a few years ago.

By phase, we mean a physically homogeneous combination of vapor and liquid; by component, we mean chemically distinct species (i.e., water and noncondensable gases). The noncondensable gases in a reactor coolant system may originate from the cover gases of various components, dissolved air in the coolant, hydrogen evolution from the metal-water reaction at high cladding temperatures, or release of fission product gases from ruptured fuel rods.

In essence, all solution methods for these applications consist of solving conservation equations for mass (of each component and each phase), for momentum, for energy (enthalpy), and for volume (pressure). The conservation equations are relatively straightforward to write, and all losses and gains can be accounted for in concept. It is the sophistication of the mathematical solution of these conservation relationships that determines the quality of the real-time solution, since the usual design code tactics of iteration and time step reduction are typically too slow for the present generation of super-minicomputers used in training simulation.

The literature abounds with numerical methods which have been developed for multiphase fluid problem solutions with digital computers. Yet very few of these are directly applicable to real-time simulation. Strongly implicit and simultaneous techniques promote realism and stability, but the order of the matrices must be low so that matrix inversion does not become the limiting element in terms of demanded duty cycle. Explicit and sequential techniques are very efficient numerically but voiding of a node or reversal of a flow direction can cause unrealistic or unstable behavior unless very small time steps are taken--at the expense of duty cycle.

Expectations on Performance

Expectations on performance of the coolant system model for nuclear reactor simulation include the representation of all phenomena that an operator trainee can monitor through instrumentation during all normal and abnormal conditions, and the representation of all responses that result from equipment failure or operator action.

Simulators being delivered today include reliable, first-principles representations of the following phenomena in each loop of a multiloop reactor system:

- Phase separation of liquid and gas including explicit representation of liquid levels in the coolant system

- Single-phase and two-phase flow capability, including flow reversal and countercurrent flow of liquid and gas phases

- Forced and natural circulation of multiphase fluids, including loss of natural circulation due to liquid traps or gas blockages

- Mixing of fluids from each inlet flow path to the vessel, including provisions for flow channeling in asymmetric cases

- A full range of single-phase and two-phase heat transfer regimes, including boiling, condensation, post-CHF, and heat flow reversal

- Core thermodynamics, including partial to full core uncovery

- Accumulation of noncondensable gases, steam voids, or saturated or subcooled liquid in the core or upper head, including their effects on natural circulation

- Reactor vessel level indications, including shrink and swell effects of pressurization and depressurization of the system

- Explicit representation of the metal heat capacity of the coolant system to facilitate thermal inputs to the fluids

- An operating range from full water solid conditions for the pressurizer or steam generator of a PWR to the empty condition

All the phenomena summarized above affect the ability of the operator trainee to control the simulated plant and cool it down to within safe limits. Therefore, these phenomena must be represented for the operator to learn how to perceive their effects and how to control them. Representations of the phenomena with "canned" (preprogrammed) models do not always permit correct response when the trainee

deviates from the emergency operating procedures, or encounters additional malfunctions during the training scenario. Therefore, the representations must be based
on first-principles models and sound technical judgments.

Fundamental Issues

The principal issue in primary coolant system modeling for real-time simulation is
achieving the necessary detail in the temporal, phenomenological, and spatial domains while maintaining real-time computation capability. The wide variation of
phenomenological regimes encountered requires very robust, broadly applicable
models. It is one thing to develop a specialized model and code for a specific heat
transfer regime or flow regime; it is quite another to develop a single model which
must handle a range from single-phase forced convection heat transfer to two-phase
regimes to post-CHF and finally vapor condensation with reverse heat transfer.
Mass inventory can vary from that of a water solid system to that of a completely
drained system.

Although the coolant system models currently being installed on training simulators
do a very good job in general, compromises must still be made in the selection of
time step size, the number of spatial nodes, and the representation of microscopic
phenomena (17). When the simulation behaves differently from the benchmark calculation or the test data, the simulation is investigated and changes are made in the
simulator modeling to match the benchmark. More confidence in the models will
result with continuing validation work. In addition, as computer technology continues to improve, the time step size and the number of nodes will continue to be
refined to provide more confidence in the predictive capabilities of the models.

Review of Methods

The basic modeling approach and conservation equations used in the models of
various simulator suppliers are similar; they stem from the FLASH work (29). All
suppliers break the real system of pipes and pots into a network of links and nodes.
The volume of the system is preserved in the nodal volumes; links or flow paths connect the nodes with proper consideration of the elevations at which flow paths intersect the system volumes. The mass and energy conservation equations are applied to
each node; the momentum conservation equation is applied to each flow link.

To the best of our knowledge, all advanced training simulator techniques are based
on the use of a global pressure and local enthalpy to evaluate fluid properties. This

global pressure is found by iteration on each time step so that conservation laws are satisfied. This simplification assumes that, during the course of a transient, density variations due to pressure changes are small; stated another way, local pressures differ little from the global pressure for the purpose of evaluating fluid properties. This assumption is essentially the same as what Porsching (31) calls the thermally expandable assumption. What is lost by this assumption is the capability to represent sonic phenomena.

By nature, the conservation equations used in the training simulation industry are considered to be very similar. Where the similarity breaks down is in the numerical solution techniques used to solve the conservation relations, and to a lesser degree, the number of conservation equations used. Wyatt, et al., (42), emphasize the importance of fully implicit, simultaneous solutions to ensure stability. The penalty of this approach is the duty cycle cost associated with inverting large matrices. On the other hand, Andersen and Fabic (1) use an explicit integration technique. Although this technique avoids the cost of inverting large matrices, it runs the risk of numerical instabilities due to the inherent asynchronous coupling of certain boundary conditions. During a rapid blowdown, the integration time step must be limited by the Courant limit, which requires that the integration time step be smaller than the nodal mass divided by the maximum flow rate in or out of the node.

The mass balance and the momentum balance may be combined (42) to make link flow the solution variable rather than nodal pressure. This technique significantly reduces the size of the original system of equations and appears to be particularly advantageous in cases where flow closes on itself (as in the primary portion of a pressurized water reactor). The order of the solution is then reduced from an NxN matrix, where N is the number of nodes, to an MxM matrix, where M is the number of independent flow loops within the network.

It is customary in today's simulation to do three mass balances and either two or three energy balances on each node, and one differential momentum balance on each flow link, for a total of either six or seven equations. All published techniques for real-time training simulation do a momentum balance on each flow link with one differential equation for total fluid flow. The total flows are decomposed by a drift flux algorithm so that the different components may, under some circumstances, have opposite signs. Mass balances are done in each node on a mixture region containing fluid and steam bubbles below the mixture level, a gas region containing vapor and entrained droplets above the mixture level, and a noncondensable gas component above the mixture level. Energy balances are done in each node on the mixture region and the entire vapor region [Deaton, et al. (15), a two-equation energy balance], or on the mixture region, the gas region, and the noncondensable gas [Wyatt, et al. (39), a three-equation energy balance]. Since Wyatt provides a separate equation for each phase and each component, his approach effectively provides a completely nonequilibrium representation of energy transport.

Validation of Models

Validation work has shown that these advanced real-time models can represent realistic behavior reliably. The earliest validation work (24,38,39,40) was based on comparison of results with best-estimate codes. More recent work (15) has compared simulation results to semiscale test results. The problem areas encountered are associated primarily with the relatively large time step requirements, particularly under conditions of low fluid inventory. Under these conditions, flow rates can move amounts equal to or greater than nodal inventories in a single time step, and transitions from one phase to another in flow links can lead to nonphysical behavior which must be detected and controlled by additional calculations in the model.

APPLICATION OF EXPERT SYSTEMS TO THE MODELING OF SINGLE-PHASE FLUID SYSTEMS

The simulation of single-phase fluid networks is the largest task facing the simulation engineer in terms of numbers of models to be developed. Approximately half of the systems modeled in a full-scope training simulator are of this class. Hence, it is not surprising that tools are being developed to aid the fluid systems modeler in the planning, coding, and documentation of mathematical models for single-phase fluid systems. One such tool has been previously described by Boire (6); this paper describes a knowledge-based software development system has been developed by Rinsma and Shemony (32).

The goals of the system of Rinsma and Shemony, called the Interactive Model Builder (IMB), are as follows:

- To standardize model practices and enhance quality control during development

- To decrease the engineering effort needed for model development and documentation

- To facilitate introduction of improvements in modeling techniques

- To facilitate addition of plant changes to existing models

- To decrease the number of coding errors and associated model deficiency reports

- To produce well-commented structured coding

The IMB is based on proven modeling guidelines that have been incorporated into a knowledge base. The modeler uses a general interactive process in which he defines the elements (check valves, heat exchangers, pumps and pump characteristics, and so forth) and links (node-to-node connections) in a flow network. The IMB processes an initialization file and defines the scope and structure of an individual model using "best techniques." Adding or subtracting nodes or nodal connections is easily accomplished by modifying the initialization data file, rebuilding the model source, and recataloging.

An IMB-generated model is based on the initialization data file, which can be generated immediately once a nodal diagram is configured from design basis information. No development of pressure, flow, or temperature nodal solution code is necessary, since they are included as modules. Since the modules are time-tested pieces of code, the time it takes to implement, debug, and integrate a model is radically reduced.

A preprocessor for the network solution uses an expert system rule base to ensure that the system matrix is not singular and to configure the system matrix for efficient inversion. The preprocessor examines the admittance relationships of each connection to each node and combines admittances where possible based on the system configuration at each time step. The node examination process is particularly useful in adding the flexibility of variable leak locations to the simulation models. For example, this process allows dynamic insertion of malfunction leaks by the instructor anywhere in the nodal network during a training scenario.

The IMB has been used successfully on six simulators delivered or under construction at the Westinghouse Instrumentation Technology and Training Center.

ELECTRICAL SYSTEMS MODELING

The electrical models for training simulators are designed to give training on correct breaker lineups, synchronization of electrical generators onto the system, operation of the diesel generator under emergency conditions, startup and shutdown of the generator following established procedures, and monitoring of equipment.

The technology of electrical systems modeling has improved in the last few years to take advantage of the greater computing power of today's super minicomputers. The changes have resulted in greater versatility in possible electrical system arrangements, and greater fidelity of system operation.

Expectations on Performance

Electrical systems modeling has not received the same level of attention as has the modeling of the core and primary systems. This trend is turning around, however, as simulator owners discover that many plant licensee event reports have their origins in electrical system faults. Therefore, recent specifications have relatively higher expectations for the electrical models.

These specifications have requested the following:

- A more complete simulation of all electrical distribution system buses and load centers

- A complete dynamic simulation of the onsite electrical system as well as the in-plant electrical distribution system, including the calculation of real and reactive load currents from startup and operation of induction motor loads, voltage, power, frequency, and synchronization phase relationships

- A complete dynamic simulation of the main generator, diesel generators, and their associated excitation systems based on the appropriate IEEE methodology for generator and excitation system transient analysis

- Backfeed of nonvital buses from the diesel generators as well as paralleled operation of the diesel generators under loss of offsite power conditions

- Simulation of generator overheating and failure due to overload and/or degraded cooling systems

- Simulation of changes in bus voltages due to transformer reactance within the distribution system as load currents change

Fundamental Issues

There are no fundamental modeling issues with this class of systems. Electrical systems can be simulated relatively easily using classical modeling techniques. The problems, if any, appear to be the tradeoffs between model fidelity (that is, cost) and training needs. Although high fidelity simulation methods can accurately represent high frequency transient behavior, the instruments available to an operator are limited in their capabilities to portray or control high-frequency behavior.

Review of Methods

The solution technique for electrical models is based on the application of Kirchoff's and Ohm's laws to a nodalized version of the electrical network being simulated. As computer power has increased, models have improved to include more scope detail and representation of real and reactive components of loads. Modelers (5) have developed efficient techniques for reconfiguration of electrical networks on line to accommodate changing breaker alignments while keeping the inversion of matrices efficient. The calculations involved may be simplified by partitioning the overall system matrix into models and submodels, where appropriate, based on plant design and obvious system boundaries. Thevenin and Norton Equivalent Circuits are used at these system boundaries.

An extension of the Thevenin and Norton Equivalent Circuit concept is used to define the connections to the in-plant distribution system and to the switchyard. Experience in using the models has shown that physical considerations of the circuit connections and arrangements can often be used to simplify the models.

Validation of Models

Validation of electrical systems models in training simulators is handled by comparisons with analytical solutions and by integrated plant systems testing.

Future Developments

The market demand for increased scope and fidelity of electrical systems models is evident. The technology and computer power to provide these models are available; however, the engineering costs of providing detailed scope models which are plant specific are relatively high. For that reason, we expect to see the development of expert systems which may be used to create plant-specific software.

INSTRUMENTATION AND CONTROL SYSTEMS MODELING

This category of power plant systems includes components for sensing, logic, signal processing, process control, and indication. The systems in this category can be classified into either a logical group or a process control group. The logical group can be represented easily in the digital computer language of the simulator; the process control group may be subdivided into either analog or digital systems. Digital control systems may be simulated directly using digital computers; simulation of analog control systems may be accomplished using standard discretization processes.

Expectations on Performance

Simulator specifications seldom, if ever, spell out in detail the performance requirements of instrumentation and control (I&C) systems. This is a reflection of the fact
that these systems are well defined and the mathematical methods for system representation are well understood and documented.

Fundamental Issues

There are no fundamental simulation issues with this category of systems in the
same sense as with the other categories, where simplifying assumptions and approximations must be made to develop a solution technique. However, elements of simulation technology for I&C systems deserve special mention: use of control system
simulation to compensate for approximations in other system models, limitations on
dynamic fidelity achievable with a fixed discrete time step, and the trend toward
smarter and more complex I&C systems (for example, sophisticated microprocessor-
based controllers).

The simulation of some thermal-hydraulic processes is not exact; consequently, the
simulated control system is often tuned during integration testing to compensate for
approximations in complex fluid models. Because of this, the actual parameters
(e.g., gains, time constants) delivered with the control system simulation may be
slightly different from the parameters in the system design documents.

If the time step size in the simulation is small with respect to the significant time
constants of the simulated control system, then the result of the simulation is stable,
straightforward, and realistic. If the time step is large, then the system may be
simulated using a quasi-static approximation where the steady-state solution is
assumed to exist at each time step. When the time step is on the same order as the
significant time constants of the system, either the time step must be changed or an
implicit solution technique must be used. Either option may be expensive in terms of
duty cycle. Fortunately, the significant time constants of power plant control systems are large enough (i.e., approximately 1 second or greater) to make simulation
time steps of 0.25 second satisfactory for the simulation of most control systems.

A relatively new challenge with control system simulation is the increased complexity that can be incorporated into microprocessor-based control systems. This
complexity, coupled with the relatively inexpensive hardware of the control system,
makes stimulation of the equipment an increasingly attractive option compared to

simulation of the control equipment, although initialization of a stimulated controller can be difficult, particularly if an integral function is required.

Review of Methods

The mathematical model for simulation of an analog I&C system is constructed by a straightforward discretization of continuous time variables into discretized time variables. This discretization can be done from either the time domain or the transfer function domain.

If the transfer function is known, the mathematical model for the simulation can be constructed in a straightforward, textbook manner using the theory of z-transforms (34). The procedure consists of replacing s wherever it appears in the transfer function by $(z-1)/\Delta t$ to obtain the z-transform form of the transfer function. The time-differenced control algorithm to be programmed digitally is simply the algebraic expansion of the z-transform with all variables multiplied by z being forward-shifted in time [that is, zI(t) becomes $I(t + \Delta t)$ where Δt is the integration time step in the simulation]. This procedure results in a control algorithm based on the one-step forward Euler algorithm (explicit differencing). The one-step backward Euler algorithm (i.e., implicit differencing) algorithm can be obtained by the same procedure except that s is replaced by $(z-1)/z\Delta t$ instead of $(z-1)/\Delta t$.

It is believed that this same basic technique is used by all manufacturers of simulators for the simulation of analog instrumentation and control systems.

Validation of Models

Validation of I&C system models in training simulators is done by comparison with analytical solutions and by integrated plant systems testing. This work is seldom published because the mathematical techniques used in modeling are standard, proven techniques.

Future Developments

If I&C systems continue to become more complex and less expensive, we expect that the stimulation option will become more attractive than the simulation option. S i m u l a t i o n, as used here refers to the mathematical modeling of the system using a digital computer. This approach minimizes hardware additions to the simulator but involves more extensive software effort. S t i m u l a t i o n, as used here, refers to representation of the I&C system by actual system hardware which is driven by signals from the simulated systems. This approach reduces the software effort but requires addition of actual I&C hardware to the simulator.

MATERIAL TRANSPORT SYSTEMS MODELING

In nuclear power plant simulation, this area of modeling deals with the transport of radioactive materials throughout all plant systems and plant buildings.

Expectations on Performance

Expectations for performance of the radiation monitoring system (RMS) model include a realistic calculation of the distribution and transport of radioactive material throughout all plant systems, and realistic responses of process and area radiation monitors.

Fundamental Issues

The fundamental issues for the transport portion of the radiation monitoring system model are maintaining real-time computation capability while providing treatment of a sufficient number of isotopes to develop realistic detector responses and treatment of a sufficient number of nodes within building areas to differentiate between detector responses.

Early models had only one "universal" pseudo-isotope. Constants were tuned for each detector to give as realistic a result as possible. However, some inconsistent alarm responses were introduced during accident conditions with this model. For example, the simplified model could not account for the high radiation transport rates of noble gases in the gaseous phase (as contrasted to the liquid phase).

Improvements have been made by using two pseudo-isotopes (predominantly liquid isotope, and predominantly gas isotope) or as many as four isotopes to model radiation transport. The inconsistent alarm response associated with the one isotope model has been, in general, eliminated.

Some attempts have been made by individual simulator owners to develop many-isotope (20- to 30-isotope) models, but satisfactory utilization of these models has not been documented. We are not aware that any of these models works satisfactorily at the present time.

Another noteworthy problem for the radiation transport calculation is the need for refined nodalizations of building spaces to represent the transport and diffusion of radioactive material throughout a building to detectors in different locations. Current technology represents an entire building as one node and it is assumed that the radioactive material is uniformly distributed throughout the building. Responses of detectors in different locations throughout the building are tuned to respond realistically based on the location of the radiation source.

Review of Methods

It is believed that all methods currently used are based on a dynamic material balance and a mass balance at each node represented in the model. Most nodal masses and flows are defined based on the solution from the fluid system models. Some additional nodes must normally be defined to represent the building spaces which are not part of the fluid system models.

Validation of Models

Validation of RMS models is difficult because of the lack of experimental data. Published data which relates RMS response to known releases simply do not exist, to our knowledge. Validation of simulator models is therefore subjective and qualitative. "Tuning" of the instrument response depends on the engineering judgment of experienced plant operating personnel.

Future Developments

Future developments in this area will continue to move toward increasing the number of radioisotopes represented in the model. Chan (9) has begun development of a concept for treating 30 to 40 isotopes important to postaccident assessment by a technique which follows the transport of a limited number of "isotope carriers." Each isotope carrier is selected to represent a particular chemical class of isotopes (for instance, noble gas, halogen, particulate, or high-temperature volatile). The transport calculations and material balance calculations at each node are based on the chemical characteristics of each carrier. Much has been learned about the behavior of the various chemical classes as a result of TMI (21). The time-dependent behavior of each isotope of a chemical class at each node is computed based on the concentration of the carrier isotope and the time-dependent relationship between each isotope in the chemical family and the carrier isotope of the chemical family.

Models like this may be used to calculate the concentrations of a large number of isotopes at various sample and detector points throughout the plant. These isotopic concentrations may be used with the emergency procedures to infer the amount of fuel failure in a major accident and thus give training in postaccident assessment.

CONCLUSIONS

The following conclusions may be drawn from the above discussion:

- Modeling technology for real-time training simulators is still evolving. Current models are able to handle training scenarios that could not be treated only a few years ago. As computer technology continues to improve, model fidelity will to continue to increase.

- Simulation models for real-time training still cannot handle many microscopic phenomena important in design and safety analysis. The models are able to give plausible (but not necessarily predictive) results.

- Simulator model dynamic fidelity today is very good for the purpose of operator testing and for training on well-understood scenarios. Fidelity improvements are still needed to permit reliable prediction of complex transient situations.

- It is time to start asking the question "How good do training simulator models need to be?" In a few more years, computer capability will outrun our needs.

- There are new frontiers of simulator modeling:

  -- Extension of simulation to new training tasks (such as postaccident assessment)

  -- Validation and verification of plant-specific models

  -- Online analysis of abnormal situations at faster than real time to assist the plant operator

  -- Incorporation of the knowledge of expert modelers into intelligent model building systems

REFERENCES

1. "American National Standard Nuclear Power Plant Simulators for Use in Operator Training," ANSI/ANS-3.5 - 1985.

2. "Analytic Simulator Qualification Methodology," NP-3873 Research Project 2054-1, Prepared for the Electric Power Research Institute, March 1985.

3. Andersen, P.S., and Fabic, S., "Theoretical Foundation of an Advanced Simulation Method for Power Plant Thermo-Hydraulics," International Conference on Simulation for Nuclear Reactor Technology, Cambridge, U.K., April 1984.

4. Avery, R., "Theory of Coupled Reactors," *Proceedings*, 2nd International Conference on Peaceful Uses of Atomic Energy, Geneva, September 1958, pp. 182-191.

5. Birsa, J.J., Personal Communication, June 1987.

6. Boire, R., "HYDRA: An Interactive Tool for Automatic Generation of FORTRAN Code for Hydraulic Circuits," *Proceedings*, Summer Simulation Conference, Society for Computer Simulation, Reno, Nevada, July 1986.

7. Chan, T.C., and Luffey, F.C., "Real Time Simulation of PWR Reactor Core System Transients," paper presented at the Summer Computer Simulation Conference, Society for Computer Simulation, Denver, Colorado, July 1982.

8. Chan, T.C., "An Advanced Simulator Reactor Core System Model," *Proceedings*, Tenth IMACS World Congress on System Simulation and Scientific Computation, Montreal, Canada, August 1982.

9. Chan, T.C., Personal Communication, June 1987.

10. Chan, T.C., "An Expansion-Based Reactor Core Model for Training Simulators," Specialists' Meeting on Training Simulators for Nuclear Power Plants, Toronto, Canada, September 1987.

11. Chao, Y.A., Suo, C.A., and Penkrot, J.A., "On the Theory of Interface Flux Nodal Method," *Proceedings*, CNS/ANS International Conference on Numerical Methods in Nuclear Engineering, Montreal, Canada, September 6-9, 1983, p. 307, Canadian Nuclear Society (1983).

12. Chao, Y.A., Hu, C.W., and Suo, C.A., "A Theory of Fuel Management via Backward Diffusion Calculation," *Nuclear Science and Engineering 93*, 78-87 (1986).

13. Chao,, Y.A., and Penkrot, J.A., "Diffusive Homogeneity--The Principle of the Superfast Multi-Dimensional Nodal Code, SUPERNOVA," paper to be presented at ANS Winter Meeting, Los Angeles, California, November 15-19, 1987.

14. Chow, S.G., Guardalben, G.V., Wight, A.L., and Chou, Q. B., "Experience in Generation and Implementation of Coupling Coefficients for a Space-Time Kinetics Model in BWR and PWR Nuclear Training Simulators," *Supplementary Proceedings*, Eastern Simulation Conference, Society for Computer Simulation, Norfolk, Virginia, March 1986.

15. Deaton, I.V., Lin, E.K., Fabic, S., and Anderson, P., "Validation of the Real-Time Advanced Core and Thermohydraulic (RETACT) Code for Nuclear Power Plant Engineering Simulation," *Proceedings*, Eastern Simulation Conference, Society for Computer Simulation, Orlando, Florida, April 1987.

16. Espinosa, R.J., Doherty, P.K., and McBeth, R.L., "Application of Thermal-Hydraulic Design Codes to Real Time PWR Simulation," Fifth Power Plant Dynamics, Control and Testing Symposium, March 21-23, 1983, Knoxville, Tennessee.

17. Fabic, S., "Thermal Hydraulics in Nuclear Power Plant Simulators," *Proceedings*, Third International Topical Meeting on Reactor Thermal Hydraulics, Newport, Rhode Island, October 1985.

18. Gregory, M.V., Aviles, B.N., and Yakura, S.J., "A Three-Dimensional Neutronics Model for Reactor Training Simulators," *Nuclear Science and Engineering 92*, 372-381 (1986)

19. Hoang, T.D., and Morin, P., "A Method of Calculation of Coupling Coefficients for a Reactor Core Nodal Model," *Proceedings*, Summer Simulation Conference, Society for Computer Simulation, Reno, Nevada, July 1986.

20. Hsu, H.H., Lea, K.C., and Shen, P., "Three-Dimensional Core Model for Nuclear Power Plant Simulation," *Proceedings*, Summer Simulation Conference, Society for Computer Simulation, Vancouver, B.C., July 1983.

21. IDCOR Technical Summary Report, "Nuclear Power Plant Response to Severe Accidents," Technology for Energy Corporation, Knoxville, Tennessee, November 1984.

22. Lin, E.K., Jen, C.L., Fabic, S., and Andersen, P.S., "Advanced Modeling of Power Plant Simulators," International Conference on Simulation for Nuclear Reactor Technology, Cambridge, U.K., April 1984.

23. Lin, E.K., Jen, C.L., Fabic, S., and Andersen, P.S., "Thermo-Hydraulic Models in Link's Advanced Power Plant Simulator," 23rd ASME/AIChE/ANS National Heat Transfer Conference, August 6-9, 1985, Denver, Colorado.

24. Luffey, F.C., Murphy, J.H., and Meyer, P.E., "Advanced Techniques for Real Time Simulation of Reactor Coolant Two-Phase Transients," paper presented at the CSNI Specialists' Meeting on Operator Training and Qualifications, Charlotte, NC, October 1981.

25. Luffey, F.C., Murphy, J.H., Garratt, C., and Wyatt, P. W., "Advanced Techniques for Real-time Simulation of Reactor Coolant System Two-Phase Transients," *Nuclear Europe*, March 1983.

26. McFadden, J.H., et al., "RETRAN 02 Program for Transient Thermal-Hydraulic Analysis of Complex Fluid Flow Systems," EPRI, NP-1850-CCM, May 1981.

27. Meyer, P.E., Spencer, A.C., and Porsching, T.A., "An Advanced Simulator Reactor Coolant System Model," *Trans. Amer. Nucl. Soc. 38*, p. 678 (June, 1981)

28. Moore, K V., and Rettig, W.H., "RELAP5 - A Computer Program for Transient Thermal-Hydraulic Analysis," ANCR-1127, Revision 1, March 1975. See also "RELAP5/MOD1 Code Manual" (Vols. 1 to 3), NUREG/CR-1826.

29. Murphy, J.H., Redfield, J.A., and Davis, V.C., "FLASH-3: A FORTRAN IV Program for the Simulation of Reactor Plant Transients in Space and Time," WAPD-TM-800, July 1968.

30. Popa, F., and Marlatt, G., "Real Time Boiling Water Reactor Core Simulation," *Proceedings*, 1987 Summer Simulation Conference, Society for Computer Simulation, Montreal, Canada, July 1987.

31. Porsching, T.A., "A Finite Difference Method for Thermally Expandable Fluid Transients," *Nuclear Science and Engineering 64*, 177-186 (1977)

32. Rinsma, R.S., and Shemony, R.A., Personal Communication, June 1987.

33. Shesler, A.T., and Wagner, S.G., "A Three-Dimensional Core Model for Personal Computers," *Proceedings*, Summer Simulation Conference, Society for Computer Simulation, Reno, Nevada, July 1986.

34. Siebert, W.M., *Circuits, Signals, and Systems*, The MIT Press, Cambridge, Massachusetts, 1986.

35. Ward, E.J., Chow, S., and Chou, Q.B., "Development of a Real-Time Spatial Simulation for a CANDU-Type Nuclear Reactor Using Modal Techniques," *Proceedings*, Eastern Simulation Conference, Society for Computer Simulation, Orlando, Florida, April 1987.

36. Wulff, W., "PWR Training Simulator, An Evaluation of the Thermohydraulic Models For Its Main Steam Supply System," BNL-NUREG-28955, September 1980.

37. Wulff, W., "BWR Training Simulator, An Evaluation of the Thermohydraulic Models For Its Main Steam Supply System," BNL-NUREG-29815, March 1981.

38. Wyatt, P.W., Kumar, S., and Luffey, F.C., "Verified Simulation of Reactor Transients in Real Time: The TMI-2 Stuck Relief Valve and the Prairie Island Steam Generator Tube Leak," WCAP-10028, Westinghouse Electric Corporation, February 1982.

39. Wyatt, P.W., and Kumar, S., "Verified Simulation of a Nuclear Reactor Transient in Real Time: Steam Generator Tube Leak," paper presented at the Summer Computer Simulation Conference, Society for Computer Simulation, Denver, Colorado, July 1982.

40. Wyatt, P.W., and Luffey, F.C., "Verified Simulation of a Nuclear Reactor Transient in Real Time: Stuck Pressurizer Relief Valve," paper presented at the 10th IMACS World Congress on System Simulation and Scientific Computation, Montreal, Canada, August 1982.

41. Wyatt, P.W., and Kumar. S., "Real Time Solutions of Multi-Phase Multi-Component Fluid Networks," *Proceedings*, Japan Society for Simulation Technology, JSST Conference on Recent Advances in Simulation of Complex Systems, Keidamren-Kaikan, Tokyo, Japan, July 15-17, 1986.

42. Wyatt, P.W., Kumar, S., and Zelnis, D.G., "Applications of Various Conservation Equations to Real Time Simulation," *Proceedings*, Eastern Simulation Conference, Society for Computer Simulation, Orlando, Florida, April 1987.

ACKNOWLEDGMENTS

This paper represents the combined efforts and contributions of many of my colleagues at Westinghouse Electric Corporation. Special recognition must be given to Dr. Paul Wyatt, whose work and text from previous publications have been freely used, and to Dr. Jack Murphy, who performed a critical review of the draft manuscript. Contributions from Frank Popa, T.C. Chan, Bill Alliston, Chuck Roman, Joe Birsa, and Henry Kwoh are acknowledged with sincere appreciation. Dr. Jean-Pierre Sursock of the Electric Power Research Institute provided many stimulating suggestions in the formulation of initial thoughts.

Concepts and Realization of the KWU Nuclear Plant Analyzer

H. MORITZ, R. HUMMEL

Kraftwerk Union AG
Dept. ST144

Hammerbacherstr. 12 + 14
D-8520 Erlangen

Abstract

The Nuclear Plant Analyzer (NPA) is a real time simulator developed from KWU computer programs for transient and safety analysis ("engineering simulator"). The NPA has no control room, the hardware consists only of commercially available data processing devices.

The KWU NPA makes available all simulator operating features such as initial conditions, free operator action and multiple malfunctions as well as freeze, snapshot, backtrack and playback, which have evolved useful training support in training simulators of all technical disciplines.

The simulation program itself is running on a large mainframe computer Control Data CYBER 176 or CYBER 990 in the KWU computing center under the interactive component INTERCOM of the operating system NOS/BE. It transmits the time dependent engineering data roughly once a second to a process computer SIEMENS 300-R30E using telecommunication by telephone. The computers are coupled by an emulation of the communication protocol Mode 4A, running on the R30 computer. To this emulation a program-to-program interface via a circular buffer on the R30 was added. In the process computer data are processed and displayed graphically on 4 colour screens (560 x 512 pixels, 8 colours) by means of the process monitoring system DISIT. All activities at the simulator, including operator actions, are performed locally by the operator at the screens by means of function keys or dialog.

1. Mode of Operation and Application of the NPA

1.1 Mode of Operation

The KWU Nuclear Plant Analyzer (NPA) is a real time engineering simulator for nuclear power plants with pressurized water reactor. The NPA is based on the principle of allowing the customer access to the KWU proprietary programs for plant transient analysis and licensing ("best estimate codes") by means of a simulator.

In contrast to the conventional training simulator (for training plant operators) the NPA has no control room, the hardware consists of commercially available data processing devices like disk drives, video display units, ink jet printer, graphic colour hardcopy device etc. The primary goal is to keep the customers financial expenditure far below that of a training simulator with comparable training capabilities despite the high computational power needed.

The NPA makes available all operator functions of a full scale simulator with nearly real time performance :

- unlimited operator action
- multiple malfunctions
- initial conditions
- starting, freezing and stopping of simulation
- snapshot,backtrack
- playback
- fast time, slow time (restricted only)

The following features were added:
- function key actuation
- analysis

Actuation by function keys accounts for the fact, that the instructor, necessary in full scope simulators, has been omitted. Off-line operation (that means, without connection to the simulation computer CYBER 990 and of course without the possibility of operator actions) is possible for playback and analysis both during a current simulator session as well as for previous sessions stored on a local magnetic disk; in this case real time performance is guaranteed. The various analysis features are provided in order to improve the understanding of the complex physical and technical processes involved in the power plant.

With the simulation frozen, all transmitted variables may be displayed on the screens as curves as a function of time as well as a function of any other variable; the same variables of previous simulator sessions can be faded in. Some expressions of variables can be formed, for example linear combinations, maximum, minimum or integrated curves.

Display of the data base at the current time is also possible containing the thousands of variables calculated but not transmitted every 1 or 2 seconds. These data can be displayed as a list or graphically.

1.2 Application of the NPA

Originally the nuclear plant analyzer of KWU was developed to be used by the vendor in plant transient and safety analysis. It will also be applied in staff training. But the most beneficial application would be the installation directly in the nuclear power plant for personnel training and operating support with the aim of higher plant availability and safety.

The fields of the application of the nuclear plant analyzer may be the following:

- rationalization in the design of nuclear power plants concerning the transient and accident behaviour;
- initial and refreshment training of plant operators partly on the NPA;
- optimization of operator guidelines and accident management;
- training onsite during plant operation;
- support in planning of plant service and upgrading;
- support during abnormal operation;
- support during slow events.

At the NPA all functions, which at the full scope simulator are fulfilled by the instructor, are attributed to the operator: He selects the initial condition - state of burnup, stationary or transient. He is able to initiate malfunctions, for example failure in the control and limitation systems. He is free in superimposing such malfunctions to get any combination and any sequence of multiple failures, and all malfunctions may be reset as well as all operator actions, irrespective of its technical feasibility in the actual plant.

Furthermore many data of the plant mechanical engineering can be changed by the operator himself without modifying the computer code (e.g. pump characteristics, load diagrams, data as setpoints of a limitation or reactor protection system).

Thus the NPA offers the user a flexibility usually characterizing an engineering computer program applied in transient and safety analysis, however in simulator environment. This flexibility and ease of use render the nuclear plant analyzer a versatile tool in plant operation and operator training.

It also can be used by the power plant staff in preparing design changes such as exchange of system components or net power uprating.

Even in upset operating conditions, the NPA proves valuable if the transient develops slowly. Then a technical supervisor or shift leader has got enough time to evaluate countermeasures on the NPA. By this the probability of a safe plant operation is increased.

The applications outlined here are mainly due to simulation accuracy, immediate onsite availability, ease of use, and economic performance of the NPA. Thus it will contribute to improve plant availibility and safety.

2. Configuration of the KWU Nuclear Plant Analyzer

For the KWU NPA we decided to follow the concept of distributed processing. Therefore two computers are involved here. The distributed tasks of the involved computers are as follows:

The simulation program itself is running on a CDC CYBER 176 or CYBER 990 in the KWU computing center under the interactive component INTERCOM of the operating system NOS/BE 1.5.

This program NLOOP is the standard (best estimate) program developed by KWU for design and licensing of PWR's. As NLOOP is normally used as a batch program, it had to be changed for interactive use in order to execute under real time conditions.

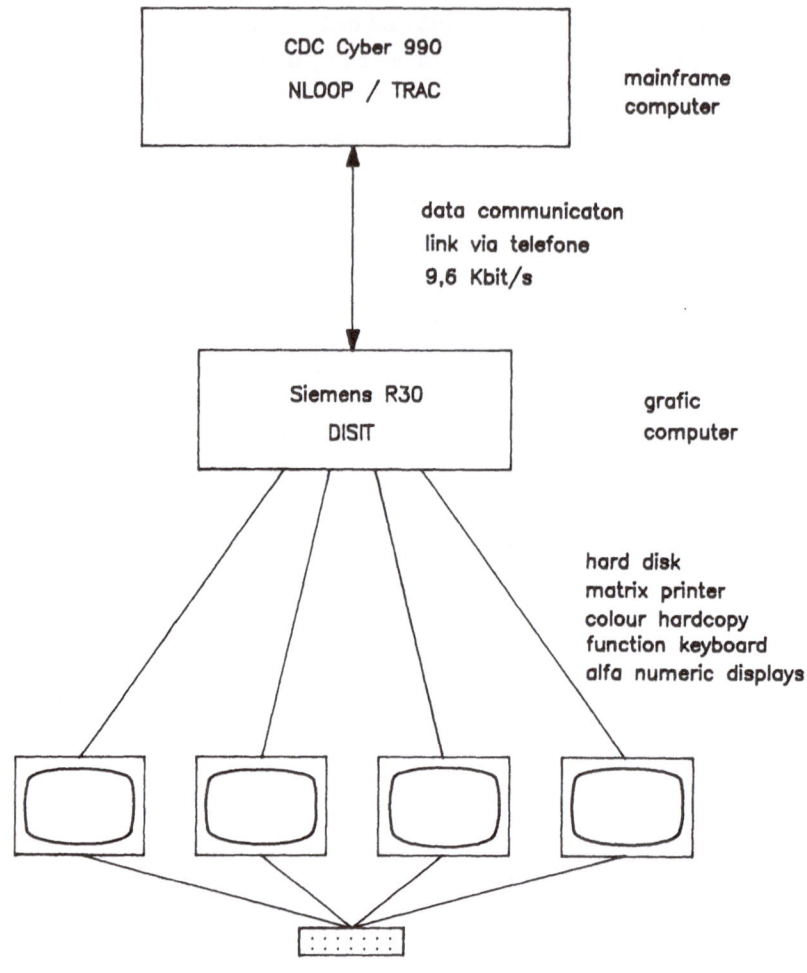

Fig. 1: Configuration of the KWU Nuclear Plant Analyzer

It transmits every n seconds (n is selectable) some 200 of its calculated data via telephone line to the so called graphic computer. This transmission is done by a standard output to a connected file in formatted manner after data compression. Thus the graphic computer acts for the CYBER computer as a single interactive terminal. An emulation of the telecommunication protocol Mode 4A is running on the graphic computer, a Siemens process computer SICOMP-R30E. (The modification we needed for this emulation and the problems which arose will be discussed later in section 4.)

The emulation writes the received data in a circular buffer and informs the other programs on the graphic computer via a semaphor, which is implemented on the process computer as an coordination counter. Then the data are fetched by the other programs running on the process computer and processed for output on 4 graphic colour screens (hardware resolution 560 x 512) and 2 alphanumeric displays. Output on the graphic screens is effected by means of the process monitoring system DISIT, a standard Siemens software system for industrial process monitoring and control. For documentation all pictures shown on any display can be printed out by a colour graphic hardcopy device. Also all data received from the simulation program can be stored on a fixed disk on the graphic computer for later use as a playback of a simulation or as comparative curves for other simulator sessions.

Vice versa data (e.g. operator actions) are sent to the mainframe computer by a similar procedure, but not at the same amount.

For better understanding we can make following images: For the CYBER 990 the whole process computer is only one terminal, and on the other side for the process computer the CYBER 990 only simulates a process which is to be monitored and controlled.

3. Involved Programs

3.1 On the CYBER

The Nuclear Plant Analyzer operates with NLOOP, the KWU standard program for transient analysis in design, licensing and operation survey of pressurized water reactor power plants. The validation of NLOOP is carried out on an ongoing basis using actual plant start-up tests and transients for comparison. Also comparison with RELAP5 calculations was done with good results.

NLOOP is a fast executing and flexible tool, which simulates plant response to a wide range from transients without loss of coolant to accidents with very small loss of coolant. Multiple assymmetric transients including flow reversal can be considered, because a number of up to 4 model loops can be chosen.

Operations:
- normal operation
- upset operating conditions

Incidents/accidents:
- reactivity disturbances
- coolant flow disturbances
- heat removal disturbances
- secondary system leaks
- accidents with very small loss of primary coolant
- anticipated transients without scram (ATWS)

The program includes models for major components of the primary and secondary side, for important auxiliary systems and for the essential control, limitation, protection and interlock systems.

The fluid in the primary system is treated as homogeneous. Temperature non-equilibrium can be considered in the pressurizer, steam generators, feedwater tank and in the reactor pressure vessel head. The one dimensional equations of mass and energy conservation are integrated by an implicit numerical method. Node / flow path networks are used to model the flow rates in the primary coolant loops and in the main steam and feedwater system.

The first version of the NPA was developed for the KWU 1300 MW standard pressurized water reactor. It can be adapted to the systems of other plants with pressurized water reactor by carrying out a small number of reprogramming tasks, in particular within the control system, limitations and reactor protection systems. This is also possible for non KWU plants with thermalhydraulic designs similar to KWU plants.

3.2 Programs on the graphic computer

While the transient code NLOOP is running on a CYBER 990, it transmits the values of the time dependent engineering variables every 1 or 2 seconds to the remote graphic processor. In the NPA, these engineering data are stored in the graphic computer before being processed and displayed on the screens.

When simulation is running, following programs are active on the graphic computer:

- emulation programs for the telecommunication protocol by which the computers CYBER 990 and Siemens R30 are coupled.

- the programs of the process monitoring system DISIT (a software product of Siemens) for output and actuation of the displays on the screens.

- the programs which enable and process the user dialog and the man machine interface (in the following called STOSIM).

With the graphic system DISIT four types of displays are available:

- semigraphic displays (symbolic process diagrams) with the process data as alphanumeric data, as analogous bars such as water level, and as graphic symbols like arrows indicating flow direction.

- Trend curves with variables displayed as a function of time. In the trend curve displays the operator deliberately can choose the variables to be displayed. The curves are distinguished by colours, and a curve point is appended each data transmission time step. If the curves reach the right hand side of the screen, the whole curves are shifted one point to the left each time step just as at the paper recorders in the control room.

- Lists containing alphanumeric data; an example is the shortened process computer protocol containing alarms and operator actions.

The tasks of the system STOSIM are:

- to initiate simulation
- to initiate computer coupling
- to supply the input data for the transient code NLOOP
- to transmit the data of operator actions to the CYBER 990

- to coordinate transmission and reception of the data
- to read the transmitted data from a buffer of the protocol emulation
- to evaluate the engineering data
- to communicate with the graphic system DISIT
- to write the trend curves
- to control the time behaviour of the simulation (real time / slow time / fast time)
- to control the whole simulation
- to effect operator actions and simulator functions
- to store the results on a local disk of the graphic computer

4. Coupling of the computers used in the NPA

4.1 Choosing the data transmission protocol

Due to the amount of data, which has to be transmitted, it would be preferable for the configuration of the NPA to use a channel coupling of the two involved computers. The concept of the NPA, however, was to offer the customer the graphic computer with its devices and run the simulation program itself on a remote CYBER 990. For this reason the coupling has to be made by telephone line.

Today on telephone lines data transfer rates up to 9.6 KB/s are state of the art (on a leased line). On CDC computers there are the protocols Mode3 and Mode4A which are supported under all operating systems, which are capable of data transmission via telephone. Synchronous and asynchronous transmission is possible on telephone line, but at higher speeds (> 9.6 KB/s) only synchronous. Mode3 is also restricted to data transfer rates up to 9.6 KB/s.

Therefore we decided to use the synchronous protocol Mode4A (also called UT200) which allows speeds up to 19.2 KB/s. Another reason for this decision was, that there already existed an emulation for this protocol, implemented by hardware, firmware and software.

4.2 Operation of the protocol Mode4A

The protocol Mode4A supports the functions

- file transfer
- dialog

File transfer is working in both directions :

Jobs are transmitted to the CYBER (e.g. read in by a card reader) and put into the input queue. Output files are transmitted to the terminal, which are printed there. Common to both operations is, that INTERCOM (the part of the operating system NOS/BE, which effects these actions) transmits only closed files released by the user program on the CYBER.

In dialog operation transmission in both directions is necessary. In direction to the CYBER single lines are transmitted, and in the other direction blocks of 12 lines (or less, if there are less in the INTERCOM output buffer). Transmission of the next block is effected by INTERCOM, after the user has triggered this by command. This is in contrast to the usual operation in interactive protocols, as in protocol Mode3, where no triggering is necessary.

4.3 Operation of the Mode4A Emulation

This emulation has been programmed in 1979 by Siemens. The program emulates the protocol by following 4 processes:

- transmit a spool-in-file (on a disk) to the CYBER (emulating the reading of punched cards)

- store printer output in a spool-out-file on a disk

- transmit one line from a keyboard to the CYBER

- display a block of lines on a screen

The last two processes emulate the dialog operation.

4.4 Concepts for Extensions to the Emulation

4.4.1 Concept of File Transfer (Disk to Disk)

The existing emulation makes possible only a file transfer from disk to disk. Rough estimate of the times for picking up data from disk and data transmission only on the side of the graphic computer lead to the result, that it is not possible to get a reasonable time behaviour with this mode of operation. Therefore it was necessary to extend the existing emulation to a data transfer from central memory to central memory. For this purpose we designed two concepts:

"engineer message" and "user mode".

4.4.2 Concept of "Engineer Message"

The protcol Mode4A uses two types of messages directed to the terminal: print message and console message.

A print message is part of the directed stream of data, which is meant to be printed on a line printer connected with the terminal.

The console message is part of the stream of data

- from the CYBER to the terminal, with the purpose to be displayed on the screen;

- from the terminal to the CYBER, with the purpose to be queued in the input queue for batch processing;

- from the terminal to the CYBER, with the purpose to be interpreted as an INTERCOM or batch command.

This concept adds to this two types of messages a third one, the "engineer message". This type of message is used to send data from the user program ("engineer program" = simulation program) on the CYBER to the graphic computer.

The extension of the Mode4A emulation would mean in this concept, to identify this engineer message and write it to a circular buffer in central memory of the graphic computer. For identification of this type of message, we studied three methods:

a. sequential: the first block after a request for data from the terminal
b. identification : on the protocol level with an ASCII control code
c. identification: covering the engineer message as a console message, starting with an identification string, which does not occur elsewhere

But to all these three methods, we found severe disadvantages:

a. The operating system NOS/BE is a batch oriented operating system with some interactive parts and it has no interrupt capabilities. An extension to it for a "request for data" from a terminal would mean a great amount of changes inside the operating system.
b. INTERCOM does not include such messages, like an engineer message, and would have to be extended. This extension would be very expensive, because INTERCOM operates on two computers (central processor and the host communication processor)
c. We did not know enough about packing of lines of data to blocks, which is done by INTERCOM, and it proved difficult to get any information about that. Therefore we had to assume, that every begin of line could be the begin of a transmitted block. So every line must contain the identification string. This leads to a higher load of the telephone line and the central processors (of both computers), but in principle it would be feasible.

4.4.3 Concept of the "User Mode"

Alternatively we made a concept realised finally, which does not imply any changes to the operating system and to the protocol. We started from the fact, that the protocol Mode4A works alternately in the modes ´dialog´ or ´file transfer´. This concept adds as a third alternate operating mode the 'user mode'. This mode has to be switched on and off by a command to the emulation and not to INTERCOM.

In this user mode the messages, which are sent to and received from the CYBER, respectively have to be written in or read from parts of the central memory, which can be accessed by the user program on the graphic computer R30. All data received from the CYBER are written in a read buffer within the central memory of the R30, and after receiving a 'POLL' from the CYBER data from a write buffer are transmitted to the CYBER. Both buffers are circular.

Realization of this concept should be easier, because no extensions or changes are necessary to the protocol itself or to the operating system. All additional tasks are attributed to the emulation extension. This leads, however, to following problems:

- all coordination has to be done by the user program on the graphic computer;
- all identification of data must also be done on the graphic computer. (This is not so easy, since all dayfile messages, INTERCOM messages etc. are received by this program and must be identified.)

4.4.4 Problems encountered in Implementation of the User Mode

Soon after beginning to implement the above mentioned concept, two problems arose, which are both attributes of the Mode4A protocol:

a. Immediately after receiving a message containing N characters, INTERCOM returns 79-N blank characters.
b. After sending 12 lines of data from the corresponding buffer, INTERCOM suspends further transmission, until it receives a message (which can even be a null string).

Both operations are due to the User Terminal 200, from which this protocol originated. After sending a message from the keyboard to INTERCOM, INTERCOM clears the rest of the corresponding line on the display (this are the 79-N characters). After transmission of a block of 12 lines the UT200 display is full, and INTERCOM waits for a receipt of the terminal operator.

Both attributes of the protocol had a fatal effect on the time behaviour of the simulator. So these problems had to be solved. This was done as follows:

a. The graphic computer is sending only full lines (79 characters) to the CYBER, by adding trailing blanks.
b. After receiving a block of data from INTERCOM the emulation
   sends back a null string to INTERCOM.

After increasing the transmission rate from 9.6 KB/s to 19.2 KB/s, we noticed, that the effective rate of transmission (that means from program to program) decreased at a factor of 1.2 to 10. With the help of a line monitor, we recognized following behaviour of the computer coupling. INTERCOM sends - in 3 different cycles - to the R30 a 'POLL'. This is a prelimi- nary to following actions: INTERCOM wants to send data (select- ing phase) or INTERCOM is ready to receive data (polling phase).

The R30 can answer this 'POLL' in 3 ways:
- sending ACK (acknowledge); thus starting the desired action
- sending NAK (not acknowledge); INTERCOM repeats the 'POLL'
- no reaction; INTERCOM repeats the 'POLL'

There are three polling cycles from INTERCOM:
- ca. 1 second
- 10-20 seconds
- > 1 minute

If all 'POLLS' are answered with ACK, the polling cycle is nearly all the time the fastest. If one NAK is answered, INTER- COM switches to the next cycle, and if one 'POLL' is not an- swered, INTERCOM switches to the slowest cycle.

With this knowledge we found an error in the emulation, which leads too often to a 'POLL' without answer. After correction of this error the double transmission rate of 19.2 KB/s at least slightly increases the effective data transmission rate. There seems to be a limitation of the hard- or software involved.

4.5 Experience in Using the NPA abroad

In November 1986 we had presented the NPA at the SNE fair in Salamanca (Spain). For this purpose we transported the graphic computer with all its devices (the same amount we offer our customers) to this place. We leased a telephone line from there to Erlangen, which was guaranteed to operate up to 9.6 KB/s. But tests we made in Spain figured out clearly, that due to the typically low line quality under exhibition conditons, time behaviour at that speed was even worse than at 4.8 KB/s . Additional tests with other transmission rates showed, that the best effective data transmission rate should be at 7.2 KB/s. Using this transmission rate, we had good results in the time behaviour of the NPA.

But after all this experience as well as two other exhibitions (in Geneva, Switzerland and in Karlsruhe, Germany) proved, that our decision for the configuration of the NPA is not only feasible but leads also to good results.

5. Conclusion

The TRAC computing program is at present being prepared within the scope of a development project for future use in the simulation of the entire loss of coolant spectrum. It will then be possible to simulate all transients involving phase separation and thermal non-equilibrium using one of the best thermal hydraulic programs worldwide available on a simulator. It is, however, to be noted that real-time simulation is not possible with TRAC (just as with RELAP) and that the computing costs will be considerably higher.

Work is in progress on a simulation model for KWU boiling water reactors on the basis of the VERENA computer code.

The realization of a Nuclear Plant Analyzer based on distributed processing was shown. It offers onsite training capability to the shift personnel on an economic base despite of the high computational capacity needed. Thus the KWU NPA will be an instrument, which improves plant availability and safety.

The Design, Development and Operation of a Compact Nuclear Power Plant Simulator

Michael F. Lynch

E. Grimm

Westinghouse Electric Corporation
P. O. Box 598
Pittsburgh, PA 15230-0598

Nordostschweizerische Kraftwerke AG
Kernkraftwerk Beznau
CH-5312 Dottingen, Switzerland

Abstract

This paper discusses the philosophy and technological considerations
necessary for constructing and utilizing a plant specific compact nuclear
power plant simulator, how it compares to full scope replica simulators,
engineering simulators, part task simulators and basic principles training
simulators. Included in this discussion are the design process, scope of
simulation, the manufacturing process, test programs and experiences with
operator training.

Items addressed include the applicability and use of a compact simulator,
how well it reproduces the actual reference plant, how well the transferal
of knowledge is accomplished and what financial considerations need to be
evaluated.

This paper will try to provide the details on just how this type of
machine was designed and developed by Westinghouse for the Swiss Utility,
Nordostschweizerische Kraftwerke (NOK) AG.

Introduction

In the United States, the market is driven by regulation toward providing
full scope replica simulators built in accordance with ANS/ANSI-3.5.
Outside the United States, the type of simulator needed for operator
training is up to the individual utility. Those utilities that do not
choose to operate a full scope simulator have available to them a complete
range of alternative simulators. At the upper end of the scale are the
full scope machines while at the lower end of the scale are the part-task
or basic principles training simulators. Between these two extremes there
are several types of machines that possess characteristics from either end
of the spectrum. An engineering simulator may appear to be a basic
principles machine but its software is more sophisticated than a full
scope machine. On the other hand, a compact simulator does not appear to
reflect the actual reference plant control room but the response of its
software is just as sophisticated as a full scope machine.

The market price for simulators ranges from 100,000 USD upward to in
excess of 10 million USD. For years, the emphasis was on physical
duplication while what is under the skin, software, has progressed at a
phenomenal rate in the ability to simulate a level of fidelity that was
not even imagined before TMI. The data base gathered since TMI and the
computing power provided by today's 32 bit super minicomputers and
intelligent I/O have allowed the software engineer almost unlimited
freedom to provide increased fidelity approaching that of reactor manufacturer's design codes. As a result, a utility procuring a power plant
simulator has a wide variety of machines to choose from with various
levels of software and hardware fidelity. This high price versus sophistication has created the need for a new machine, a compact power plant
simulator. A compact simulator is one having the capabilities of a full
scope simulator but packaged so that it can be located in an area the size
of a classroom. A compact nuclear power plant simulator combines the
technology of full scope replica simulators, engineering simulators, and
basic training simulators into an extremely powerful machine capable of
the most demanding tasks. This type of machine has been developed to fill
the gap between the extremes of the market and to provide a cost effective
compromise sought by many utilities.

The Design Process

Control Boards

Since the compact simulator does not have to duplicate the reference plant control boards, the task of designing the control boards can become somewhat difficult. It is easier to duplicate than to design. The first objective was to determine what was an acceptable size for a compact simulator taking into consideration the facility in which it was to be used. In the case of the NOK Compact Simulator an area of approximately 36 square meters was available for the simulator. An additional requirement was that the simulator should be sized for operation by one individual. Two basic designs were considered. One, a benchboard type operated from a standing position and the other, a console type design operated from a sitting position. The standing design would have the controls positioned so that an operator would have to reach down to manipulate the controls while having to observe resultant control responses of meters and CRT displays. The CRTs would then have had to be placed in an upward angle, a very undesirable situation due to inherent reflection of ambient light. The alternate design was for a sitdown control board having a three part winged configuration (Figure 1) typical of many full size control rooms and one that reinforces the feeling of being surrounded by an actual power plant control room. All controls are located within recommended reach based upon human factor guidelines. The CRTs are angled slightly downward toward a sitting operator. After proper dimensioning, it became apparent that this design allows operation from both a sitting or standing position as well as providing the extra benefit of having a usable area for placing training materials and procedures used during operation. In order to establish a relationship to the actual plant control boards, it was decided to duplicate in color and design the component (pump, heat exchanger, piping) mimics used in the reference plant as well as using meter scale ranges and engraving identical to the reference plant. Although difficult to achieve on such reduced sized control boards, the transferal of learning is greatly enhanced. Even though the simulator is physically small, experienced operators feel quite at ease and are immediately capable of operating the simulator due to the cognitive relationship built into the design.

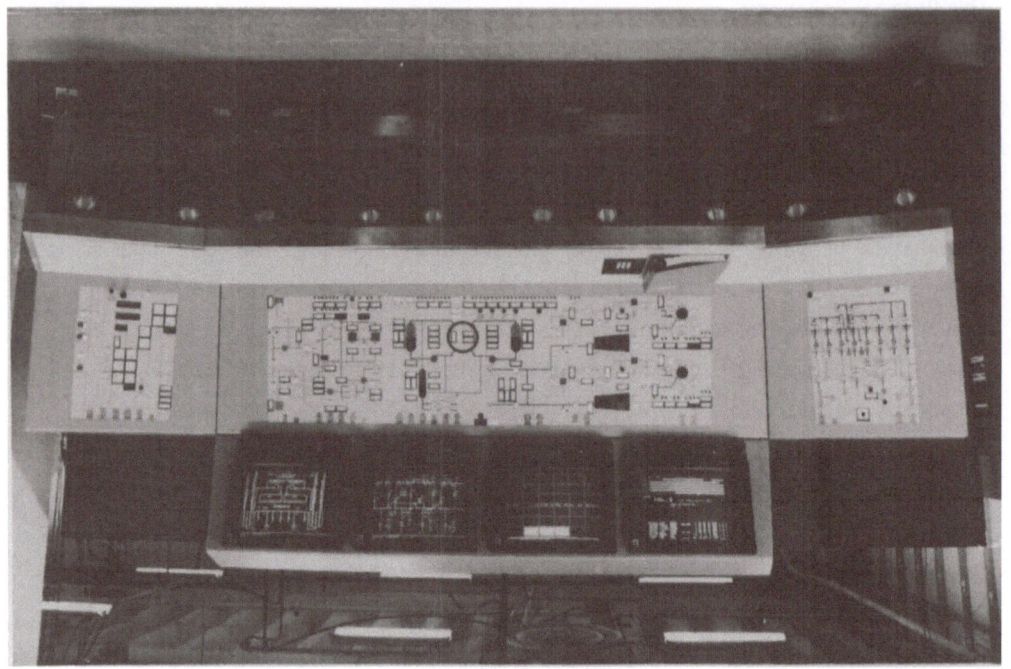

"Figure 1. The Compact Simulator"

<u>Instruments and Controls</u>

When one thinks of the various sources for meters, control switches, and
status lights available from today's manufacturers, you might assume that
the designers' options are endless. This may be true if you have
sufficient project lead time to mix and match the best each manufacturer
has to offer. In reality, very few manufacturers have a complete line of
products available to construct a complete control system. One manufacturer may specialize in analog meters, another in switches, another in
digital displays, and even fewer in miniature control room components
needed for a compact design. The customer specified requirement for a
grid-tile (mosaic) type of construction combined with a manufacturing
requirement for a metal substructure narrowed the selection to only two
qualified vendors. All components making up the entire control portion of
the boards, including mimic layout and engraving, were ultimately obtained
from a single supplier thus reducing lead times associated with having to
deal with multiple suppliers and greatly facilitating component integration.
The basic building block for a grid is a 50 mm square. Each 50 mm square
is capable of housing any combination of up to two meters on four control
switches. Each wing is made up of 88 squares arranged in an 8 x 11 matrix.

The center section is 418 squares arranged in a 38 x 11 matrix. Most power plant designs can be accommodated in this amount of space. For consistency, all levels and valve positions are indicated on vertical meters and all flows, temperatures, and pressures are indicated on horizontal meters. Where possible, all meters and controls are placed on or within mimic lines and symbols that reflect the actual layout of the systems.

Computer System

In order to maintain uniformity and standardization, the choice of a computer system was a GOULD Concept 32/97, an overwhelming choice of most full scope replica simulator manufacturers. Westinghouse, as a supplier and user/owner of full scope and engineering simulators, made the choice to use the same type of computer system for its compact simulator. The advantages are complete portability of software, consistent use of the GOULD MPX operating system, and the familiarity gained through past and current application and use. To try and introduce a new and different computer system would have required an extended learning period. The customer advantages include the ability to interface with owners or operators of full scope and engineering simulators in areas of common interests such as computer system maintenance, modeling, and operation. In fact, during the manufacture of the compact simulator, use was made of both in-house engineering simulator and full scope simulator computer systems for compact simulator model development.

Scope of Simulation

The scope of simulation of the compact simulator includes all major systems necessary to perform all normal, abnormal, and emergency operations. The NOK specification and subsequent system "tracing" of plant system diagrams by a combined NOK/Westinghouse project team defined the final scope of simulation. The compact simulator models 27 major plant systems to the same standards of fidelity and performance as a full scope simulator. This includes multi-phase Reactor Coolant System models and complete simulation of Plant Control and Protection Systems. All models are plant specific and are based upon NOK Beznau Unit 1, a 2-Loop Westinghouse Pressurized Water Reactor with dual BBC Turbines having a generating capacity of 350 MW(e).

In order to maintain its compact size, only single channel indications of process parameters (temperatures, pressures, flows) are provided by control board mounted instruments. Since redundant equipment trains and channels are necessary for proper plant control and protection system operation, the models are identical to those found on full scope simulators. Redundant channels or train indications, typical of nuclear power plants, can be accessed on any of four CRTs located above the control portion as well as on a dedicated and independently located instructor graphic CRT, one of two located on a separate instructor's console.

Two different graphic packages are provided as a means of presenting all the operator required information. One set of graphic displays duplicates those available to the operator in the actual control room, further enhancing the transition between the simulator and the actual plant as well as the transferal of knowledge. The second set of graphic displays is based upon simulated plant systems. This second set of graphic displays serves two purposes. It is available to the trainee/operator to provide a more complete picture of the system or process as well as being available to the instructor. Also, the instructor CRT is equipped with a touch screen for manipulating remote equipment necessary for plant operation but not available to the control room operator, such as the operation of local manual valves or pumps.

The Manufacturing Process

The manufacturing process for full scope simulators now averages approximately 30 months with control board manufacture taking 10 to 12 months and software integration and testing taking 18 to 20 months. By comparison, customers procuring an engineering or compact simulator expect to have their machine within 12 to 18 months. This expectation places a severe strain on the entire manufacturing process. Everything is critical path. Basic principles training simulators or a generic engineering simulator can be delivered to customers in 12 months or less. In the case of a "plant specific" compact simulator, the short schedule combined with the plant specific modeling require a high degree of time dependent actions that include control board manufacture concurrent with modeling effort starting from the date of the contract award or sooner. Manufacture of a plant specific compact simulator requires a risky head start by the manufacturer. The actual physical manufacture of smaller control boards does take less time, is a known variable, and is not the limiting factor, assuming acceptable and existing designs are available. The software effort is the limiting item. On a full scope simulator project, the first few months are spent defining the scope of model simulation. However, in order to meet an 18 month versus a 30 month schedule, the scope of model simulation for a plant specific compact simulator should essentially be complete by the time of contract award. This effort has to be an integral part of the specification/proposal/negotiation chain of events. There is no time to "come up to speed" on such a project. This entails early use of in-house computer systems, setting up the software DATAPOOL structure, and establishing software protocol and interfaces. The use of a single supplier for all instruments and controls and the use of standard I/O gear and computer systems equivalent to those for a full scope simulator enabled Westinghouse to manufacture the plant specific NOK Compact Simulator from the ground up in eighteen months. Plant specific compact simulators are not tuned copies of a similar plant, and therefore the system modeling effort approaches that of a full scope simulator. Most simulator manufacturers have well established data bases for various reactor plant designs that provide a good starting point for critical mathematical models. However, the secondary plant systems such as turbines, turbine controls, condensate, feedwater and, without question, the electrical plant systems are always unique to the reference plant if the compact simulator is to be considered truly plant specific. The secondary plant models thus became the critical path items for such a project. The use of the grid-tile (mosaic) system of I&C mounted in a standardized control board provides the hardware flexibility required for accommodating various plant designs.

The Test Programs

The increased sophistication of model software brought about by the corresponding increase in computing capability has resulted in simulator test programs that approach, and in many cases, exceed actual plant startup test programs, particularly with respect to older power plants. The impetus for simulator testing has for years been directed at proving a simulator meets or exceeds the requirements of ANS/ANSI-3.5. Many simulator manufacturers augment ANS/ANSI-3.5 recommended tests with power plant vendor startup, hot functional, and core physics tests. Over the years, the simulator owner/operators have added surveillance tests, tests to verify emergency operating procedures, and stand alone system tests. In today's full scope simulator market, it is not unusual to have a defined six month test program separated into two phases, a vendor test program followed by a customer test program. In this respect, the NOK Compact Simulator test programs were not very different than programs conducted on full scope simulators. The reasoning for two separate test programs is due, unfortunately in large part to manufacturer's performance; it seems there is never enough time to complete all the software work and preliminary tests before the start of formal factory acceptance testing.

Simulator customers now require some period of time for testing or retesting the simulator after completion of a preliminary test program. The industry norm for a total simulator test program during the period of 1980 to 1985 was three months of at least two shifts of testing a day. The conduct of test programs ultimately led to customer/vendor conflicts; the vendor desiring to ship the simulator, the customer wanting to have all software and hardware discrepancies corrected before shipment. Rather than continue to create conflicts, simulator vendors have established two-part simulator factory acceptance test programs, a preliminary or vendor acceptance test, and a final or customer acceptance test. The latter being anywhere from a subset to a complete repeat of the former. Most full scope simulator test programs now take three to six months to complete due to the increased number of tests. The test program for the NOK Compact Simulator was essentially the same as most full scope acceptance test programs. The preliminary factory acceptance test program was scheduled for two months, the final test program for one month, with the final test program being a subset of the most significant tests conducted during the preliminary program. Even with fewer simulated systems, the acceptance test programs required the entire three month period.

Experience With Operator Training

Compact simulator operational experience gained to date includes Westinghouse and NOK operations personnel as well as project management personnel, all of whom have operations backgrounds. It has been learned that the reduced size of compact simulator control boards requires the operator to be much more attentive to the control board indications. This is a little more difficult since the meters are 50 x 25 mm in size. For example, narrow range analog instruments become much more meaningful on a compact simulator, i.e., a 275 to 325°C meter is more useful than a 0 to 325°C meter. Status lights and alarms become the primary indicators of transient conditions, and they should be as long as they immediately direct the operator to the appropriate controls. It is quite important to place alarm annunciators as close as possible to related instruments and controls, which is not always the case or even possible in actual control rooms. The operation of a compact simulator is enhanced and supported by graphic displays. Main control board instruments provide indication of primary or critical parameters but the ability to completely diagnose the cause of a particular transient is most often confirmed or reinforced by the information displayed on graphic CRTs. Graphic displays have proven to be a significant factor in operating the compact simulator. The use of graphic trending and bar chart displays also provide the operator with immediate feedback. Therefore, it is very important that the compact simulator operator have easy access to graphic displays. On the NOK Compact Simulator this was achieved by providing the same man-machine interface that exists in the actual control room, a numeric entry keypad of the same design for selecting displays, a further case of reinforcing operator response. Further examples of "helping" the operator operate include having status light engraving correspond to the lighted condition. For example, a trip is active when the status light is on or a valve is open when its status light is on. Another example is having pump start/stop pushbuttons flash when the associated pump is tripped or having valve open/close pushbuttons flash when the associated valve or valves are opening and closing. This is facilitated by designing the simulator with only momentary contact switches. This operating logic also helps reduce the number of required alarm annunciators on the control board. The existing hardware alarms are augmented by a CRT based alarm system. If for example, a low level alarm condition is indicated on the control board, a review of the CRT alarm system will identify exactly which channel is causing the alarm. This is made possible because the models are full scope models by design.

Applicability and Use

The applicability and use of a plant specific compact simulator varies depending upon regulatory requirements and the desires or ultimate goals of the utility. In the United States, the regulatory pressure and requirements dictate the use of full scope replica simulators for operator training. Even this should not preclude the use of compact simulators for initial operator training, emergency team training, or procedure verification. Given the required amount of annual retraining required by regulation, many multiple unit utilities are finding it increasingly more difficult to train operating staffs on a single full scope simulator. Many multiple unit plant sites with a full scope simulator are now using the simulator 24 hours a day. The use of a compact simulator in conjunction with full scope simulator training can significantly reduce this burden.

The differences between similar units also become transparent when using a compact simulator. This can also be further minimized by having software compensate for differences between the units. Similar units at a common site always operate with different fuel cycles and sometimes have completely dissimilar secondary plant operating characteristics. Use of a compact simulator combined with plant specific software can reduce such conflicts. Different fuel cycles can be accommodated by simply having a separate disk with only the reactor core model initialized to reflect the desired reference plant. Differences in secondary plant operating characteristics due to different steam generators, feedwater control systems, or turbine control systems can also be easily demonstrated through the use of plant specific models without having to change the physical layout of the control boards.

Outside the United States, where the regulatory process does not require a plant specific full scope replica simulator and where there is considerably more plant standardization, a single full scope simulator may have to accommodate operators from as many as six units. There is just not enough simulator time available. With a compact simulator a crew can practice and train on the compact simulator and be examined on the full scope simulator.

Financial Considerations

The cost of a plant specific compact simulator is approximately 25 percent of a full scope replica simulator. For a utility buying or now operating a relatively new full scope machine, the cost of obtaining a compact simulator to support the full scope machine can be substantially less due to software portability. It is the author's opinion that 80 percent of required training objectives can be accomplished on a compact simulator which represents a significant return on the investment. The ability to reduce the competition for precious full scope simulator time and having the ability to train non-operator personnel require more than a casual analysis of the usefulness of a compact simulator. Furthermore, given the 40 year life expectancy of nuclear power plants and the fact that many were built in the late 1960's and early 1970's for less than 100M USD, the thought of having to invest another 10+M USD in a full scope replica simulator is hard to imagine and even harder to justify. Indeed, even today a plant built in 1970 and which is expected to operate until 2010 can be almost impossible to duplicate exactly from a hardware point of view since many control board components are no longer manufactured. Full scope simulator manufacturers are now finding it increasingly difficult to locate switches and meters still in use in the reference plant, a condition that will escalate and one that can significantly increase the cost of a full scope simulator. A plant specific compact simulator using the latest available hardware is not affected by material obsolescence and becomes a financially attractive alternative.

Summary

The plant specific compact simulator is a machine that bridges the gap between a basic principles training device and a full scope replica simulator. Its size does not place an additional burden upon the utility to have a large facility, is every bit as sophisticated from a modeling standpoint, has the potential for creating a phased approach to plant operator training, and allows non-operator training of the support staff. The compact simulator will never replace the quality of training that is achieved on a full scope machine but is intended to augment and support the operation of the power plant. For those utilities building new plants, a compact simulator can be an attractive alternative for initial operator training until the plant design is finalized.

Design and Analysis of Nuclear Processes with the Apros

M. Hänninen*, E. K. Puska*, P. Nyström**

*   Technical Research Centre of Finland
    Nuclear Engineering Laboratory
    P.O.B. 169
    SF-00181 Helsinki, Finland

** Imatran Voima Oy
    Electrotechnical Department
    P.O.B. 138
    SF-00101 Helsinki, Finland

Summary

APROS is the product being developed in the Process Simulators
project of Imatran Voima Co. and Technical Research Centre of
Finland. The aim is to design and construct an efficient and
easy to use computer simulation system for process and
automation system design, evaluation, analysis, testing and
training purposes. At halfway of this project a working system
exists with a large number of proven routines and models.
However, a lot of development is still foreseen before the
project will be finished. This article gives an overview of the
APROS in general and of the nuclear features in particular. The
calculational capabilities of the system are presented with the
help of one example.

1. Introduction

The APROS (acronym for Advanced PROcess Simulator) is the product being developed in the Process Simulators project of
Imatran Voima (IVO) and Technical Research Centre of Finland
(VTT). This three-year co-operation project was started in
1986. The aim is to design and construct an efficient and easy
to use computer simulation system for process and automation
system design, evaluation, analysis, testing and training
purposes.

In building the APROS the flexibility of the system and
easeness of use are emphasized. For most applications little or
no expertise in computer programming, operating systems or
simulation techniques and solution algorithms is required.

2. System Overview

2.1 General

In Figure 1 the composition of the APROS is shown. Four main
elements can be distinquished: the workstation system, the
simulator executive system, the model programs and the application specific files. The simulator executive system can be
considered as the core of the system interpreting system commands, managing system data bases and controlling the simulation. Application specific files contain the description of
the specified process to be simulated. Model program files
include the generic component models, solution algorithms and
fast access material property tables. The workstation provides
the user with a graphics I/O system which includes a commercial
CAD-system for input preparation and several modules for data
output and user communication.

2.2 Software

All programs in the system are written using the standard
Fortran F77 language. In all programs the principle of strict
modularity is emphasized. Due to the modularity and good documentation the program entity is comprehensible. It is easy to
replace small parts of the code without a thorough knowledge of
the rest of the system.

Concerning nuclear applications the models of thermohydraulics
and reactor kinetics rely much on the work done at the Nuclear
Engineering Laboratory of VTT. The mathematical solution
algorithms used in the system as well as the fast access material property tables for steam and water origin from the work
done at the Electrical Engineering Laboratory of VTT.

2.3 Hardware

The basic development of the APROS has been done using the
VAX-computers (VAX 8650, VAX 785) of IVO. At present the system

is used also in two micro-VAX II computers reserved solely for this purpose. Thus the present versions of the APROS are for the supermini-class computer systems. However, in future the system will be available for a wide range of computers from the PC's for small applications to multiprocessor minisuper-class computers for the largest applications.

3. Design Aspects

3.1 Definition of a Process

The APROS contains three component definition levels shown in Figure 2. The highest is the so-called process level. This is composed using the elements of the lower level, the so-called generic component level. The generic components are composed of the lowest, the so-called calculational level, components. The designer works most often at the definition level two using the generic components included in the system library. A typical generic process component is a pump, a valve, a heat exchanger, a pipe, or a tank. Correspondingly, in automation system there exist such generic components as sequence program, change over automatic, PID-controller and so on.

The designer builds the system to be simulated using the generic components from the library. The desired component is selected and its connecting points to other components must be defined. The system inquires also for other parameters needed to specify the component. In addition, a user can indicate the accuracy level of the simulation by choosing one of three flow models.

If a desired generic component is not available in the library a user can define the process using the components of the level one, i.e. branches and nodes. Once process/subprocess has been created this new entity can be stored in the system with a specific name and it is ready for use in any later application.

3.2 Simulation

After a designer has built the process on the screen of the
workstation, the preparation system of the simulator forms the
calculation network of the elementary components i.e. nodes and
branches. During the preparation phase also the initialization
of the simulated process is performed and the system is ready
for simulation.

If the designer wants to alter some component/components in the
system this is realized with simple commands like "add", "de-
lete" and "modify". After the alterations the whole system is
again ready for simulation calculations.

In terms of the workstation capabilities the APROS can be considered also as a computer aided design (CAD) tool for process
and automation system design containing efficient features for
dynamic testing and analysis of the systems to be designed. In
Figure 3 the plant design data flow is represented as an
example of the APROS as a design tool.

It is also foreseen that the APROS application data base can be
linked with other CAD-systems to continue with the total design
process, e.g. to produce lay-out drawings, component and material lists, wiring diagrams etc.

4.   Specific Models

4.1 Thermohydraulics

In the calculation the hydrodynamic system is divided into
nodes which are connected to each other with flow branches. The
static properties like pressures, enthalpies and densities are
calculated in the nodes while flow properties are calculated in
connecting branches. A branch begins always from one node and
ends to another. The system allows several branches to be
connected to each node. The same node and branch structure is
applied also in the heat conduction solution.

The equations for conservation of mass and momentum can be written for a node in the following form /1/:

$$\frac{\partial \alpha_k \rho_k}{\partial t} + \frac{\partial \alpha_k \rho_k u_k}{\partial z} = \Gamma_k \tag{1}$$

$$\frac{\partial \alpha_k \rho_k u_k}{\partial t} + \frac{\partial \alpha_k \rho_k u_k^2}{\partial z} + \alpha_k \frac{\partial p}{\partial z} = \Gamma_k u_{ik} + \alpha_k \rho_k \vec{g} + F_{wk} + F_{ik} \tag{2}$$

where α_k is the volume fraction, ρ_k is the density, u_k is the velocity, Γ_k is the mass transfer rate, p is the pressure, u_{ik} is the velocity of the interface, g^+ is the component of gravitational acceleration in the pipe direction, F_{wk} is the wall friction and F_{ik} is interfacial friction. The subscript k holds for the liquid (l) and for the gas (g).

The total energy equation is written in the form /1/:

$$\frac{\partial \alpha_k \rho_k h_k^*}{\partial t} + \frac{\partial \alpha_k \rho_k u_k h_k^*}{\partial z} =$$

$$\alpha_k \frac{\partial p}{\partial t} + \Gamma_k h_{ik}^* + q_{ik} + q_{wk} + F_{wk} u_k + F_{ik} u_i + \alpha_k \rho_k \vec{u}_k g \tag{3}$$

where h_k^* is the total enthalpy = $h_k + u_k^2/2$, h_{ik}^* is the total enthalpy at the interface, q_{ik} is the heat transfer rate at the interface and q_{wk} is the heat transfer rate from the walls. In the homogeneous flow the mass, momentum and energy equations of the mixture are employed.

For the thermohydraulic calculation in the APROS a one dimensional homogenous two-phase flow model and a one-dimensional heat conduction model for solid structures are currently used. In the present method the pressure equation is formed using the discretized and linearized mass and momentum continuity equations. After the pressures have been solved from the pressure equation, the mass flows can be computed from the momentum equations. Then, using new pressures and mass flows the mixture enthalpies can be solved from the discretized energy equations. Due to non-linearities an iteration procedure must be applied. When calculating pressures and enthalpies the linear equation groups having as many equations as nodes to be simulated must be solved. In order to get fast solution of the linear equations the sparse matrix technique is applied. The density and other steam and water properties are calculated with the aid of the material property tables using the pressure and enthalpy as arguments. The material property tables for water and steam have a very fast access and they have been specially tailored for use in the APROS. For tanks and other large vertical volumes the homogeneity assumption is abandoned and the water and steam are considered fully separated.

The solution for the separated two-phase flow is performed according to the same methods as in the homogenous two-phase flow. The separated two phase flow is solved with a six-equation model. In the model the matrices for pressures, void fractions and for phasial enthalpies are solved. The new phasial velocities are calculated using new pressures and old velocities.

For heat conduction calculation in solid structures a one-dimensional heat conduction model in cartesian, cylidrical or spherical coordinate system is available. The boundaries of the heat structures are either at constant temperature or isolated or they are limited to a hydrodynamic node. Also heat generation inside a heat structure can be described. When a heat structure is connected to a hydrodynamic node a heat transfer correlation package is used to calculate heat flows between heat structures and fluid. Temperatures are calculated in each node using the material properties, geometric dimensions, boundary conditions and heat generation rates in the zones left and right to that particular node. The heat conduction is considered in the radial direction in each heat structure. Axial heat conduction in the structure is not calculated. The heat structures are connected to the hydrodynamic nodes via the heat transfer correlations. In most applications the axial division of a heat structure is

the same as the nodal disretization of the hydrodynamic model
where it is connected to. However, the axial length of the
structure can be different from that of the hydrodynamic
node.

4.2 Reactor Kinetics

Concerning the reactor the most straightforward approach for
the user is to omit the actual kinetics analysis by inputting
the thermal power produced by each axial fuel section of the
core. This option is useful in cases where the role of the
reactor is to function solely as an auxiliary device
producing the required amount of heat.

The point kinetics approach of the APROS for neutron flux and
delayed neutron concentration starts from the well-known
equations (e.g. /2/)

$$\frac{d\phi(t)}{dt} = \frac{\rho(t)-\beta}{\Lambda} \phi(t) + \sum_{i=1}^{N} \lambda_i C_i(t) + S \qquad (4)$$

$$\frac{dC_i(t)}{dt} = \frac{\beta}{\Lambda} f_i \phi(t) - \lambda_i C(t), \quad i = 1, N \qquad (5)$$

where ϕ is neutron flux, ρ is reactivity, β is effective
delayed neutron fraction, Λ is prompt neutron generation
time, λ_i is decay constant of group i, f_i is fraction of
delayed neutrons in group i, C_i is concentration of delayed
neutron group i and t is time.

Six delayed neutron groups are used in APROS. The equations
are discretized in respect of time and the neutron flux and
delayed group concentrations are obtained readily from the
discretized form of the equations. In the calculation of
reactivity the changes versus stationary state are described.
The total reactivity is obtained as a sum of the components
due to control rods, boron, coolant density, fuel temperature
and coolant void fraction.

The iodine and xenon behaviour are also described with the wellknown equations (e.g. /2/)

$$\frac{dX}{dt} = -\lambda_x X - \sigma_x \phi X + \lambda_I I + \gamma_x \Sigma_f \phi \qquad (6)$$

$$\frac{dI}{dt} = -\lambda_I I - \sigma_I \phi I + \gamma_I \Sigma_f \phi \qquad (7)$$

where X is xenon concentration, I is iodine concentration, λ_x is xenon decay constant, λ_I is iodine decay constant, σ_x is xenon absorption cross section, σ_I is iodine absorption cross section, γ_x is xenon effective fission yield, γ_F is iodine effective fission yield, Σ_f is fission cross section, ϕ is neutron flux and t is time. The equations are again discretized versus time and the solution for iodine and xenon is readily obtained.

In the APROS the reactor core can be divided axially into n sections which can be of various length. For each of these sections a point kinetics calculation is performed taking into account the average conditions in that heat structure section and in the thermohydraulic section associated with it.

The point kinetics approach is suitable for several applications of the APROS. However, for discussion of reactor transients more detailed methods are required. The one-dimensional neutron flux model describing the axial behaviour of the flux is composed on the basis of one-group diffusion equation. In the method used in APROS the second derivative of the neutron flux is approximated with the help of finite differences. The method results in a tridiagonal matrix which is solved with a tridiagonal matrix solver especially tailored for APROS.

4.3 Automation System

Automation system can be divied into three parts; measurement system, group of elementary modules of the control system and the interface to the controlled devices.

Measurement system consists of analog and binary measurements. All the information from proces to control system is coming through this system. Measurement system generates a big group of analog and binary signals for the other control system components.

The heart of the control system consists of different kind of elementary components, which are connected to each other with analog and binary signals. These elementary components are for example controllers, adders, nonlinear curves, MAX/MIN-selectors, different logical elements, sequence programs and so on. Most of the external signals to the control system components are coming from measurement system, but some of them are generated by the user of the system.

The interface to the controlled devices consists of the device controllers for four types of devices:

- continious devices
- shut-off valves
- on/off-devices with state feedback
- on/off-devices without state feedback

The device controller gets its input signals from protection or interlocking systems or from controllers. Also the manual commands are connected to the device controller.

For the reasons of calculation it is useful to make difference between components that handle binary signals from those handling only analog signals. The calculation of these module groups is made separately.

The analog part of the control system is calculated together with the process in the same iteration loop. The binary part of the control system is calculated just after the analog calculation procedure has converged. Also the binary calculation can include its own iteration loop, if so wanted.

The components of the analog part of the control system can
be described by algebraic and differential equations. All
differential equations are discretized with time and so the
total system is converted to the group of algebraic equations
which should be solved in each time step. Because of the
discontinuous components of the control system (for example
analog switches, MIN/MAX-selectors, dead band and hysteresis
elements) the solutionof that equation group is not
calculated in matrix form, but one equation after another.

The convergence speed of the calculation depends on the
calculation order of these equations. The equations are
arranged automatically in such an order that all the
variables in the right side of an equation are, if possible,
calculated earlier during that time step. The inside
feedbacks in control system will for example slow up the
speed of convergence.

The diskretization of the equations is carried out in such a
way that the change of the time step is possible during the
simulation. Automatic control of the time step is possible.

The components of the binary part of the control system can
be described by logical equations which can contain time
delays. These equations are calculated by the same method as
analog equations. It is also very important to handle these
equations in correct order to be sure that the calculation
will converge.

5. Simulation Example

The example process describes the primary and secondary
circuit of a pressurized water reactor. In the primary
circuit the following components are described: reactor,
pressurizer, two cooling water circuits and two steam
generators. In the secondary circuit the following components
are described: waterpipes from feedwater tank to steam
generators and steampipes from steam generators to turbine

control valve. Heat structure is described in the reactor
core, in pressurizer and in the heat exchangers. In the
example the number of thermohydraulic nodes is 36 and the
number of branches is 38. In the heat structures there are
altogether 67 nodes and 54 branches. The example process is
shown in Figure 4. In this case reactor power was given as an
input.

The simulated event is a leak in the primary circuit cold
leg. In the system the leak is caused by opening the valve in
the branch B38 which leads to the external node N36 being at
pressure of 1 bar. At the early phases of the simulation the
reactor is at full power. When the pressure in the primary
circuit has fallen to approximately 100 bars the reactor
scram was modeled by changing the heating from 100 % to 3 %.
In the time scale this corresponds to approximately 150
seconds from the beginning of simulation. The total
simulation time of the event was 500 seconds. The simulation
runs were performed using time-step sizes 0.5, 1.0, 2.0 and
10 seconds.

In Fig. 5 the calculated pressures in the nodes N14, N22 and
N28, the discharge mass flow in the branch B38, the void
fraction in the node N7 and the water level in the node N22
have been presented. The physical behaviour of the system was
as expected. The main purpose of the example was to indicate
that the implicit calculation model stays stable even with
long time steps.

The total calculation time using VAX-8650 was 250 s with a
time step of 0.5 s and 26 s with a time step of 10 seconds.

6.   Further Developments

6.1 Physical Models and Solution Methods

Concerning the thermohydraulics the separated two-phase flow
model is well in the test phase. After finishing the tests
successfully it will join in the standard models of the

APROS. Development of mathematical methods two-dimensional flow calculation is also in progress. One application of this twodimensional flow calculation is the flow in the reactor pressure vessel downcomer.

Tests for the one-dimensional neutron flux model are going on. As further development in the reactor kinetics a three dimensional model is foreseen. The model will benefit the proven principles and techniques of the neutronics model of the finnish transient analysis code TRAWA /3/. In this code the basis of the neutron kinetic model is the two-group neutron diffusion equation.

The mathematical methods used in the solution of the APROS, especially the sparse matrix techniques, are well proven /4/. The entire set of physical models for flow calculation and heat transfer are unified into a network that will be solved simultaneously. Also this task is already in test phase. In the area of the fast access material property tables further development is expected only upon specific needs.

The code in the APROS has been written taking into account the possibility for vectorized parallel processing. However, this capability has not yet been tested since the present APROS environment does not contain a vector-machine.

6.2 Executive System and Interfaces

The executive system interpreting system commands, managing system data bases and controlling the simulation has already found its final framework in APROS. However, further development will take place upon specific needs, as in all living systems. For the communication of the user with the system and for the output of results properly working procedures exist. However, in these areas further development is expeced, since these systems are under the constant critical test of all the users of the APROS.

7. Concluding Remarks

At present the APROS contains all the main features of the final product. The simulator executive system has its final appearance. The fast access material property tables for water and steam have been developed. The fast implicit solution methods are working well. All the main models of the nuclear application of the system exist. In the thermo-hydraulic area the separated two-phase flow model is in the test phase. In the reactor kinetics area the one-dimensional flux model is in test phase and a three-dimensional model will be developed. The user interfaces, system display and output routines are working. However, these will still be subject of further development. The test phase of the physical models has been initiated.

References

1. Siikonen, T.: A Numerical Method for One-dimensional Two-phase flow. To be published in Numerical Heat Transfer.

2. Bell, G. I., Glasstone, S.: Nuclear Reactor Theory, New York: Van Nostrand Reinhold Company 1970.

3. Rajamäki, M.: TRAWA, A Transient Analysis Code for Water Reactors. Technical Research Centre of Finland, Nuclear Engineering Laboratory, Report 24, 1976.

4. Juslin, K., Silvennoinen, E.: Real Time Solution Approach for Sparse Network Equations. Technical Research Centre of Finland, Electrical Engineering Laboratory, Research Notes 615, 1986.

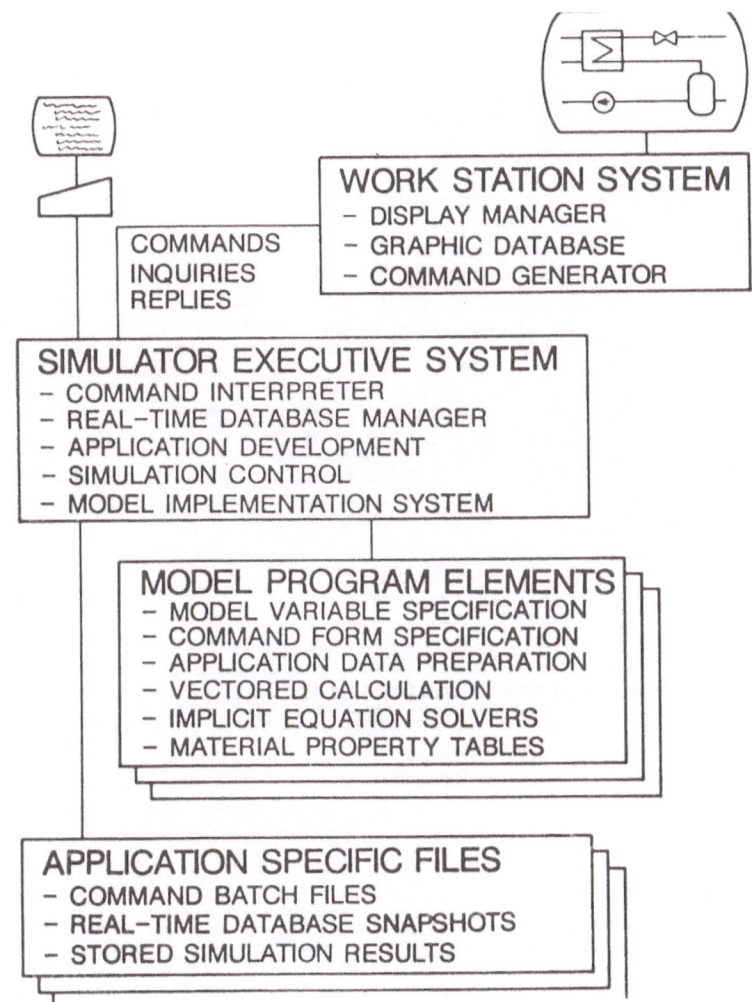

Figure 1.   The composition of the APROS.

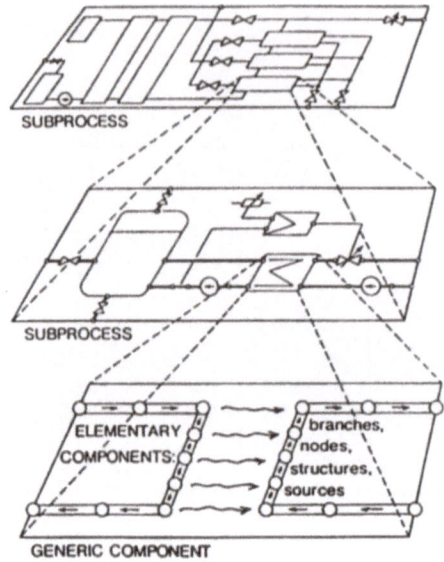

Figure 2. The component definition levels of the APROS.

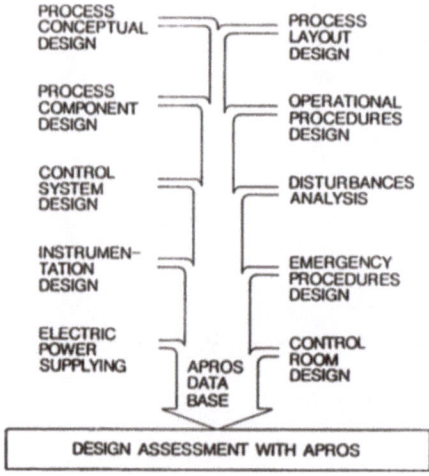

Figure 3. Plant design data flow for APROS as a design tool.

203

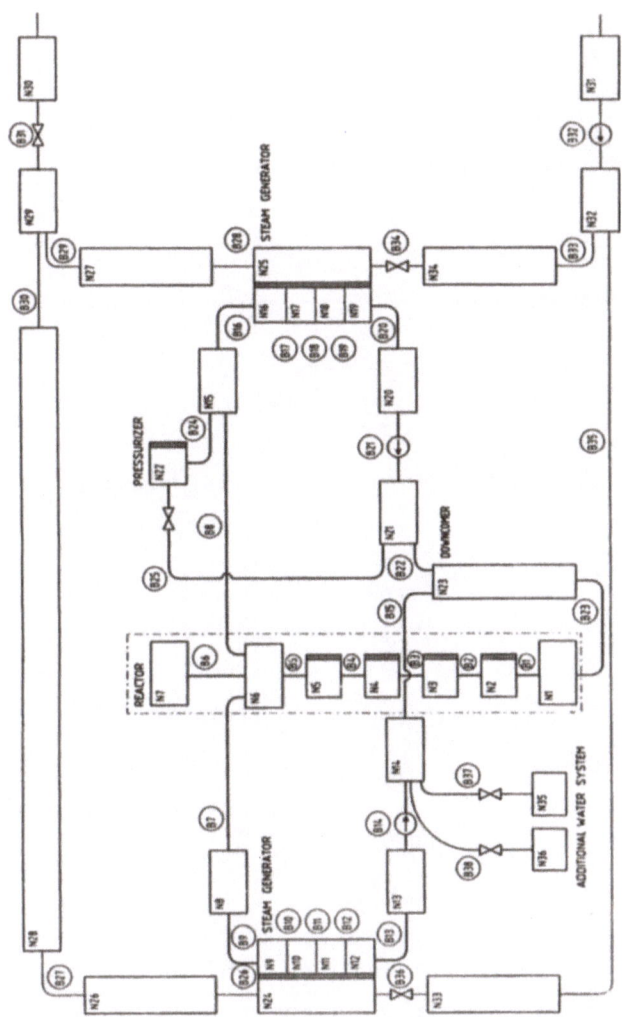

Figure 4. Example process describing a pressurised water reactor primary and secondary circuits.

204

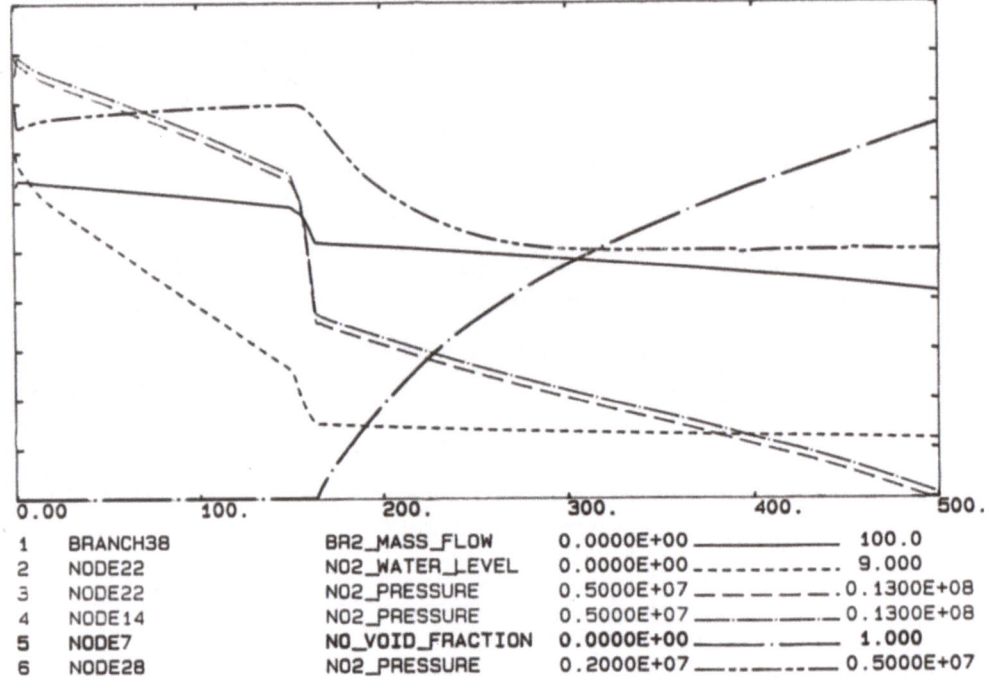

1	BRANCH38	BR2_MASS_FLOW	0.0000E+00 ————————	100.0
2	NODE22	NO2_WATER_LEVEL	0.0000E+00 -----------	9.000
3	NODE22	NO2_PRESSURE	0.5000E+07 ———————	0.1300E+08
4	NODE14	NO2_PRESSURE	0.5000E+07 ———————	0.1300E+08
5	NODE7	NO_VOID_FRACTION	0.0000E+00 —— · ——	1.000
6	NODE28	NO2_PRESSURE	0.2000E+07 —— ·· —— ·	0.5000E+07

Figure 5.   APROS simulation results. The pressures are in Pa,
the mass flow in kg/s and the water level in
meters.

The Role of Simulation in Control System Design/Modification

SEN-I CHANG SHIH-JEN WANG MIN-SONG LIN

Institute of Nuclear Energy Research
P.O. BOX 3, LUNGTAN. TAIWAN 325
REPUBLIC OF CHINA

Abstract
Due to the discrepancy between design and actual plant data,
controller tuning is required during power test of new plant.
Furthermore, after a period of operation time, the aging effect
of the sensors and components will cause the system performance
to change. And with the improvement of control system hardware,
better control algorithm can be implemented to assure the safety
of the system operation. Control system tuning/modification is
necessary to keep the system at its best performance. So, it is
not once in a system life time job.

Simulation plays an important role in control system design and
modification. Because the reactor itself has a very restricted
operation regulation, we can't afford to modify/tune the control
system by applying the "trial-and- error" method directly on
actual power plant. Instead, physical plant is modeled in terms
of set of mathematical equations in advance. Then control system
design and analysis are performed by computer simulation.

Besides the computer simulation, we can also verify and try the
newly developed algorithm on experimental reactor system. The
Taiwan Research Reactor power regulating system modification work
is taken as an example to demonstrate the role of simulation. In
this work, the model of Taiwan Research Reactor and its power
regulating system were set up and analyzed by simulation. Parameter optimization and tuning map technique is applied to optimize the control system performance. Analysis results will be
verified on Taiwan Research Reactor. These procedures can also be
applied to commercial nuclear power plants.

1. Introduction

One can always finds the discrepancies between design and actual

plant data during power test of new plant. Furthermore, after a

period of operation time, the aging effect of the sensors and

components will cause the system performance to change. And with

the improvement of control system hardware, better control algo-

rithm can be implemented to assure the safety of the system

operation. Hence, the control system tuning/modification is necessary to keep the system at its best performance. In view of these facts, it can be sure that all these tuning/modification activities will be needed from time to time, during the system life operation.

Simulation plays an important role in control system design and modification. Because the reactor itself has a very restricted operation regulation, we can't afford to modify/tune the control system by applying the "trial-and-error" method directly on actual power plant. Instead, physical plant is modeled in terms of set of mathematical equations in advance. Then control system design and analysis are performed by computer simulation.

With the goal to modify control system performance for nuclear power plant, it is good for engineer to be familiar with the modification procedure and analysis technique in the research reactor. The Taiwan Research Reactor (TRR) is a heavy water type research reactor located in Taiwan, R.O.C. During current TRR power operation, large overshoot in neutron power is observed during change in thermal power demand (this will be discussed more detail in section 3). Hence, the dynamic performance of the TRR power regulating system is degraded. But in current TRR power increasing procedure, it is divided into many steps and each step with small power demand. Thus, this degraded control system has no serious impact on the normal system operation. However, it is obvious that for large step increase of power demand during large transient tests may cause reactor trip. From the control system point of view, it is worthwhile to find the cause of the degraded control system and tune the corresponding controller setting to achieve better performance. So, the TRR power regulating system modification work will be taken as an example to demonstrate that the simulation is not limited to the computer only. The research reactor, TRR itself, also plays an important role in simulation.

The analysis work is broken into a number of sections. First, the TRR system and its main power regulating system are described briefly in section 2. In section 3, a comparison of control system performance between the designed system and current system is performed. The main cause of the large undershoot is found through the identification of the transfer function between neutron power and thermal power in section 4. In section 5, control system modification is performed based on the new transfer function. Simulation and parameter optimization are applied to get the new controller settings. The important results of this analysis are summarized in section 6.

2. Description of Taiwan Research Reactor

The Taiwan Research Reactor is a 40 MWt heavy water research reactor (1). It is built for material testing, irradiation, and neutron flux related experiments. Its main control system is the power regulating system. The function of this control system is to regulate reactor power to demand by adjusting the level of moderator, i.e. D_2O. The block diagram of this control system is shown in Figure 1. Power demand and the corresponding feedback signals are fed into the control system. The control system generates a valve demand signal to the moderator system. The moderator system contains a control valve, calandria tank, storage tank, and pump as shown in Figure 2. The calandria tank contains hundreds of flow tubes. Inside the flow tube is flow channel and fuel rod. Outside the flow tubes is filled with moderator. Constant pumping rate is flowing into the calandria tank, while the flow rate leaving the calandria tank is controlled by the control valve. By varying the control valve position, the moderator level is changed and so is the reactivity.

At high power operation, the system has two control mode. One is linear control mode and the other is thermal control mode. In linear control mode, the neutron power signal measured by neutron detector is compared with power demand. The generated error signal is used to adjust the control valve position, moderator level, and neutron power. In thermal control mode, the thermal power signal that is determined by the coolant temperature rise across the reactor core and coolant flow rate is compared with

the power demand signal. The generated error signal is used to
adjust the control valve position, and thus the moderator level,
neutron power and thermal power. This sequence forms a closed
loop. A rate compensation term include K and T is used to provide anticipatory action to avert overshoots of neutron power.

During power operation, fission power generated in the fuel rods
is removed to the cooling tower via the heat exchanger. The
coolant flow is pumped at constant flow rate. The thermal power
is measured as the temperature rise in the reactor core and
coolant flow rate. At the inlet and outlet header of the reactor,
there are two thermowells and the temperature sensors are located
outside the thermowell.   The transfer function of thermal power
to neutron power, G(S) is simplified as

$$G(S) = \frac{e^{-T_d S}}{(1 + T_1 S)(1 + T_2 S)} \tag{1}$$

Where:

S = Laplace transfer complex variable
T1 = dominant time constant of heat transfer from fuel to
      coolant
T2 = thermal time constant of the outlet temperature sensor
Td = coolant transport time from center fuel to the place
      where the outlet temperature sensor located

The parameter that has significant influence on the stability in
thermal control mode is the total delay time of T1, T2, and Td.
These are defined as delay times between neutron power and thermal power. From the analysis, we found that the delay time between neutron power and thermal power is an important parameter
for control system performance of the thermal control loop. The
thermal loop of TRR power regulating system is designed based on
13 seconds delay time between thermal power and neutron power
(2). The design values of these three delay times are shown in
Table 1.

3. Performance of TRR Power Regulating System

Since the delay times between neutron power and thermal power has strong influence on the performance of power regulating system. Thus, in the following control analysis, we will focus on the response of neutron power to thermal power demand.

After the detailed models of the control block diagram is implemented by using Advanced Continuous Simulation Language (3). The transient response of TRR during linear control mode and thermal control mode are performed. The original controller settings are used to analyze the designed control system performance. Figure 3 shows the response of neutron power for a step decrease of 5 percent of full power demand in linear control mode. Figure 4 shows the response of neutron power for a step decrease of 5 percent of full power demand in thermal control mode with 13 seconds delay time. From these two Figures, it is found that the system is designed with maximum undershoot (overshoot) no more than 3 percent of step demand. The response of thermal mode is slower than linear mode due to the delay of thermal power.

In order to check the designed control system performance, two tests with the same initial condition and perturbation in Figure 3 and 4 are performed. Figure 5 shows the transient responses of neutron power in linear mode and Figure 6 shows the transient responses in thermal mode. From these Figures, it is clearly shown that the system performance of linear control loop is quite close to the design performance. However, the system performance of thermal control loop is very different from the design case. Large undershoot of about 24 percent of step demand and the lagged responses are observed in test data.

From these comparison, it clearly indicated that these three delay times are different from design value. We believe that after long operation time, the system transfer function may vary due to both fouling in the primary system and the aging effect of the temperature sensors.

4. Identification of System Parameters

Due to the discrepancy as described in the previous section, both test data of neutron power and thermal power during power transient are used to identify the parameters shown in the transfer function of equation 1. Three delay times are identified by using Advanced Continuous Simulation Language (ACSL) and the parameter optimization code, OPTIM (4). The identified scheme is shown in Figure 7. At first, neutron power signal from TRR is fed into the system model. The generated thermal power signal is compared with the actual TRR thermal power signal and the square of the error signal is integrated to generate a performance index. Optimization method is used to adjust these three delay times by minimizing the performance index. The simplex algorithm of optimization method is applied in this analysis. These identified delay times are shown in Table 2. The total delay time is 27 seconds which is 14 seconds more lagged than the design data. Figure 8 shows the identified result and the response with the designed delay times. The identified system response is very close to the test data.

With these new delay times and original controller settings, the system is perturbed by a step decrease of 5 percent full power in thermal power demand. Simulation result of neutron power is shown in Figure 9. Large undershoot is found which is similar to the test result shown in Figure 6. This Comparison strongly indicated that the main cause of the degraded control system performance is due to the discrepancy of three delay time between the design data and the actual one.

In order to have better control system performance with these three identified delay times, it is necessary to tune the controller settings in thermal control loop.

5. Control System Modification by Simulation

For the performance of the control system is not changed much when operated in linear control mode. Plus, we found that the main cause of large undershoot is due to the deviation of the

designed delay time between neutron power and thermal power from
the actual data. So, only the controller settings in the thermal
control loop are required to be tuned.

The system with the designed parameter values is perturbed by a
step change in thermal power demand. The important time domain
performance characteristics of neutron power, such as rise time,
settling time, and maximum overshoot of neutron power are recorded in Table 3 as designed performance specifications.

The rate compensation term which contains rate feedback gain, K,
and time constant, T, in the thermal control loop as shown in
Figure 2 is used to provide anticipatory action to avert overshoots (or undershoots) of neutron power on approaching power
demand. With the delay times changed, the rate feedback gain, K,
and time constant, T, should be tuned. The goal is to tune these
two controller settings in such a way that the control system
performance approaches the designed performance specifications.

Tuning map is a map which shows a family of constant control
system performance characteristics with respect to the controller
settings. With this map, the designer can chose proper controller
settings according to the performance specifications. This technique is applied in the following control system analysis. The
system with the identified delay times is then perturbed with
step change in thermal power demand. The time domain simulation
and parameter optimization method are applied to construct the
time domain tuning map. The block diagram of constructing time
domain tuning map is shown in Figure 10. The parameter optimization with one variable is used. With controller setting , T,
changed each time in the range of interest, the controller setting, K, is obtained through the optimization procedure. The
corresponding tuning map for maximum overshoot, rise time, and
settling time with respect to K and T is shown in Figure 11.

From the TRR operation point of view, the maximum overshoot is of
primary concern. Too large overshoot might cause plant trip

during high power operation. In this study case, the desired maximum overshoot is 2.37 percent of step demand as shown in Table 3. In Figure 11, controller settings are chosen along the line with 2.37 percent maximum overshoot. At the same time, the other two performance characteristics, the rise time and settling time, should not change much from the values shown in Table 3. Finally, the new performance characteristics are chosen from the tuning map as shown in Table 4. The corresponding new controller settings for K and T are shown in Table 5.

By using these new controller settings, simulations are performed. A comparison of the control system performance of new settings with the old settings is shown in Figure 12. With the old controller settings, the neutron power experienced a large overshoot of about 24 percent of step demand. However, the overshoot of the neutron power with new controller settings is reduced to 2.37 percent of step demand. Also, frequency domain analysis is performed to check the relative stability of this control loop. Comparison result is shown in Table 6. Both gain margin and phase margin of the modified system are larger than the old system. From the above analysis, it is evident that the performance of the modified system with the tuned K and T is better than the old system.

6. Conclusions and Recommendations

The parameter optimization code has been developed and verified from the identification of three delay time between neutron power and thermal power. Large undershoot of neutron power in thermal mode operation is found due to the discrepancy of these three delay times from the design value.

With the new delay time, time domain tuning map for the thermal control loop is constructed with the aid of simulation and optimization method of one variable. The controller settings are chosen easily from this tuning map according to the design specifications. After that, simulation results verified the performance of the new controller settings. We believe that the result of this analysis with simulation and optimization method could be applied to the actual TRR control system modification. We strongly recommended that hardware modification be taken to confirm the simulation results and up-grade the control system performance.

Due to the aging effects, the performance of nuclear power plant would be degraded. Control system modification work is necessary and it is not once in a system life time job. The TRR, itself, is taken as a "simulation" for control system modification in this study. The analysis work of TRR control system modification has been successfully completed. After engineers go through these entire procedures and familiar with these techniques, we are convinced that it will help a great deal in the control system modification work of commercial nuclear power plant in the future.

7.Acknowledgements

The authors would like to thank nuclear engineering division and TRR personnel of Institute of Nuclear Energy Research for their supporting of this work.

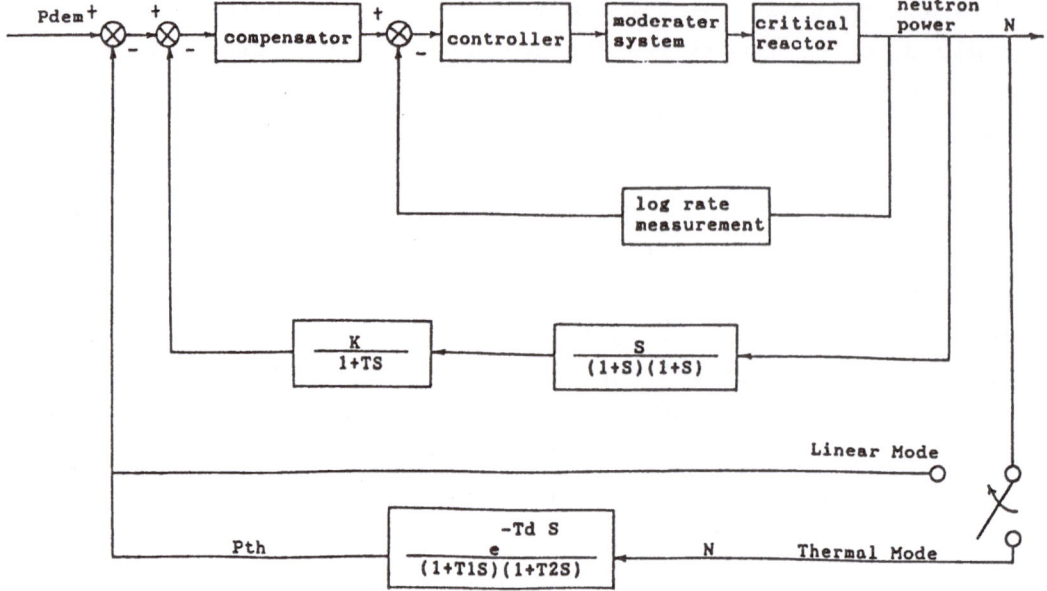

Figure 1. Block diagram of power regulating system

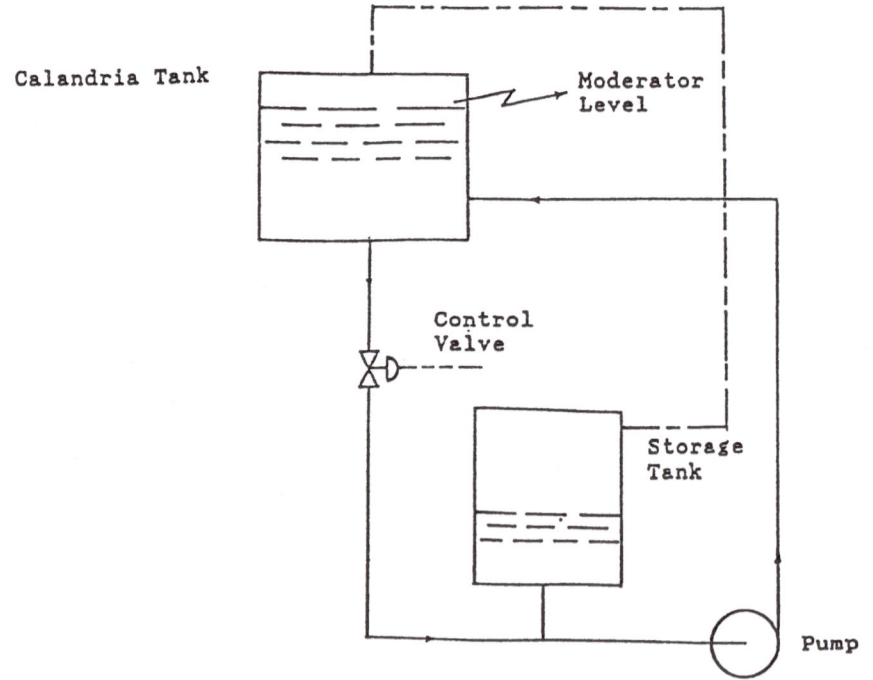

Figure 2. Block diagram of moderator system

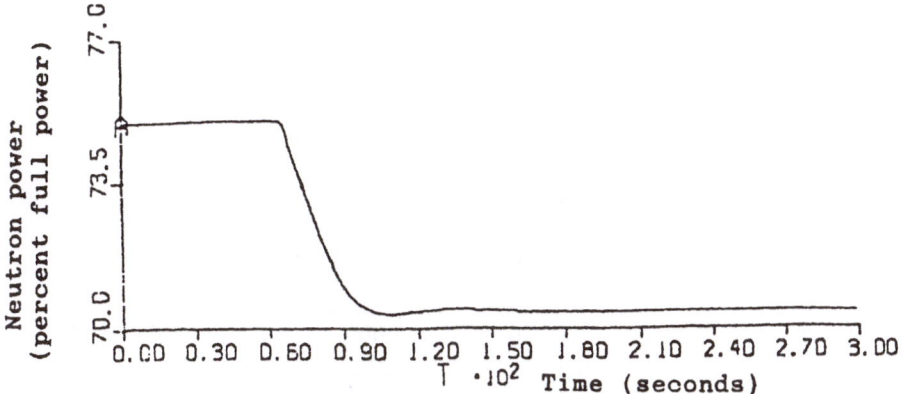

Figure 3. Response of design system to a step decrease of
5% in neutron power demand

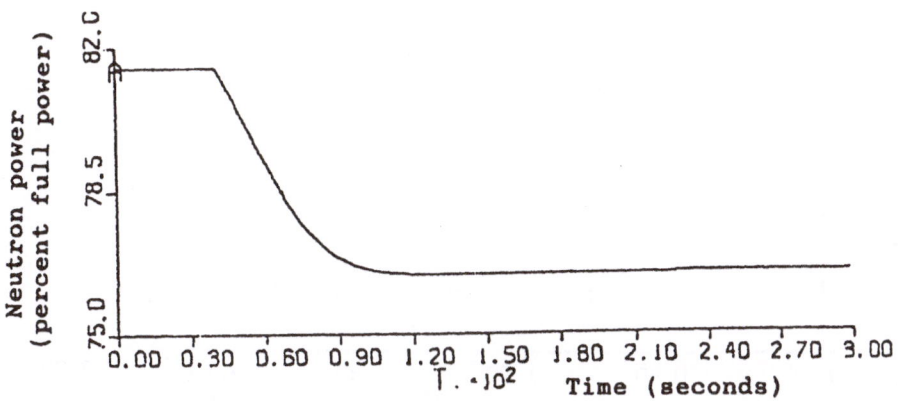

Figure 4. Response of design system to a step decrease of
5% in thermal power demand

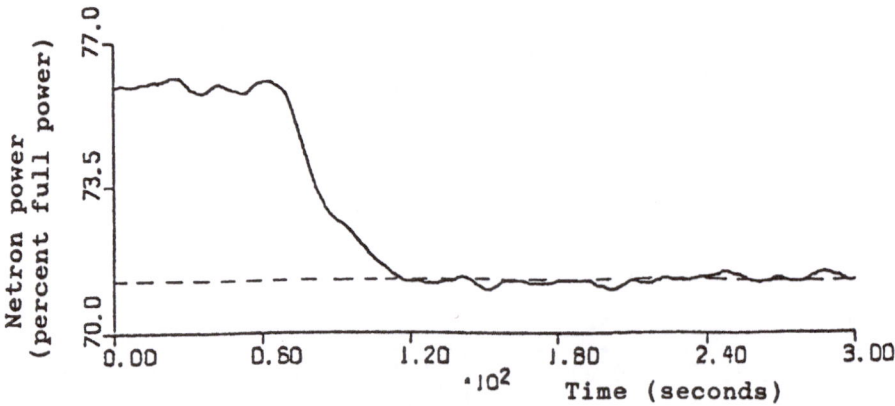

Figure 5. Response of neutron power to a step decrease of
5% in neutron power demand

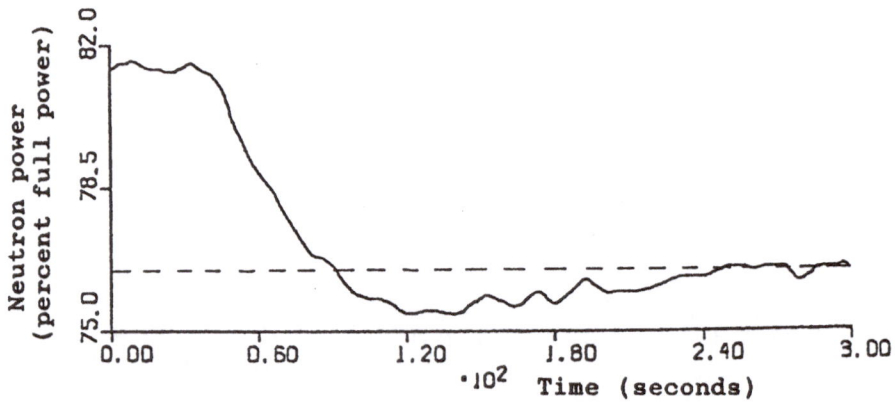

Figure 6. Response of neutron power to a step decrease of
5% in thermal power demand

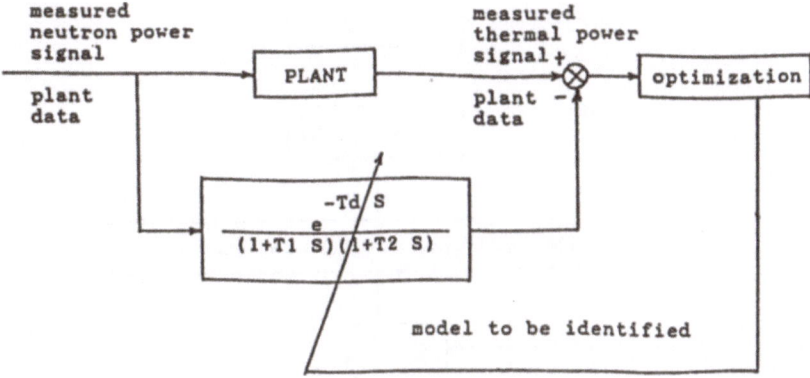

Figure 7. Block diagram of identification of three delay time

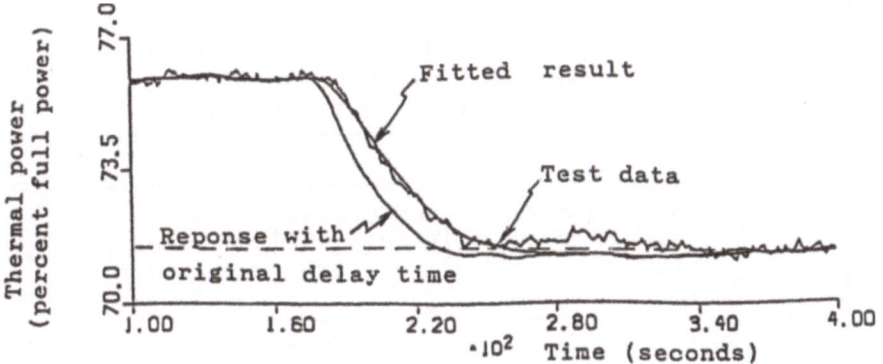

Figure 8. Fitted result and the response with original delay time

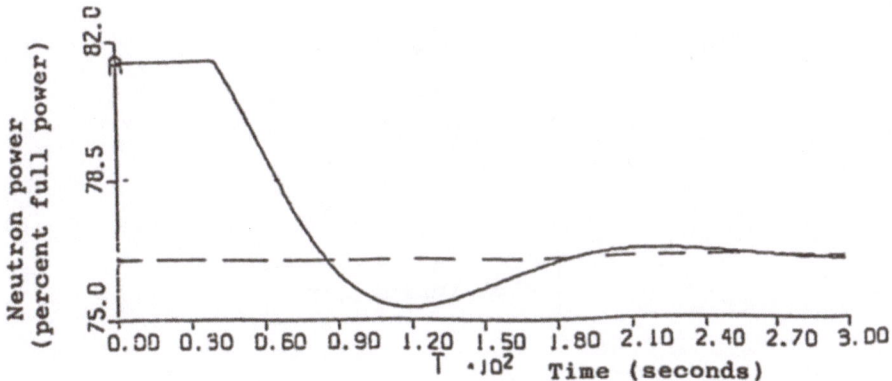

Figure 9. Response of neutron power to a step decrease of
5% in thermal power demand with new delay time

218

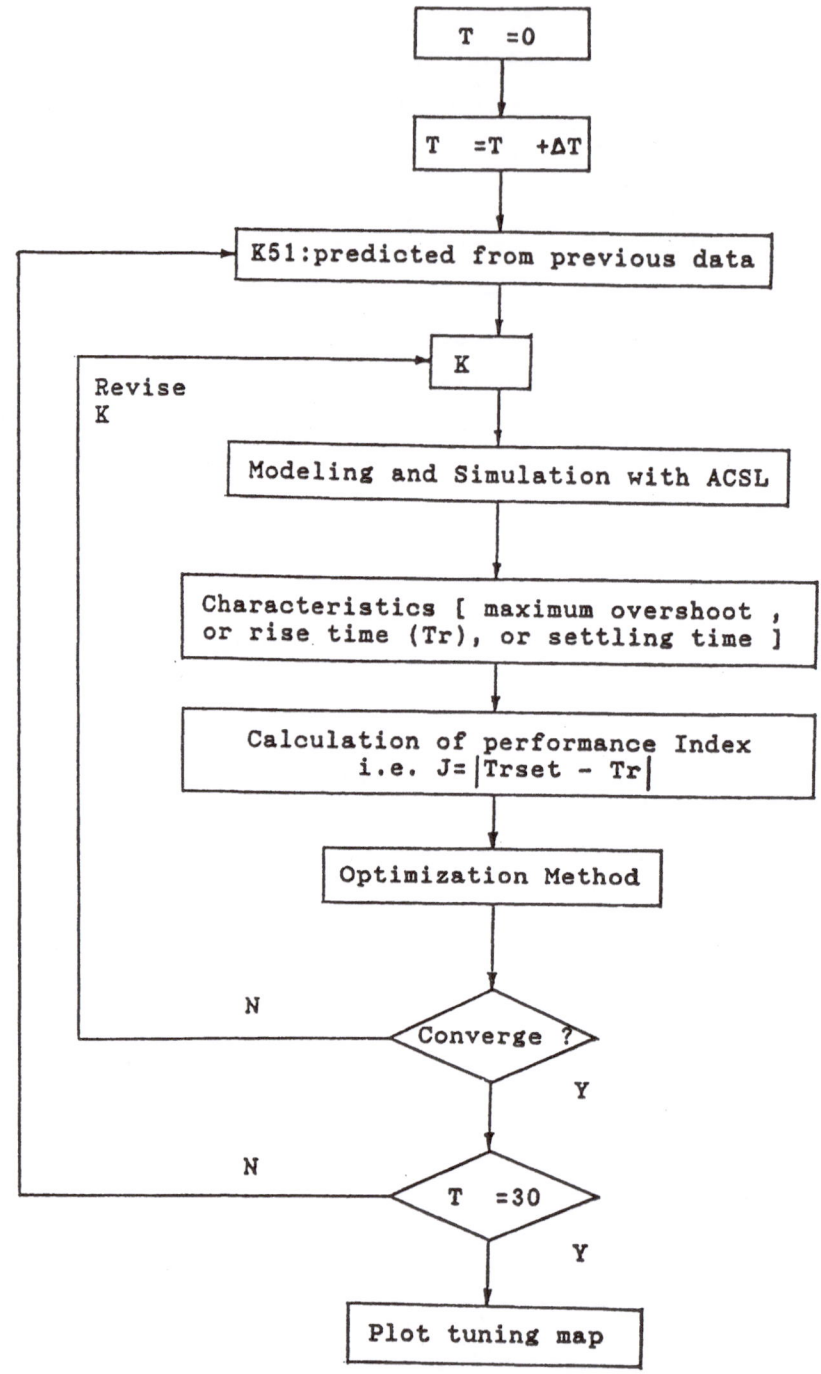

Figure 10. Block diagram of constructing time domain tuning map

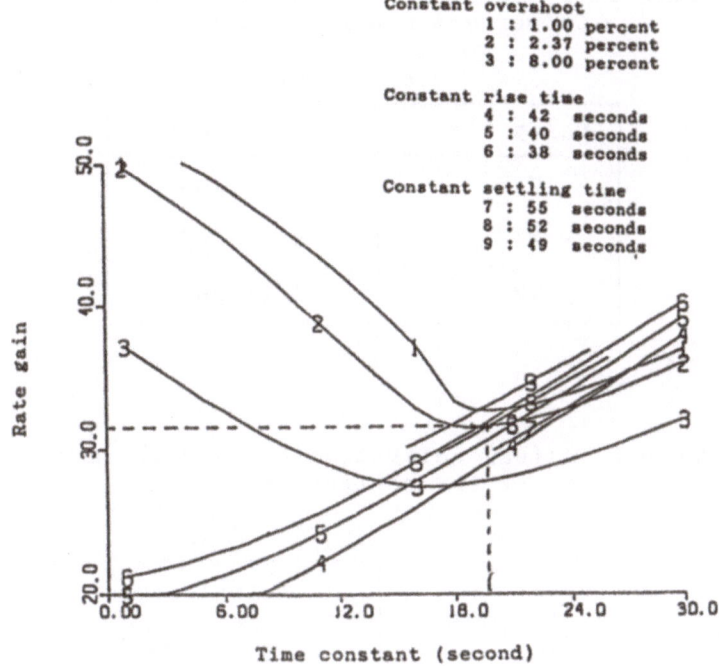

Figure 11. Time domain tuning map of thermal N loop

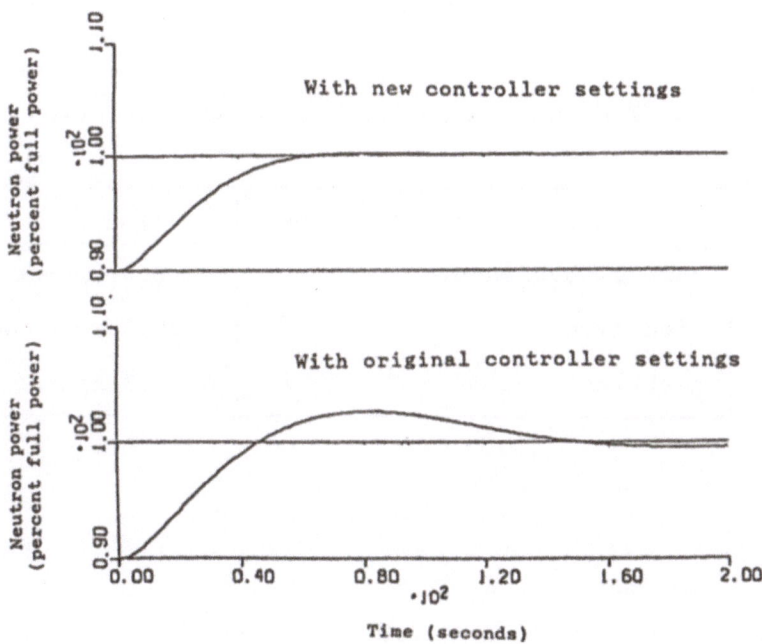

Figure 12. Comparison of power control system performace

Table 1. Designed delay time between thermal
power and neutron power

Delay time	Design data
Td	2 second
T1	6 second
T2	5 second

Table 2. Actual delay time between thermal
power and neutron power

Delay time	Actual data
Td	3.5 second
T1	6.2 second
T2	17.3 second

Table 3. Time domain performance characteristics

performance characteristics	value
maximum overshoot	2.37 percent
rise time	39.2 seconds
settling time	53.3 seconds

Table 4. Performance characteristics with new
controller settings

performance characteristics	value
maximum overshoot	2.37 percent
rise time	39.7 seconds
settling time	53.8 seconds

Table 5. Controller settings

controller	new setting	original setting
rate feedback gain K51	31.5	17.
time constant T51	19.	10.

Table 6. Comparison of relative stability

performance characteristics	new setting	old setting
gain margin	11.6 DB	10.2 DB
phase margin	66.4 degree	50.5 degree

222

8. References

1) "Final Safety Analysis Report of Taiwan Research Reactor," Chalk River, Ontario, Canada, November 1972.

2) "Design manual: TRRP power regulating system," TRR-681-1014 page 92, (1972).

3) Advanced Continuous Simulation Language Users' Guide/Reference Manual, Mitchell and Ganthier, Associate, INC.   (1975)

4) Min-Song Lin, Shih-Jen Wang, "ACSL based parameter optimization module for control system analysis," Internal report to be published

USNRC's Nuclear Plant Analyzer: Engineering Simulation Capabilities in the 1990's

E. T. Laats

Idaho National Engineering Laboratory
EG&G Idaho, Inc.
P. O. Box 1625
Idaho Falls, Idaho 83415

Summary

The Nuclear Plant Analyzer (NPA) is the U.S. Nuclear Regulatory Commission's (NRC's) state-of-the-art nuclear reactor simulation capability. This computer software package integrates high fidelity nuclear reactor simulation codes such as the TRAC and RELAP5 series of codes with color graphics display techniques and advanced workstation hardware. The NPA first became operational at the Idaho National Engineering Laboratory (INEL) in 1983. Since then, the NPA system has been used for a number of key reactor safety-related tasks ranging from plant operator guidelines evaluation to emergency preparedness training.

The NPA system is seen by the NRC as their vehicle to maintain modern, state-of-the-art simulation capabilities for use into the 1990s. System advancements are envisioned in two areas: first, software improvements to existing and evolving plant simulation codes utilized by the NPA through the use of such techniques as parallel and vector processing and artificial intelligence expert systems, and second, advanced hardware implementations using combinations of super-, minisuper-, supermini-, and supermicrocomputer systems and satellite data communications networks for high flexibility and greatly increased NPA system performance.

---

*Work supported by the U.S. Nuclear Regulatory Commission, Office of Nuclear Regulatory Research, Under Department of Energy contract No. DE-AC07-76ID01570.

Introduction

For the past three decades, the U. S. Nuclear Regulatory Commission (NRC) and its predecessor organizations have sponsored the development of large, state-of-the-art nuclear reactor systems simulation computer codes for use in resolution of safety related issues. Hundreds of staff-years of effort have been expended to arrive at the current simulation capabilities represented by the RELAP5[1], TRAC-PWR[2,3], and TRAC-BWR[4] codes.

When first developed, these simulation codes were meant to be used in batch rather than interactive mode. Since the March 1979 accident at Three Mile Island-Unit 2, greatly increased emphasis and value have been placed on using these complex codes as interactive engineering simulators. The development of a system now called the Nuclear Plant Analyzer (NPA)[5-7], arose from this need. The NRC has sponsored its development and application at the Idaho National Engineering Laboratory (INEL) and the Los Alamos National Laboratory[8] since 1983, and the INEL is currently chartered as sole proprietor of the NPA system for the NRC. The NPA now serves as the executive processor to the large NRC-sponsored reactor simulation codes (RELAP5, TRAC-PWR, TRAC-BWR), providing the capability to redirect interactively the course of the simulation while it progresses, and then provides the capability to view the output through a very high resolution color graphics workstation, as the output is generated.

The NPA engineering simulator has been used for a number of applications that address the NRC's legislatively mandated mission of ensuring the public's safety. These applications cover the areas of evaluation of safe power plant performance, training of the NRC's emergency response personnel for a crisis situation, evaluation of power plant operator guidelines, analytical testing of advanced reactor concepts, and orientation training for engineers, scientists, and management personnel. Due to the generalized nature of post-processing data for display on the workstation, analysts have also been able to display in an after-the-fact manner, the data obtained from complex thermal-hydraulic test facilities at the INEL. (See Reference 9 for more details.)

The NRC envisions the NPA system, in conjunction with its reactor simulation codes, as the means to provide state-of-the-art engineering simulation capabilities into the 1990s. The NPA system will take advantage of rapidly advancing technologies in the areas of software and hardware.

This paper includes an overview of the NPA system, discusses the envisioned enhancements in the software and hardware areas, and end with conclusions and completes the discussion.

Overview Of The Nuclear Plant Analyzer

A depiction of the major components of the NPA system is shown on Figure 1. The heart of this engineering simulator is the Executive Processor, which is the user-friendly interface to all other components. The key components are the simulation codes such as RELAP5 and TRAC, and the graphical display functions that convert the data to the desired output format. The other components on the figure represent the numerous support data bases (for storage of input decks for simulation codes, storage of past calculations, storage of experimental data[10]) and utilities that enhance the overall capabilities of the system. An important NPA feature is the capability to allow an analyst to redirect a code calculation as it progresses through its simulated scenario. The analyst can have the same power plant control capabilities as the operator of the actual plant.[11,12]

The NPA currently resides on the Control Data Corporation (CDC) CYBER 176 mainframe computer at the INEL, the Cray-1S computer at the LANL, and soon on the Cray-X/MP computer at the INEL. Users access the NPA through high speed communications lines and a workstation that consists of a color graphics device (Tektronix 4115B or 4125) to display the data, and an alpha-numeric terminal to enter the control commands. An example display that may be seen on the color graphics device is shown on Figure 2. The raw data have been converted to colors to depict coolant water state, component status (open or closed valves, for example), water levels, fuel rod stored energy, or to depict flow direction (arrows), or X-Y plot data points, or just digital data. These examples represent only a few of the

techniques that have been employed for the NPA. All of these NPA software
and hardware features taken collectively make it a most flexible true
engineering simulator.

The simulation codes currently used with the NPA engineering simulator are
the RELAP5/MOD2, TRAC-PF1/MOD1, and TRAC-BF1/MOD1 codes. The NPA also has
the capability to replay data through the color graphics display system
that have been generated by simulation codes run in "batch" mode. The
"replay" capability has been used to display data from the SCDAP/RELAP5,
the CRAC-II, and the SCDAP codes.[13-15]

The three primary codes used in the NPA engineering simulator have been
utilized to model both pressurized water reactors (PWRs) and boiling water
reactors (BWRs). The particular nuclear power plants that have been
modeled to date now include Oconee-1, Davis-Besse, Calvert Cliffs,
Arkansas Nuclear One-Unit 2, H. B. Robinson-2, Browns Ferry, Three Mile
Island-Unit 1, and Westinghouse reference plant RESAR-3S.

Future Software Enhancements

The nuclear reactor simulation codes have reached a high level of
maturity. Very few major advancements are now envisioned on modeling the
basic physics principles that are present when simulating power plant
behavior. Instead, future enhancements will take advantage of bigger,
faster, more sophisticated computing hardware and emerging, maturing
software techniques to integrate more of the NRC's large simulation codes
into a common system and enhance their ease of use. The NPA is mechanism
to integrate these codes that provide the NRC with state-of-the-art
engineering simulation capabilities at least for the next decade.

The developers of the RELAP5, TRAC-PWR, and TRAC-BWR simulation codes are
just now starting to take advantage of the vector and parallel processing
capabilities of both large and small computers. These efforts have barely
begun and the potential for computational speed increases is tremendous.
A first attempt[16] at utilizing parallel processing with the RELAP5 code
on a Cray X-MP computer resulted in a computational speed increase of a
factor of 2. Much more work in this area is envisioned.

Some areas of simulation that are likely additions to the NPA system include severe reactor damage (SCDAP/RELAP5 code), containment behavior (CONTAIN code[17]), radionuclide dispersion (MACCS code[18]), and overall safety of advanced reactor system designs and concepts (ATHENA code[19]). Some of these codes have already been partially implemented into the NPA by using the "replay" capability mentioned earlier.

A key area of growth to the NPA system is the integration of artificial intelligence technologies in the form of expert systems. These systems can efficiently operate in parallel with the simulation codes on multiprocessor computers. Expert systems will, as a minimum, help create code input models from libraries of plant specific data, aid the codes in efficient problem execution, and help explain results as they are generated or later replayed.

The first and last of these activities are well underway at the INEL as projects outside the NPA program. The ATHENA Aid[20] uses expert system knowledge to guide the user through code input model development and utilize the knowledge of experts, for example, to properly nodalize the model in compliance with established, tested, and documented procedures for the ATHENA advanced simulation code. This same expert system has been used to generate RELAP5 input models. In the results interpretation area, the rule bases of the NRC's Reactor Safety Assessment System (RSAS)[21] are presently being tested for use in the NRC Operations Center. This expert system is designed to track and enunciate the major events occurring in an off-normal nuclear power plant event. This system currently resides on a stand-alone Xerox 1186 workstation, but has been designed to easily interact with the NPA. A manual link between the NPA and the RSAS has already been demonstrated. An expert system, whether it is the RSAS or a similar system, is expected to be integrated into the NPA system to perform the same functions that provide rapid discernment of simulation code output.

Other expert systems are in various stages of development at the INEL. An alarm annunciator handling expert system is presently being developed as part of the control room upgrade of the Advanced Test Reactor. Another concept being discussed is an expert system that discerns plant status and advises the operations staff regarding the utilization of emergency procedure guidelines. The NPA is an ideal testbed for developing the evaluating these and many other expert systems of this type.

Future Hardware Implementations

The rapid advancement of computational hardware will provide the opportunity to implement the NPA on several efficient and cost effective configurations that meet a much wider variety of needs. Presently, the NPA resides on mainframe computers that reflect 10 to 15 year old technologies.[20] Within the next five years, NPA implementations will likely utilize computing hardware that range from personal desktop computers of the IBM-PC vintage, to modern mainframe computers of the Cray X-MP (and beyond) vintage. Also, computer networking technologies will be an integral part of design strategies.

The NPA system designs will likely accommodate six computing hardware configurations.

(a) Stand-alone personal computers (PCs) that are intended for a single user and likely have partial NPA capabilities. A single simulation code will reside on the PC and specialized NPA features for that code and input model will be available. The RELAP5 code has already been demonstrated to run acceptably fast (1/30 of the speed of a CDC CYBER 176) for a simple analysis problem.[21]

(b) Several PCs with the capabilities of (a) mentioned above, but networked directly to a central super-minicomputer that has full NPA capabilities and acts as the repository and server to all its PCs.

(c) Stand-alone super-minicomputer with full NPA capabilities, including the simulation codes, that acts as a single user computational device for multiple NPA workstations of the current type of workstation configuration (i.e., workstations with minimal local capabilities).

(d) Same configuration as (c) except using a mini-supercomputer as the server machine.

(e) Current, older generation mainframe computers that enable a multiuser environment for serving either the PC-based workstations or some version of the present type of workstation.

(f) Same configuration as (e) except using a modern Class VI (or newer) computer.

Options (a) and (b) will provide a very favorable environment to develop and check-out simulation models, but real-time simulations of power plant behavior is very unlikely to be possible with the RELAP5 and TRAC series of codes. Real-time operation is much more likely with options (c), (d), and (e), and faster than real-time performance on the Class VI computers used in option (f).

The more computationally powerful systems such as options (d) and (f) provide a flexible simulation environment for development, concepts testing, and real-time simulation during production use. When combined with high speed land or satellite communications capabilities, even remote workstations worldwide may utilize the NPA system.

Another area where hardware advancements are progressing very rapidly is equipment utilized for the HPA as workstations. The current workstation configuration is centered around a Tektronix 4125 color graphics device. The technology of this machine, though continually being upgraded, is still not nearly as advanced as the newest product releases from many hardware vendors. Many of the graphics functions that were custom built some 5 years ago for the NPA system may now be duplicated within the firmware of these latest products. Also, other color graphics devices have recently been released that duplicate nearly all of the important capabilities of the important capabilities of the present workstation, but these products cost 1/2 to 1/4 the price of the current workstation hardware. Thus, it is apparent that the NPA workstation hardware will need to change to take advantage of these hardware advancements and those to come later.

Conclusions

The NPA system is an engineering analysis simulation tool developed by the
NRC. The future of the NPA system will be based on expanding its power
plant simulation capabilities and implementing the NPA system on several
hardware configurations. The software additions will be comprised of more
power plant simulation codes in the severe accident area, faster versions
of the presently implemented reactor systems codes, and supporting expert
systems for enhanced code usability.

Computing hardware ranging from personal computers to the most modern
mainframe computers will provide a cost effective environment for a
multitude of users. Many more NPA functions will be moved to the NPA
workstations that will be networked to a central computer system with full
NPA capabilities.

Acknowledgments

The author is the project manager of the Nuclear Plant Analyzer program.
He wishes to gratefully acknowledge the many efforts of the people whose
work is being reported in this paper and referenced below. In addition,
the following people are acknowledged for their invaluable contributions:
R. F. Audette, R. J. Beelman, N. Bonicelli, J. D. Burtt, W. H. Grush,
R. N. Hagen, M. A. Lintner, H. Makowitz, G. A. Mortensen, and J. E. Tolli.

Notice

References

1. V. H. Ransom, R. J. Wagner, <u>RELAP5/MOD2 Code Manual Volume 1: Code Structure, Systems Models and Solution Methods</u>, EGG-SAAM-6377, April 1984.

2. D. R. Liles, J. H. Mahaffy (Principal Investigators), <u>TRAC-PF1/MOD1: An Advanced Best Estimate Computer Program For Pressurized Water Reactor Thermal-Hydraulic Analysis</u>, Los Alamos National Laboratory Draft Report, February 1984.

3. B. E. Boyack et al., <u>TRAC User's Guide</u>, NUREG/CR-4442, November 1985.

4. TRAC-BD1 Code Development Group, <u>TRAC-BD1/MOD1: An Advanced Best Estimate Computer Code For Boiling Water Reactor Transient Analysis</u>, NUREG/CR-3633, EGG-2294, April 1984.

5. K. D. Russell et al., <u>Nuclear Plant Analyzer and Data Bank Common User Interface: Functional Requirements, Conceptual Design, and Hardware Considerations</u>, EGG-SAAM-6419, September 1983.

6. H. D. Stewart et al., <u>NECTAR (NPA) Program and Reference Manual</u>, EGG-IS-6825, January 1985.

7. E. T. Laats et al., <u>User's Manual For The U.S. Nuclear Regulatory Commission's Nuclear Plant Analyzer</u>, EGG-RST-7044, September 1985.

8. R. G. Steinke et al., <u>The Nuclear Plant Analyzer: An Interactive TRAC/RELAP Power-Plant Simulation Program</u>, Los Alamos National Laboratory Report LA-UR-84-2953 (1984).

9. E. T. Laats, R. J. Beelman, "U.S. Nuclear Regulatory Commission's Nuclear Plant Analyzer," <u>Transactions of ANS/ENS Thermal Reactor Safety Meeting</u>, San Diego, CA, February 2-6, 1986.

10. H. A. Hardy, E. T. Laats, <u>Introductory User's Manual For The U.S. Nuclear Regulatory Commission Reactor Safety Research Data Bank</u>, NUREG/CR-2531 Rev. 3, EGG-2164, March 1985.

11. R. J. Beelman, D. G. Hall, <u>Operator Guideline Evaluation Tool</u>, EGG-NTAP-6287, May 1983.

12. R.J. Beelman, B.D. Stitt, <u>Application Of RELAP5 To Evaluation of Babcock & Wilcox (B&W) Abnormal Transient Operating Guidelines (ATOG) For Portions Of Selected Small Break Scenarios In Oconee-1</u>, EGG-SAAM-6508, July 1984.

13. G. A. Berna et al., <u>RELAP5/SCDAP/MODO Code Manual, Volume 1: Code Structure, System Models, and Solution Methods, Draft Preliminary Report</u>, EGG-RTH-7051, September 1985.

14. L. T. Ritchie et al., <u>Calculations Of Reactor Accident Consequences Version 2: CRAC2 Model Description</u>, NUREG/CR-2552, December 1982.

15. G. A. Berna et al., <u>SCDAP/MODO: A Computer Code for the Analysis of LWR Fuel Bundle Behavior during Severe Accident Transients</u>, WR-NSMD-078, December 1982.

16. H. Makowitz, "<u>Numerical Experiments in Concurrent Multiprocessing with the RELAP5 Nuclear Reactor Systems Code,</u>" Proc. 11th Int. Mtg. Advances in Nuclear Engineering Computational Methods, Knoxville, Tennessee, April 9-11, 1985.

17. K. D. Bergeron et al., <u>User's Manual for CONTAIN 1.0, A Computer Code for Severe Nuclear Reactor Accident Containment Analysis</u>, NUREG/CR-4085, May 1985.

18. D. J. Alpert, <u>MELCOR Accident Consequence Calculation Code System</u>, NUREG/CR-4691, To Be Published.

19. H. Chow et al., (Principal Investigators), <u>ATHENA Code Manual Volume 1: Code Structure, Systems Models, Solution Methods, and Input Data Requirements, DRAFT</u>, EGG-RST-7034, September 1985.

20. R. K. Fink et al., "Expert Systems Approach To Modeling Systems With The ATHENA Thermal-Hydraulics Code," <u>Proceedings Of The Fourth Symposium On Space Nuclear Power Systems</u>, Albuquerque, New Mexico, January 12-16, 1987.

21. M. A. Bray et al., "Reactor Safety Assessment System - A Situation Assessment Aid For USNRC Emergency Response," <u>IEEE Proceedings of the Expert Systems In Government Symposium</u>, McLean, Virginia, October 24-25, 1985.

22. J. Schefter, "Supercomputer: Incredible Cray-1 Cruises At 80 Million Operations a Second," <u>Popular Science</u>, June 1979.

23. A. R. Edwards, and F. P. O'Brien, "Studies of Phenomena Connected with the Depressurization of Water Reactors," <u>Journal of the British Nuclear Energy Society</u>, 9, April 1970, pp. 125-135.

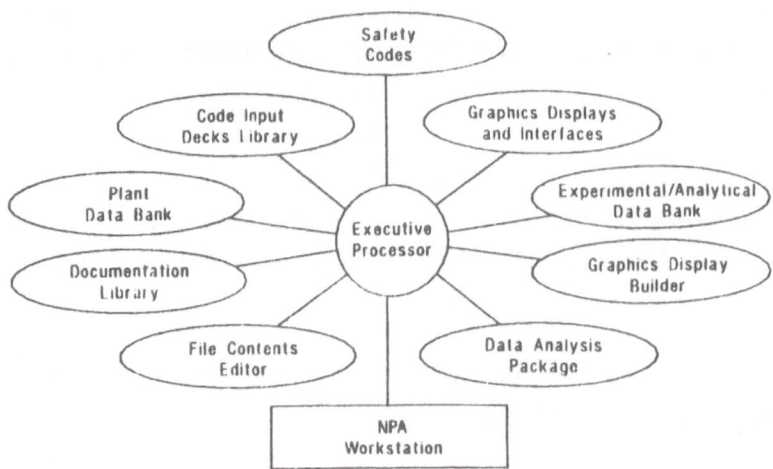

Figure 1. Major components of the Nuclear Plant Analyzer.

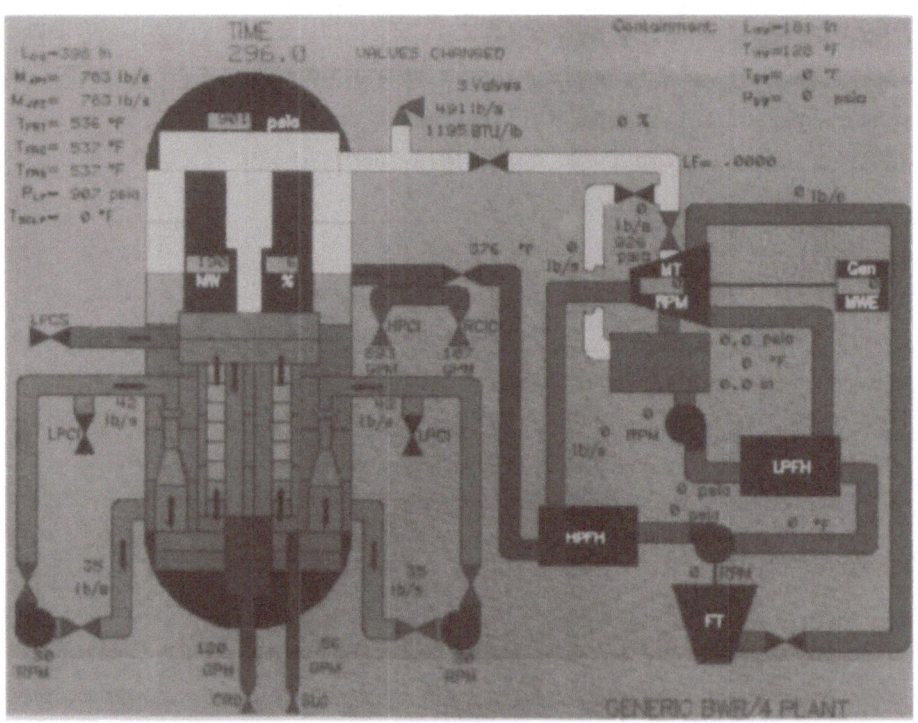

Figure 2. Example color graphics display for generic
boiling water reactor plant

CASMO-3/SIMULATE-3 Core Follow Calculations on Oskarshamn 3

K EKBERG

S LUNDBERG

STUDSVIK ENERGITEKNIK AB

S-611 82 Nyköping, Sweden

Summary

The new Studsvik core analysis package, comprising CASMO-3 and
SIMULATE-3 as the main components, has been used to perform core
follow calculations on the Oskarshamn 3 ASEA-ATOM BWR-3000
reactor.

The initial and the second cycle have been followed. The average
k-effective value of cycle 1 is 1.0005 with a standard deviation
of 0.0005 and of cycle 2 1.0018 with a standard deviation of
0.0009.

At some of the reference points the calculated distributions
have been compared to TIP traces. The calculated axial traces
agree very well with the measured traces. Both neutron TIP and
gamma TIP traces have been analysed. The radial power is predicted within 3-4 % (rms).

1   INTRODUCTION

The new Studsvik core analysis package consisting of the lattice
code CASMO-3, the linkage code TABLES-3 and the advanced nodal
reactor code SIMULATE-3 has been benchmarked against core follow
data for the Oskarshamn 3 BWR, owned and operated by OKG AB.

The purpose of this benchmark was to show that the new ICFM code
package is fully capable of calculating the reactivity level and
the power distribution in a large BWR with a complex initial
core without using any adjustment parameters.

2   THE CODES

A schematic graph of the codes and data used in this benchmark
is shown in figure 1.

2.1   CASMO-3

CASMO is a multi-group two-dimensional transport theory code for
burnup calculations on BWR and PWR assemblies or simple pin
cells. The code handles a geometry consisting of cylindrical
fuel rods of varying composition in a square pitch array with
allowance for fuel rods loaded with gadolinium, burnable absor-
ber rods, cluster control rods, in-core instrument channels,
water gaps, and cruciform control rods in the regions separating
fuel assemblies. Typical fuel storage rack geometries can also
be handled.

All bundle calculations to produce homogenized 2-group cross
sections and discontinuity factors for SIMULATE-3 were performed
with the most recent version CASMO-3 (1), which was released in
1987.

Nuclear cross section data were taken from the CASMO-3 40-group
library based on ENDF/B-IV with selected data from ENDF/B-V (2).

CASMO-3 also contains a two-dimensional gamma transport calculation module which was used to calculate gamma TIP detector
constants. Gamma cross sections were taken from the Studsvik
10-group CASMO-3 gamma library (3).

Effective GD cross sections for fuel containing Gadolinium were
calculated with the stand-alone MICBURN-3Z code (4, 5).
This code can be used for both homogeneously distributed Gd and
the ASEA-ATOM heterogeneous gray Gd concept, in which fuel
pellets with Gd alternate with Gd-free pellets.

MICBURN-3Z utilizes a discrete ordinates transport theory module
in r,z geometry for this calculation.

2.2   TABLES-3

The linking of data from CASMO to SIMULATE was performed with
the TABLES-3 code (7, 6). The CASMO microscopic, macroscopic
data, fission product data and ADF's are resolved internally in
TABLES into tables of base values as functions of the independent variables of importance and additive series of partial
cross section functions. The cross sections of interest in
SIMULATE-3 are then developed as a summation of several partials
from the base value

$$\Sigma_i = \Sigma_o + \sum_{\ell} \Sigma \Delta \Sigma_{i\ell}$$

The partial cross section can be expressed as a separable
function of three variables as

$$\Delta \Sigma_{i\ell} = F_{i\ell}(x,y,z) F$$

Microscopic cross section tables are used for xenon and samarium.

TABLES-3 produces a binary master library for SIMULATE-3.

2.3   SIMULATE-3

SIMULATE-3 (8) is an advanced two-group nodal code for analysis
of both PWR and BWR cores. It has been developed at the Studsvik
subsidiary Studsvik of America. Earlier versions were developed
by Yankee Atomic Electric Co. sponsored by EPRI (9).

The neutronic part of SIMULATE-3 consists of the QPANDA model
(10, 11). QPANDA is a full two-group neutronics model method
using fourth order polynomials to represent the intranodal
fluxes in the fast and thermal group. The nodal equations are
solved by utilizing a non-linear iteration technique (12).

Fuel heterogeneities are accounted for by using advanced spatial
homogenization techniques (discontinuity factors, ADF) (13,14).

In SIMULATE-3 the reflector is treated explicitly by using
reflector cross sections and ADF's from one-dimensional CASMO
calculations. The reflector can be treated as a homogenized
region and the strong gradients at the reflector/core boundary
are accurately accounted for by the ADF's. This eliminates
entirely the need of separately calculated empirical albedos.

One of the advantages with an advanced modern nodal code such as
SIMULATE-3 is that normalization to higher order fine mesh codes
or measurements is not required. Also, there is no need for
adjustable factors such as a-factors, thermal leakage corrections, edge corrections or node size corrections etc.

The hydraulic calculations in SIMULATE-3 (15) determine the
relative water density of each node (void values) by using the
EPRI void correlation (16).

A reactor heat balance is available to calculate the core inlet
subcooling. The channel coolant flow distribution can be calculated either by using a core pressure drop model or an empirical
power/flow fit. Presence of bypass region voids can be treated
by an optional void fraction correlation.

3   DESCRIPTION OF THE REACTOR AND THE FUEL

3.1   The core

The Oskarshamn 3 reactor is a modern ASEA-ATOM BWR with internal
pumps that went into commercial operation in 1985. The reactor
core consists of 700 fuel channels. Control is provided by means
of 169 cruciform B_4C control rods and the core flow circulation
system. The flow window at full power extends from 10300 kg/s
(82 %) to 12500 kg/s (100 %) giving a reactivity swing of about
450 pcm. Operation at up to 13100 kg/s has recently been allowed
by the authorities.

The flow is measured in 8 incore locations thus providing accurate in- channel coolant flows.

There are 37 strings for the TIP detectors. Each detector string has 4 fixed LPRM neutron detectors.

3.2   Fuel types

The initial core uses three different bundle types of the ASEA-ATOM 8x8 design. Two of these are enriched to 2.63 w/o U235 and the third is enriched to 1.27 w/o U235.

The low enriched bundle has one Gd rod with 1.0 w/o Gd_2O_3 homogeneously distributed. This bundle also has 4 central water holes.

One of the high enriched bundle types does not contain any Gd rods. This bundle has a low enriched top zone (1.21 w/o U235). The second high enriched bundle type has 3 rods with nominally 5.5 w/o Gd_2O_3 axially distributed. The bottom node has no Gd and the top node has low enrichment (1.30 w/o).

The reload fuel for the second cycle is of the ASEA-ATOM SVEA design with 2.85 % average enrichment. Five Gd rods with nominally 2.55 w/o Gd_2O_3 (axially distributed) are used except in the top and bottom node.

In cycle 2 also four demo bundles of the new ASEA-ATOM SVEA-100 design were loaded. This assembly has top and bottom natural uranium blankets and Gd loading of 8 rods with 4.4 w/o Gd_2O_3. The average enrichment is 2.71 w/o U235.

The axial distribution of fuel enrichment and Gd poisoning in cycles 1 and 2 is shown in fig. 2.

3.3   Core loadings

The initial core low enriched fuel was loaded in control cell modules. The core periphery was loaded with the high enriched non-Gd fuel. The loading of the initial core is shown in figure 3.

Cycle 1 was ended before the core reached zero potential reactivity. The reactivity content in the core would have been sufficient for an additional 3.5-4 MWd/kgU full power operation. Therefore only 12 fresh bundles (4 8x8 high enrichment, 4 SVEA and 4 DEMO) were loaded into the cycle 2 core. Additional reactivity for a cycle of about 5.4 MWd/kgU length was gained from an extensive shuffling. The low enriched initial core bundles were maintained in control cell modules. Low enriched assemblies were also moved to the core periphery. The cycle 2 core loading is shown in figure 4.

These core loadings result in regions, between which there are large discontinuities in fuel enrichment and Gadolinium content, well suited for demonstrating the advantages of an advanced nodal model.

4   PREPARATION CALCULATIONS

All the CASMO-3 depletion calculations were performed by OKG and kindly put at Studsvik's disposal.

Data for the explicit reflector tratment in SIMULATE-3 were calculated using the special reflector option in CASMO-3. The reflector data calculated in this way are very insensitive to the properties of the adjoining fuel nodes.

As the control rod history in SIMULATE-3 is defined in an alternate way compared to the POLCA code used by OKG, the CASMO runs had to be completed with rodded depletion vectors and branches with the rod withdrawn. The control history of a node is defined in SIMULATE-3 as the exposure weighted control fraction. The rodded depletion calculations were performed at 0 % void.

The cross section library for use in SIMULATE-3 was generated with the TABLES-3 code. The TABLES runs generated macroscopic and microscopic data, discontinuity factors and detector data.

In order to facilitate the SIMULATE-3 calculations the pressure drop was approximated to occur at the inlet. Thus the inlet orificing coefficients were adjusted to give the same pressure drop as resulted from POLCA calculations performed by OKG. Also the dependence of pressure with coolant flow was adjusted to the POLCA values. The steam dome pressure was 70 bars in all calculations.

The subcooling was calculated with the heat balance model in SIMULATE-3.

5   CYCLE DEPLETION

All the input steps to the depletion calculations were provided by OKG (17). TIP reference cases for calculation of the reference k-effective and the TIP response were inserted at various exposures throughout the cycles.

5.1   Cycle 1

Cycle 1 was depleted with steps of an approximate length of 0.24 MWd/kgU equivalent to 10 days of operation. Eleven reference points were analysed in cycle 1. The k-effective values of these runs were corrected for the neutron absorption in the inconel spacers (sec 5.3).
Figure 5 shows the resulting k-effective values.

Excluding the first point because the samarium transient was not modelled gives an average k-effective of 1.0005 with a standard deviation of 0.0005 pcm. Full power was reached at the exposure 2.7 MWd/kg as shown in figure 6, showing the power, the coolant flow and control rod percentage inserted into the core. The reactor power of the reference points was 100 % with coolant flows from 82 to 99 %.

The results of cycle 1 indicate that the CASMO-3 code using effective Gd cross sections from MICBURN-3 well predict the depletion of the Gd in the bundles.

5.2   Cycle 2

Cycle 2 was depleted until mid February 1987 using depletion steps of about 2 weeks length. 17 reference state points were included in cycle 2, all at 100 % reactor power and with coolant flows between 82 and 93 %.

Figure 7 shows the k-effective values of the reference points after spacer correction. The average value is 1.0018 with a standard deviation of 0.0009.

The core power, the coolant flow and the control rod percentage inserted into the core during cycle 2 depletion is shown in figure 8.

For the operation after February 19 a predictive calculation was performed. This calculation indicated a start of power coast down operation on March 12 using the last reference k-effective value. The true date was March 13 obtained with a 7 $^{\circ}$C feedwater temperature reduction just before coast down. This result shows that the k-effective reference value is well predicted. As the cycle proceeds in coast down mode further analysis will be performed and compared to predictive calculations.

5.3   Spacer correction

CASMO data produced for the POLCA system does not include the neutron absorption in the spacers. In POLCA the spacers are represented with an absorption correction in the correct axial node.

In SIMULATE-3 a model to represent the spacers is not yet available. In order to study the influence of spacers on the k-infinity values the high enriched cycle 1 assembly with and without Gd was depleted with and without spacers. The spacer reactivity worth decreases with exposure (after the Gd has been burnt out) due to the change of spectrum with exposure. The correction over cycle 1 and cycle 2 varies between 480 pcm and 430 pcm.

6   TIP TRACE ANALYSIS

TIP traces corresponding to some of the reference points were provided by OKG (17). Comparisons to calculated traces have been performed. Core average neutron TIP comparisons for cycle 1 are shown in figures 9-10 and for cycle 2 in figure 11.

Neutron TIP detectors were used until late October 1986 (cycle 2). At that point gamma sensitive TIP detectors were installed in the reactor. The gamma TIP traces were analysed for uncontrolled detector strings only, due to lack of CASMO data, and the result is shown in figure 12.

In cycle 1 the power shape is bottom-peaked while it is more peaked towards the top in cycle 2. The calculated power is slightly overpredicted in the bottom part and underpredicted in the top nodes.

A comparison between calculated and measured neutron TIP traces shows about 4 % rms deviation radially and about 5 % axially. The agreement with gamma TIP traces is better, with rms values of about 3 % both axially and radially.

The larger spread of data in the neutron TIP traces is attributed to the much larger sensitivity to geometric dislocation of these detectors.

7   CONCLUSIONS

These benchmark calculations on the complex initial and second
core of Oskarshamn 3 show a constant k-effective level. Also the
calculated TIP responses are in good agreement with measured
values. This is especially true for the gamma TIP traces.

The conclusion that can be drawn from this, yet limited material, is that the user friendly STUDSVIK CASMO-3/ SIMULATE-3 core
analysis package is well capable of predicting the behaviour of
a BWR core.

8   ACKNOWLEDGEMENTS

This work could not have been performed without the kind and
helpful support of the staff at OKG AB. Also we are grateful to
the staff of Studsvik of America for their helpful advise with
this benchmark.

245

10   REFERENCES

1   M Edenius and Å Ahlin
            CASMO-3. A Fuel Assembly Burnup Code. User's Manual
            STUDSVIK/NFA-86/7 (1986)

2   H Häggblom
            The CASMO-3 Nuclear Data Library
            STUDSVIK/NFA-86/12

3   A Ahlin et al
            Integral Transport Computations of In-core Gamma
            Effects with CASMO/CPM
            Trans. Am. Nucl. Soc. 47, 434 (1984)

4   M Edenius et al
            MICBURN-3. User's Manual STUDSVIK/NFA-86/26
            MICBURN-3. Methodology STUDSVIK/NFA-86/28

5   C Grägg
            MICBURN-3Z. Microscopic Burnup Radially and Axially in
            Burnable Absorber Rods. User's Manual
            STUDSVIK/NFA-86/27

6   D M Ver Planck
            TABLES-2. A Cross Section Library Preparation Code for
            SIMULATE
            STUDSVIK (internal report)

7   D M Ver Planck
            TABLES-3. User's Manual
            STUDSVIK report to be published

8   D M Ver Planck
            SIMULATE-3. User's Manual
            STUDSVIK report to be published

9   D M Ver Planck
            SIMULATE-E: A Nodal Code Analysis Program for Light
            Water Reactors
            EPRI-NP 2792 CCM (1983)

10   K S Smith et al
            QPANDA: An Advanced Nodal Method for LWR Analyses
            Trans. Am. Nucl. Soc. 50, 532 (1985)

11   M Edenius et al
            Recent developments in the MICBURN/CASMO/SIMULATE LWR
            Analyses Package
            ANS Topical meeting on Advances in Fuel Management
            Pinehurst, NC (1986)

12   K S Smith
         An Analytic Nodal Method for Solving the Two-group,
         Multidimensional, Static and Transient Neutron Diffu-
         sion Equations
         Engineers Thesis, Dept. of Nucl. Eng. MIT
         Cambridge MA (1979)

13   K S Smith
         Spatial Homogenization Methods for Light Water Reactor
         Analysis
         PhD Thesis, Dept. of Nucl. Eng. MIT
         Cambridge, MA (1980)

14   K S Smith
         Assembly Homogenization Techniques for Light Water
         Reactor Analysis
         Progr. Nucl. Energy Vol 17, No 3 (1986)

15   D M Ver Planck
         Methods of the Analysis of BWR. Steady State Core
         Physics' YEAC-1238 (1981)

16   Lellouche et al
         Mechanistic Model for Predicting Two Phase Void Frac-
         tion for Water in Vertical Tubes, Channels and Rod
         Bundles
         EPRI-NP-2246-SR

17   K Adielsson et al, OKG AB, Private communication

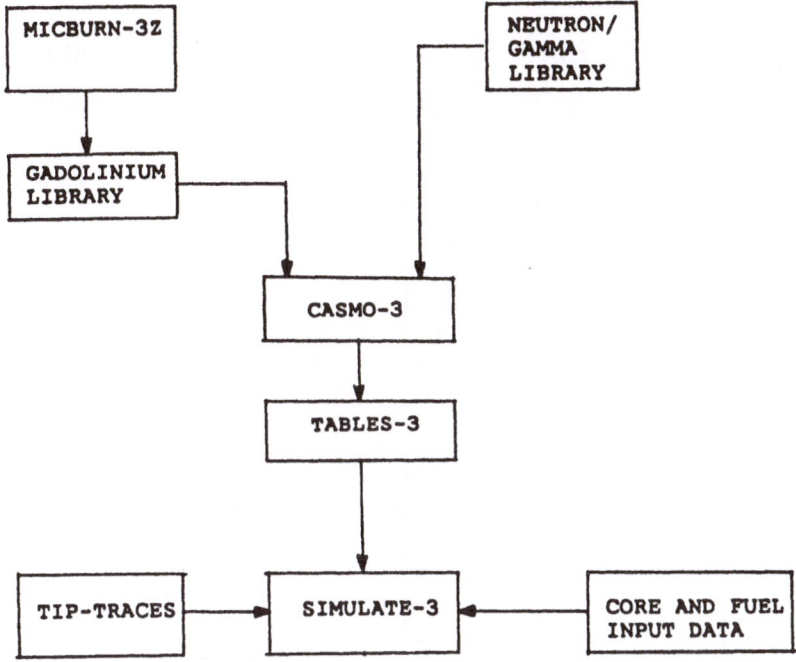

Fig. 1   The STUDSVIK ICFM Code Package

248

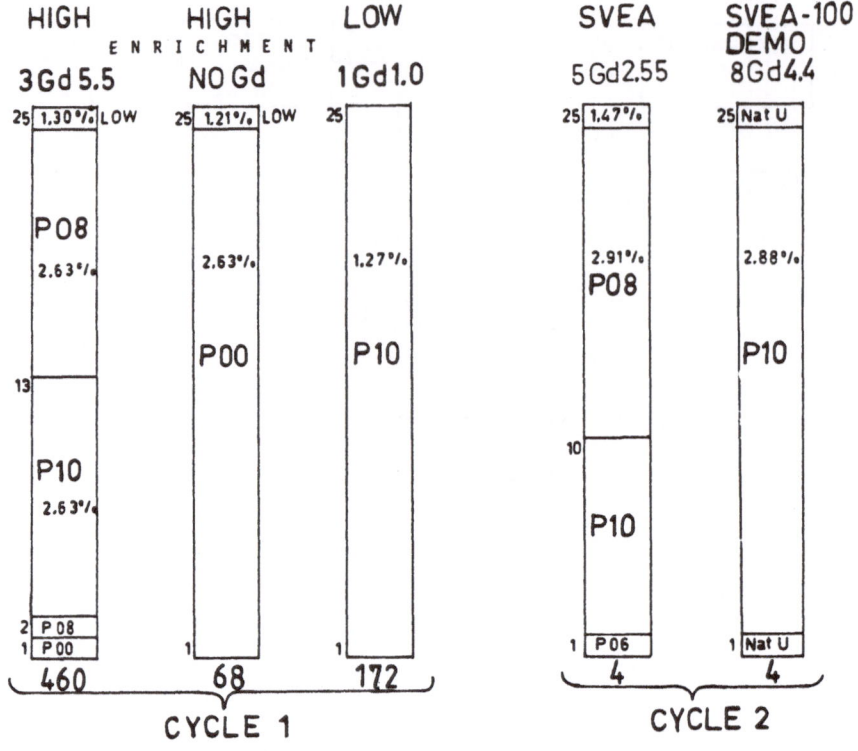

Fig. 2   Oskarshamn-3. Axial fuel enrichment and
         Gd design

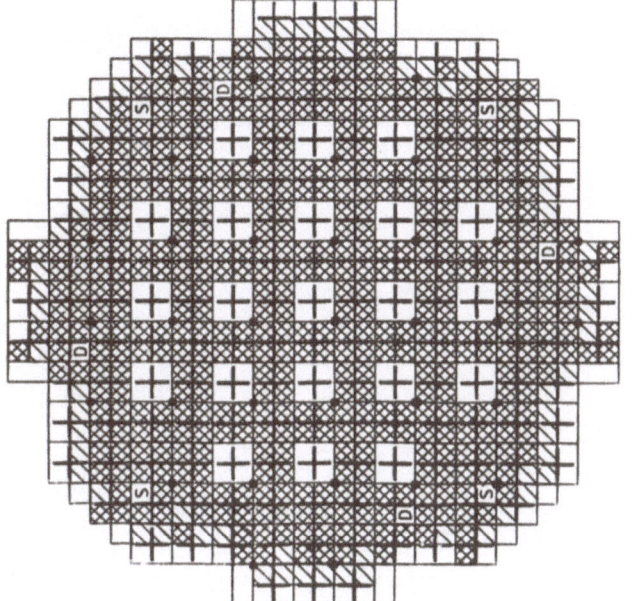

☐ Low enrichment with Gd
▨ High enrichment with Gd
▨ High enrichment without Gd
Ⓢ SVEA with Gd
Ⓓ SVEA−100 DEMO with Gd and nat. U blankets

Fig. 4 Cycle 2 Core Layout

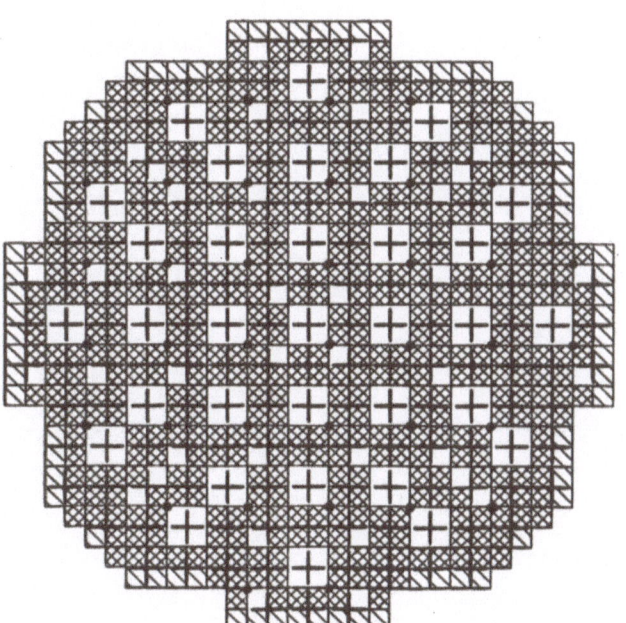

☐ Low enrichment with Gd
▨ High enrichment with Gd
▨ High enrichment without Gd

Fig. 3 Cycle 1 Core Layout

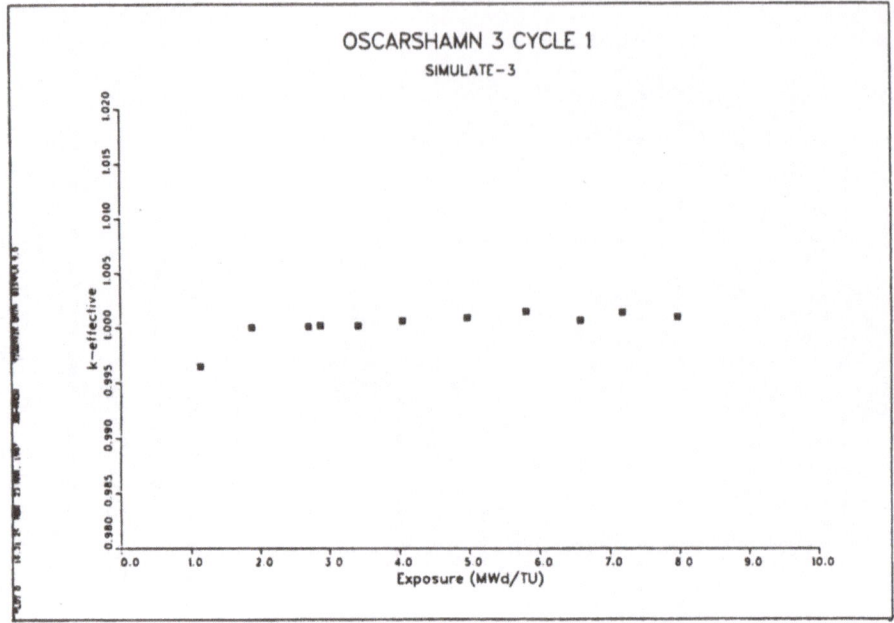

Fig. 5 Cycle 1. K-effective reference points

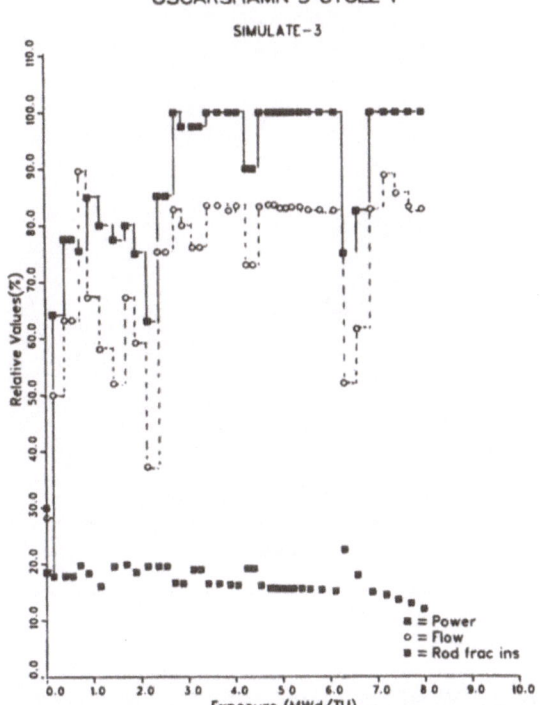

Fig. 6   Cycle 1. Reactor power, Core flow and Control
Rod fraction inserted

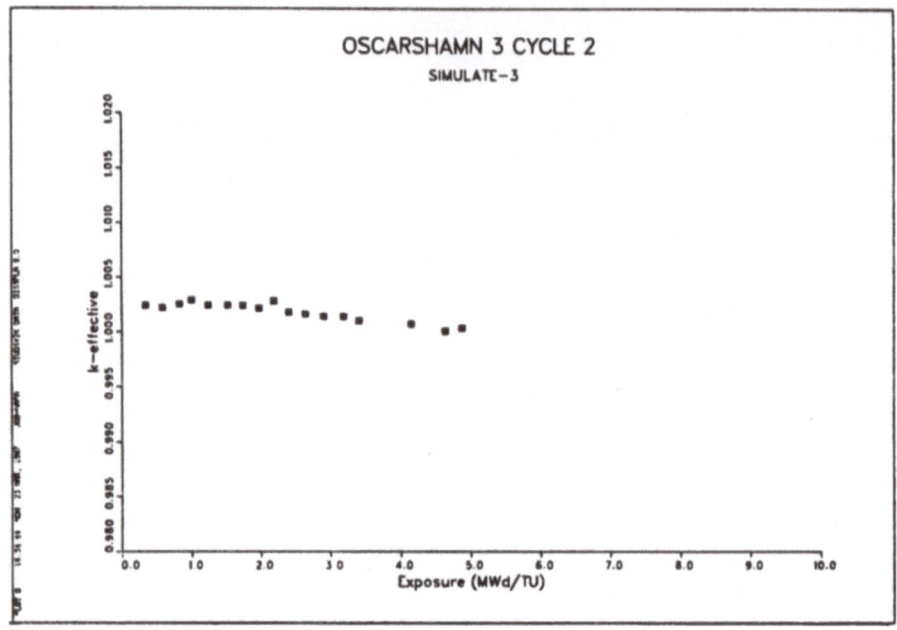

Fig. 7   Cycle 2. K-effective reference points

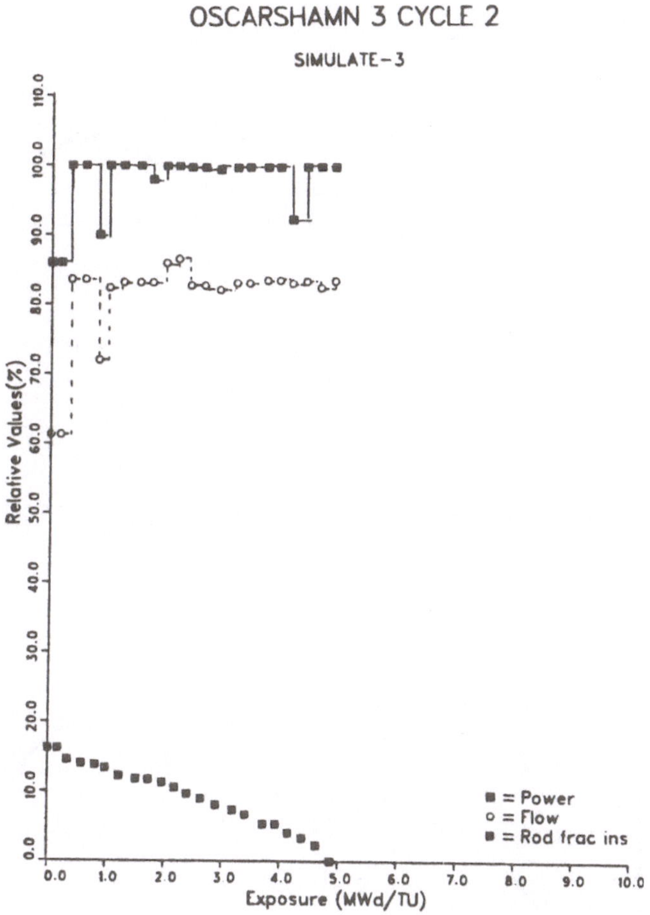

Fig. 8 Cycle 2. Reactor power, Core flow and Control
Rod fraction inserted

C1 TIP ANALYSES

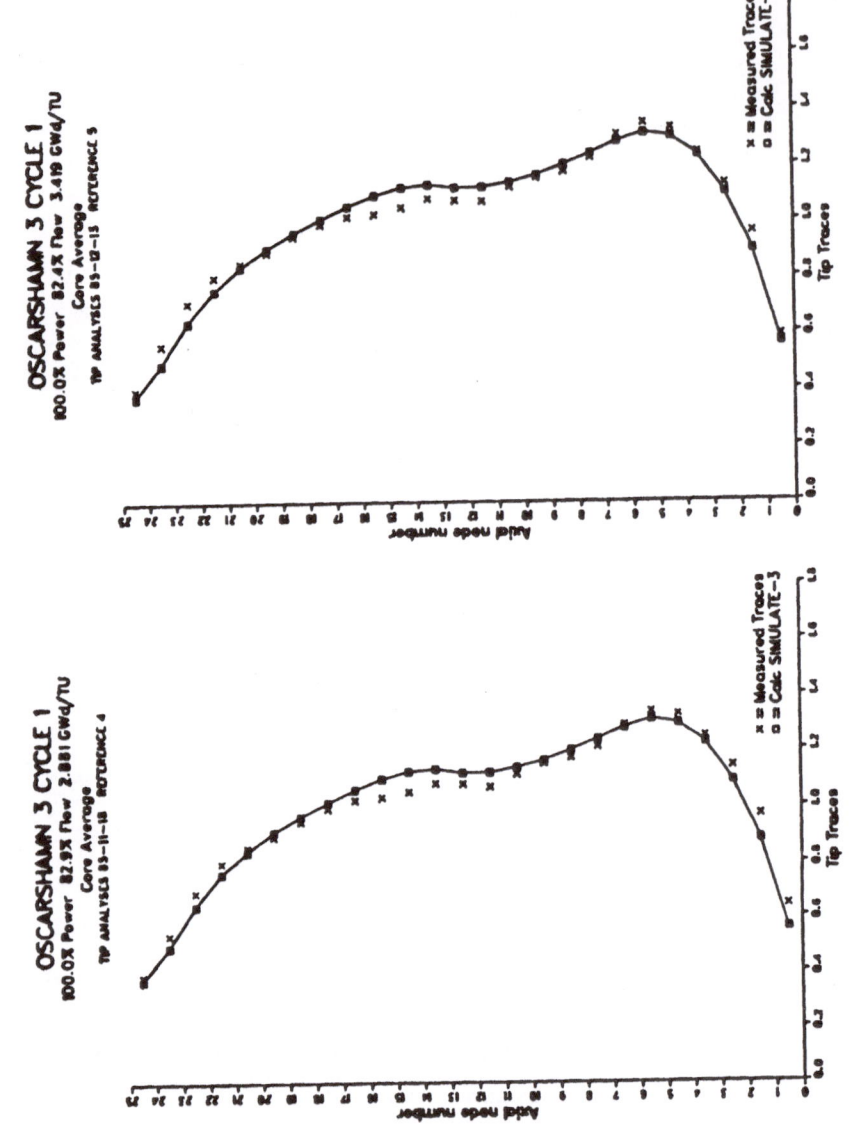

OSCARSHAMN 3 CYCLE 1
100.0% Power 82.4% Flow 3.419 GWd/TU
Core Average
TIP ANALYSES 85-12-13 REFERENCE 5

x = Measured Traces
o = Calc SIMULATE-3

OSCARSHAMN 3 CYCLE 1
100.0% Power 82.9% Flow 2.881 GWd/TU
Core Average
TIP ANALYSES 85-11-18 REFERENCE 4

x = Measured Traces
o = Calc SIMULATE-3

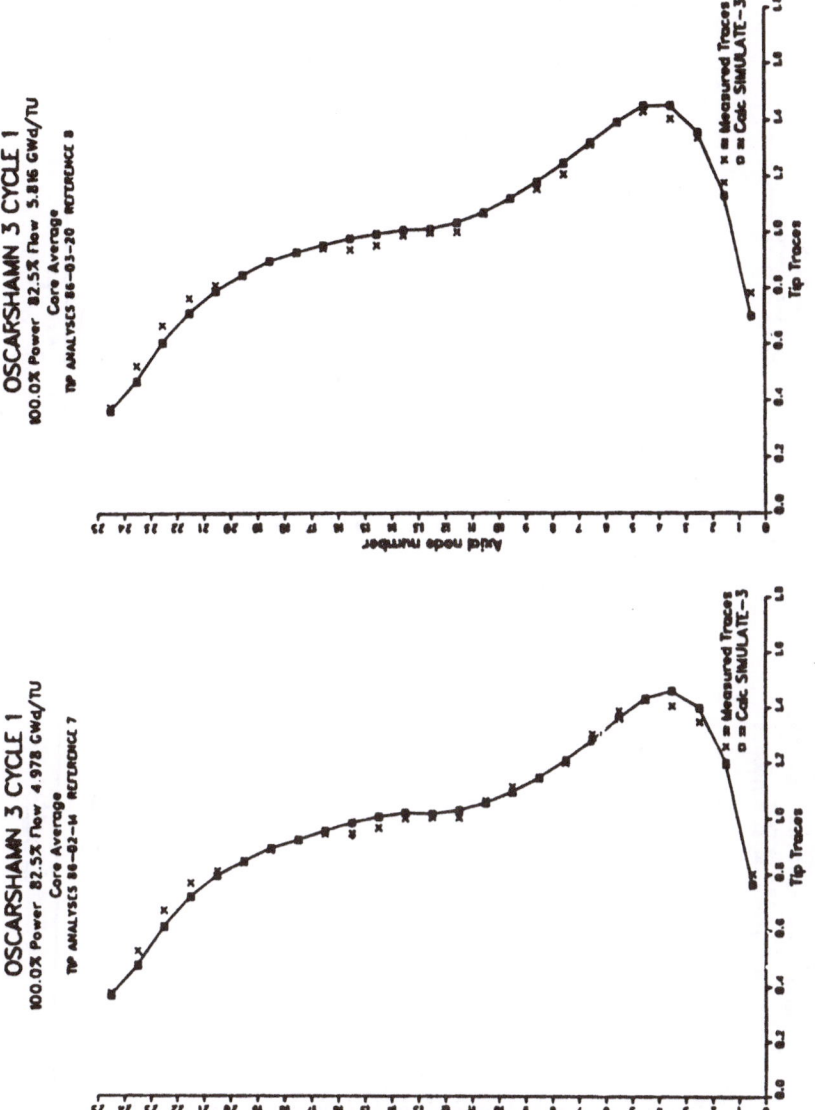

Fig. 9 Cycle 1. TIP traces (neutron)

C1 TIP ANALYSES

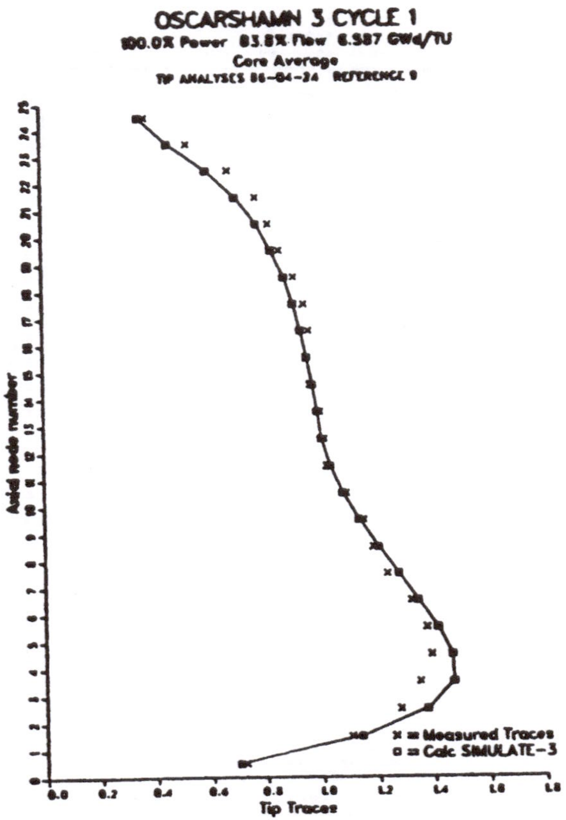

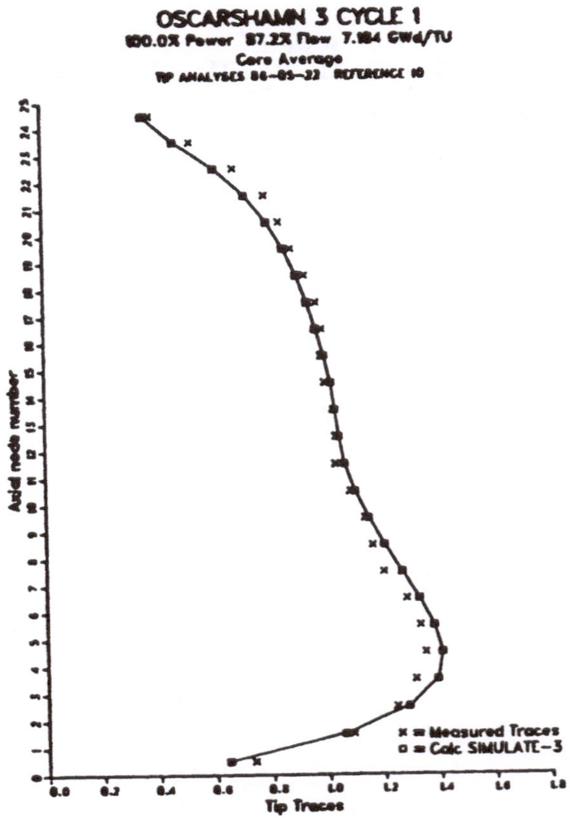

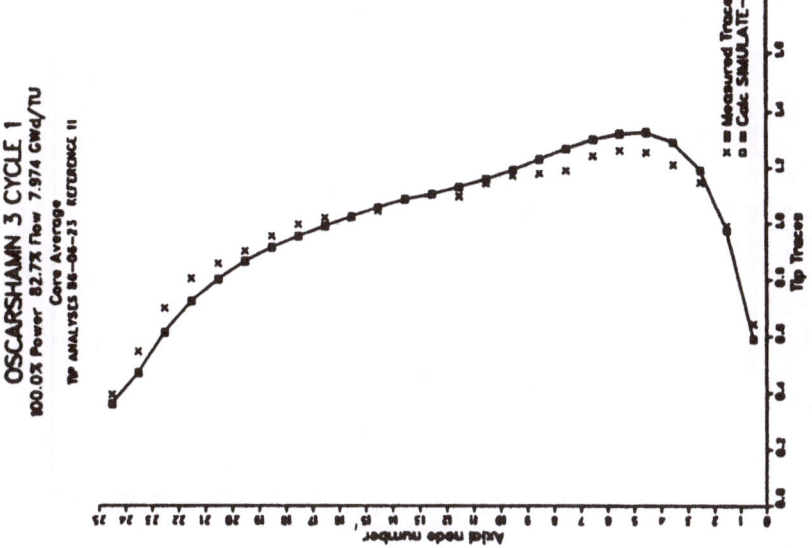

Fig. 10 Cycle 1. TIP traces (neutron)

C2 TIP ANALYSES

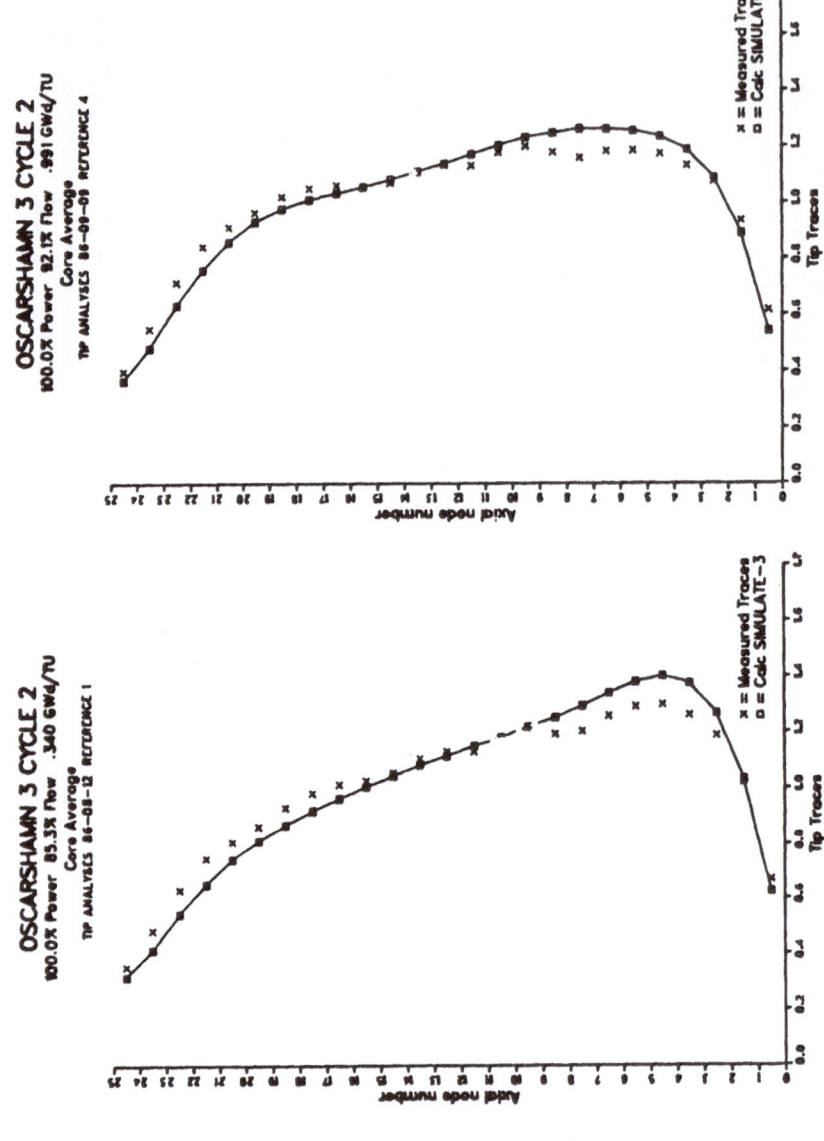

259

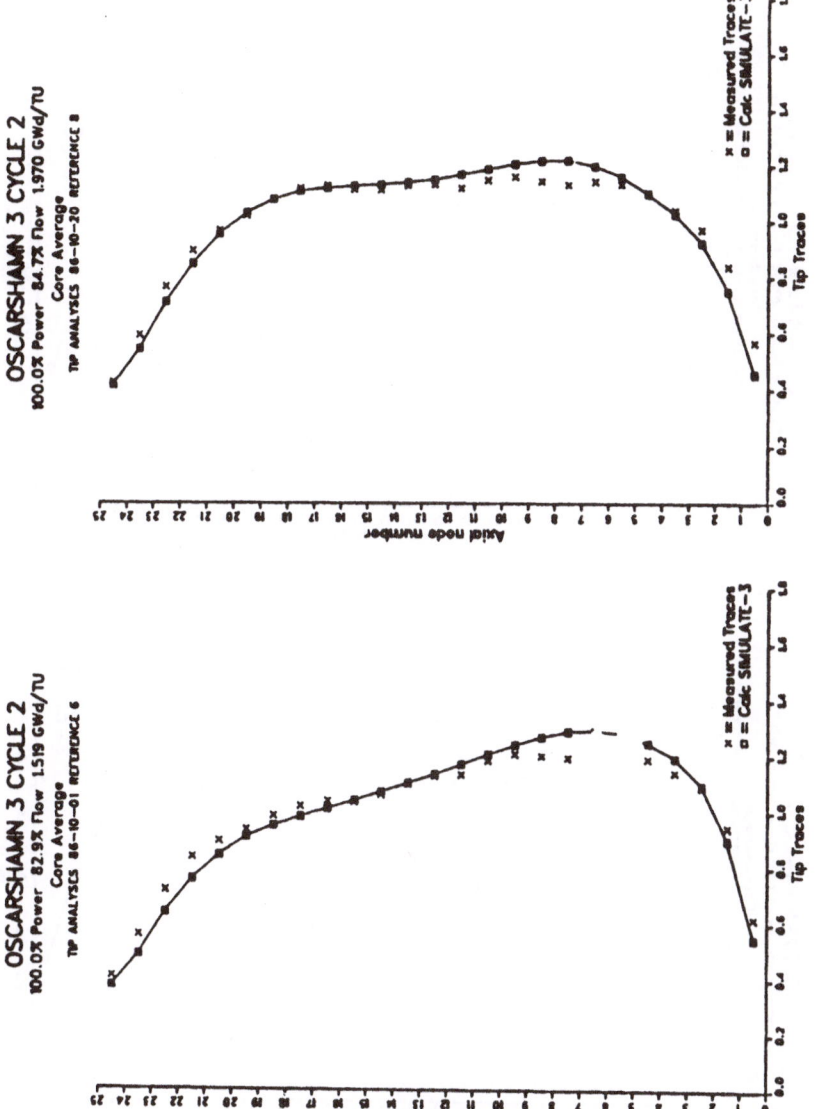

Fig. 11 Cycle 2. TIP traces (neutron)

C2 GAMMA TIP ANALYSES

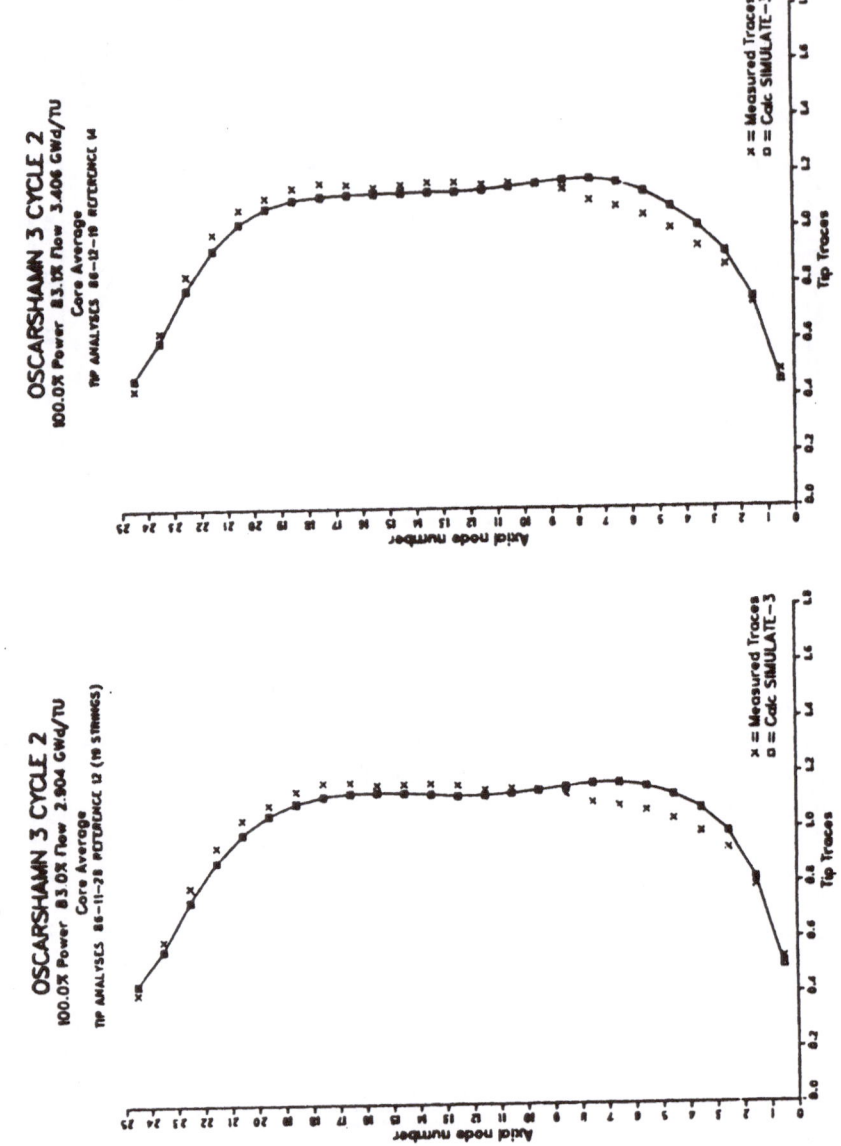

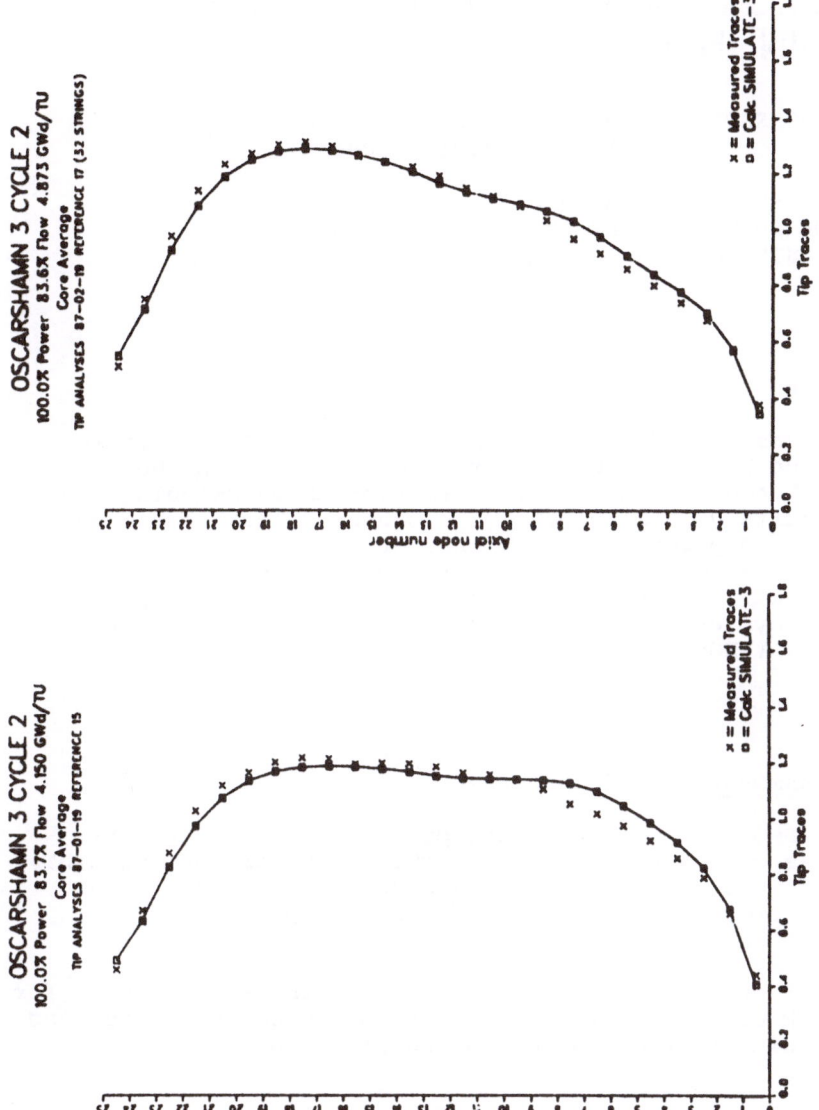

Fig. 12 Cycle 2. TIP traces (gamma)

CASMO-3/MBS Benchmark Calculations on RINGHALS PWR

E.B. JONSSON   Studsvik Energiteknik AB
                S-611 82 Nyköping, Sweden

M. ERIKSSON   Swedish State Power Board
C. HOLMLUND   S-162 87 Vällingby, Sweden
G. NORSTRÖM

Summary.

The Studsvik code package CASMO-3/MBS for PWR analysis
has been benchmarked against several cycles in the
three Ringhals PWR. The results from these calculations
show quite good agreement with measured boron let down
curves and power distributions. Cores with different
degrees of leakage loading pattern show the same good
agreement. The overall RMS deviation in assembly power
over 4 cycles is 1.05%, with single cycles going down
to 0.71 % RMS.

The CASMO-3/MBS code package.

The CASMO-3/MBS (1,2,3) code package is used for ICFM
studies on PWR reactor cores. The CASMO cell and assembly
code is used for production and homogenization of fuel
bundle cross sections as function of exposure, boron
content and temperatures.

The MBS code is a fine mesh diffusion theory code used for
2-D PWR core analysis of critical boron concentration and
power distribution on bundle and pin level.

The multigroup CASMO-3 cell code

The bundle properties are evaluated with the multi-group 2-D
transport theory code CASMO. Neutron spectrum calculations
are performed in 40 or 70 energy groups for all different pin
cells in the assembly. The nuclear data library is based on
ENDF/B-4 and B-5 data (4).

Special treament is provided for highly absorbing pins. With
the MICBURN code (5) effective microscopic cross sections
can be evaluated for Gd poisoned rods.

The homogenized 2-group cross sections for the bundle are
produced as function of exposure, boron concentration and
moderator temperature for the controlled and uncontrolled
bundle. The influence of fuel temperature and xenon is
accounted for.

Extensive testing of the CASMO-3 produced 2-group data has been made against many different experiments and operating reactor cores (4,6).

The MBS 2-D diffusion theory code.

The MBS code (3) has been developed for fast and accurate solution of the diffusion equation in 2-D. The code is very general and can handle 2-4 energy groups as well as non-symmetric assemblies and core layouts. The main approach for PWR analysis is, however, to use the code in a x,y-geometry with bundle homogenized 2-group cross sections. Core layout and fuel handling are on assembly level which means that loading,shuffling as well as saving fuel in a pool is easy. Baffle and reflector are treated with separate cross section data.

Depletion is normally done with critical boron search with the predictor/corrector method allowing rather large exposure steps without loosing accuracy. Xenon and Sm are normally in equilibrium but time/power dependence can be explicitly accounted for. The boron let down curve is calculated during depletion and pin powers are determined by superimposing the CASMO calculated pin power distribution on the homogeneous MBS flux solution.

The linking code

The cross section library for MBS is processed by the MBSLINK code (7) from the CASMO formatted output. Homogenized bundle data are used in a multidimensional setup with cross section dependence on depletion,boron content,moderator temperature and control rod presence. Data for missing state points are interpolated and the tables are completely filled.
The linking code has an option for homogenisation with flux discontinuity factors (ADFs) which has been shown to improve the power sharing calculation significantly in bundles with strong absorbers. The tables are used with linear inter-polation.

The fuel temperature and Xe (Sm) effects are represented with polynomials.

In addition to the cross sections, also the CASMO calculated pin power distributions as function of exposure and boron concentration are transferred to the MBS library. These are used in a special mode to superimpose the detailed CASMO pin power distribution on the homogeneous flux solution. The detector signals are calculated at the same time and a special edit gives the analytic constants for the INCORE core surveillance code.

The baffle/reflector model

The flux gradients in the baffle/reflector regions require special attention in the diffusion code. The strong thermal absorption in the baffle can be quite accurately described with the CASMO baffle/reflector model (8),which directly gives homogenized cross sections with flux discontinuity factors. This treatment gives a baffle/reflector cross section set that is only dependent on boron and water temperature but insensitive to the fuel type and the exposure. The same reflector data can therefore be used both in low and high leakage core layouts.

The material layout

The MBS code subdivides each assembly in 4x4 material regions which then are updated during power iterations and depletion. The material regions are normally further subdivided with one more flux mesh at least at the core periphery. The core layout, shuffling and unloading of fuel are all done on assembly level which facilitates normal core follow over several cycles. Full,half and quarter core symmetry options are available as also rotation of assemblies.

Unloading of fuel is done to a special 'POOL' file where besides the exposure distribution some history effects are saved.

The pin power calculation and the analytic constants

The flux solution is saved for each depletion point and it can be used for an automatic creation of analytic constants for the INCORE core surveillance code and for pin power determination.

A pin power reconstruction technique is used where the CASMO calculated assembly pin power distribution is superimposed on the homogeneous MBS flux solution.
The MBS method with few material mesh in the core layout and superimposing the pin power is quite computer efficient and certainly very easy to use.

Results from core follow on the RINGHALS PWR reactors

The RINGHALS 2,3,4 are all Westinghouse three loop PWRs with 157 assembly cores. RINGHALS 2 is the oldest with 15*15 pin assemblies while RINGHALS 3 and 4 have the 17*17 pin assembly design.

The cycles analyzed are for

     RINGHALS 2   cycles 10,11   high leakage

     RINGHALS 4   cycles 3 and 4   high and low leakage

which cover typical both out,in and in,out loading schemes.

Cycle 4 of RINGHALS 4 contains 8 fresh assemblies with 5%
Gd-pins with the unconventional design of Gd-pins (with
depleted uranium) placed in peripheral positions in the bundle.
The reactor power has been at nominal 100 % level except for
R2 where the power was reduced to 80% due to steam generator
problems in the latter part of cycle 10 and during cycle 11.

Modeling the RINGHALS cores

The MBS model of the RINGHALS cores is a standard layout with
the assemblies subdivided into 4*4 material meshes and with
one extra flux mesh in gradient areas. Standard baffle and
reflector data have been calculated and used for all cores
and cycles.

Burnup dependent local buckling is used for all cycles but no
corrections has been made for partly inserted control rods.

Fuel temperature has been described with a linear dependence
on specific power and no exposure dependence.

Results from critical boron search

The boron let-down curves for R4 cycles 3 and 4
are shown in figures 1 and 2

In R2 the power decrease in cycle 10 was not evaluated with
the Sm transient option and therefore a slight deviation
between calculated and measured boron content can be seen in
figure 3 during this transient. The R2 cycle 11 shows good
agreement at 80 % power (figure 4).

Comparisons with measured assembly powers

During the depletion several comparisons have been made
with measured (symmetrically corrected) power distributions.
A summary of the RMS (Root Mean Square) deviations for the
different cycles is given in table 1.

Table 1. Summary of RMS differences in calculated and
measured assembly powers over four cycles.

Case	Cycle depl. MWd/tU	RMS % (MBS-MEAS)	Case	Cycle depl MWd/tU	RMS % (MBS-MEAS)
R2	330	1.35	R4	210	0.90
cy 10	1500	1.15	cy 3	2500	0.86
	3650	0.90		3640	0.68
	5790	0.64		4840 *	0.56
	6920 *	0.62		5980	0.63
	7790	0.83		7110	0.46
R2	0	1.98	R4	100	0.67
cy 11	1050	1.53	cy 4	1510	1.74
	2140	1.05		2610	1.43
	2720	1.09		3510	1.30
	3745 *	0.74		4730	0.99
	4290	0.78		5810 *	0.87
	5165	0.82			

Typical examples of assembly power distributions are shown
in figures 5 to 8. The distributions represent the depletion
values marked with * in Table 1 above. In all cases MBS
calculated values are compared with symmetricized INCORE
evaluated power distributions. In Ringhals 2, Figs. 5
and 6, both figures represent 80 % power. In Ringhals 4,
Fig. 7 represents the low leakage core of cycle 3 and
Fig. 8 the high leakage core of cycle 4.

Conclusions.

The CASMO-3/MBS code package is a computer efficient system
to be used for PWR in-core fuel management. All types of
steady state core calculations of interest can be performed
with a minimum of manpower and computer time without any
sacrifice of accuracy.

References

1. M Edenius,Å Ahlin,"CASMO-3. New Features,Benchmarking and Advanced Applications",Intern.Topical Meeting on Advances in Reactor Physics,Mathematics and Computation, Paris (1987)

2. M Edenius, et al,"Recent developments in the MICBURN/CASMO/ SIMULATE LWR analysis package",Topical Meeting on Advances in Fuel Management, Pinehurst N.C. (1986)

3. O Norinder, et al "MBS, A 2-D Core Analysis Diffusion Theory Code", STUDSVIK/NR-85/80 (1985)

4. P Jernberg,"CASMO-3 Benchmarks against Criticals", STUDVIK/NFA-86/11 (1986)

5. Å Ahlin,M Edenius, "MICBURN-CASMO/CPM for analysis of Assemblies with Gadolinium",Trans.Am.Nucl.Soc.,41,594 (1982)

6. K Ekberg,S Lundberg,"CASMO-3/SIMULATE-3 Core Follow Calculations on OSKARSHAMN 3",Paper submitted to this Symposium.

7. R Håkansson,"MBSLINK, Program Description and User's Manual" STUDVIK/NR-85/101 (1985)

8. K Smith,"Assembly Homogenization Techniques for Light Water Reactor Analysis",Prog. Nucl. Energy.,Vol. 17,Number 3, Pergamon Press,Oxford,England, (1986)

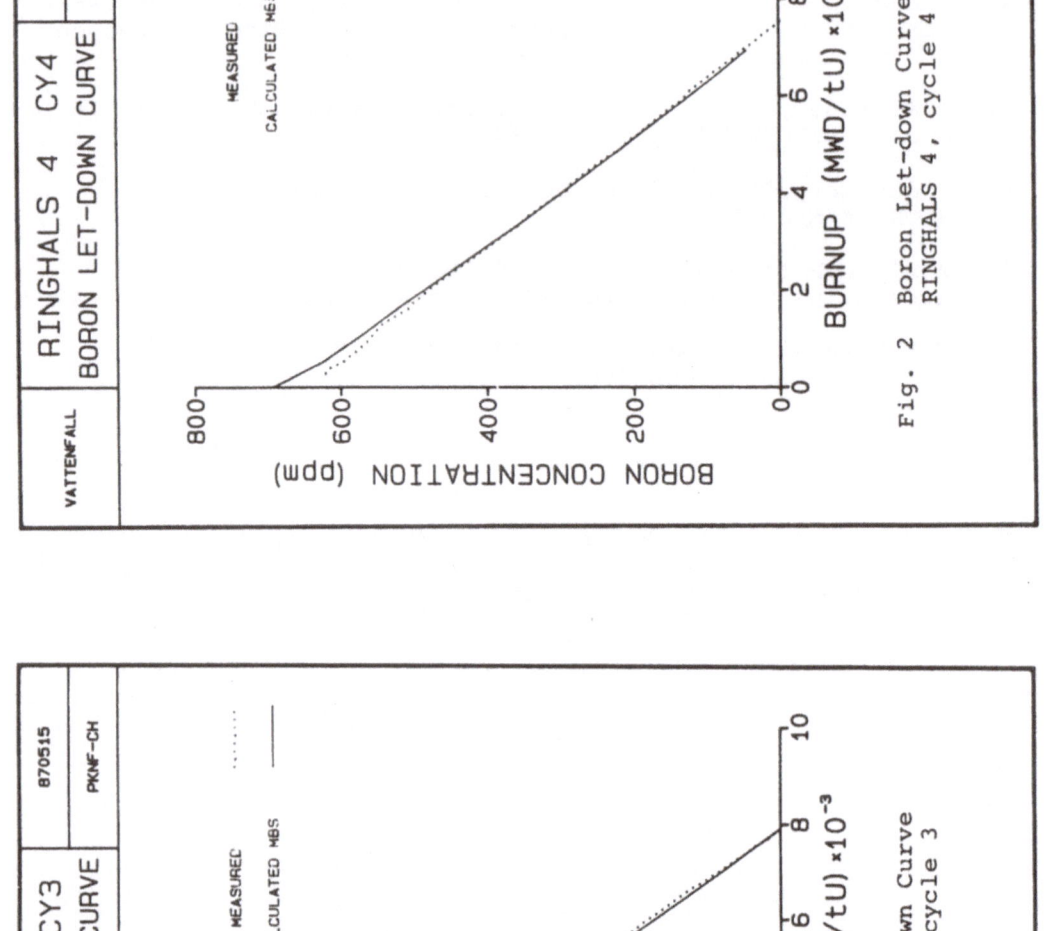

Fig. 1   Boron Let-down Curve
         RINGHALS 4, cycle 3

Fig. 2   Boron Let-down Curve
         RINGHALS 4, cycle 4

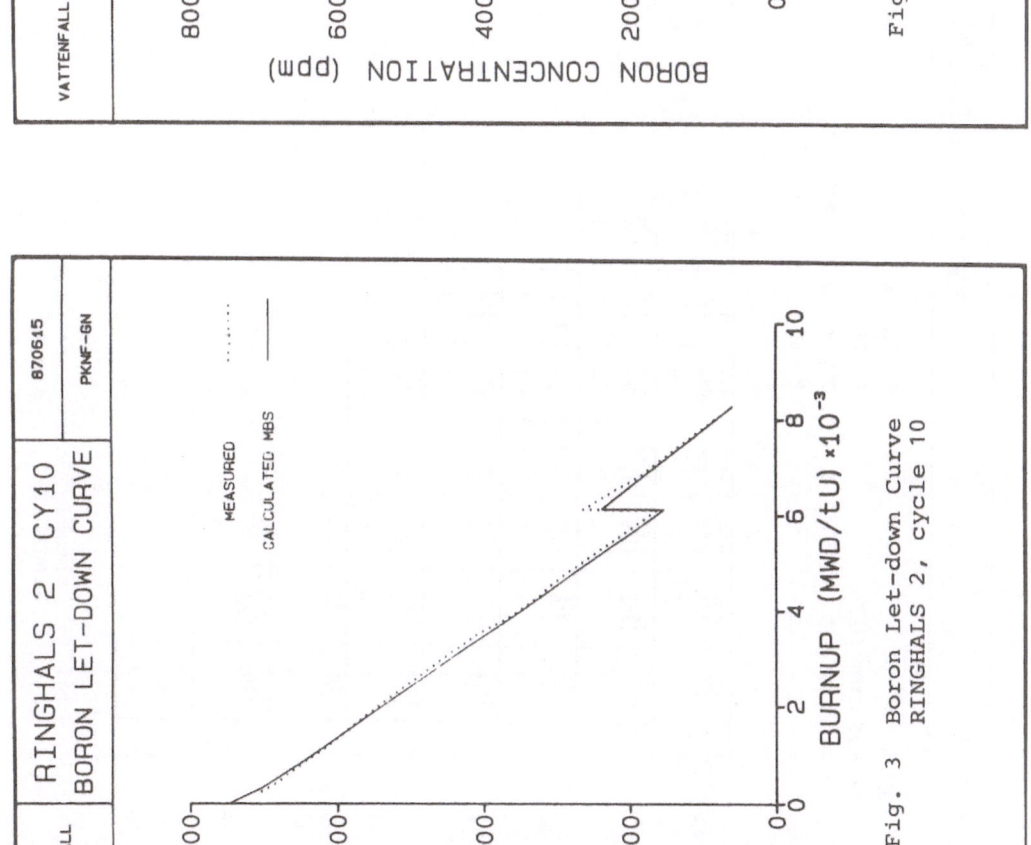

Fig. 4 Boron Let-down Curve
RINGHALS 2, cycle 11

Fig. 3 Boron Let-down Curve
RINGHALS 2, cycle 10

Fig. 5 — RINGHALS 2 CYCLE 10

VATTENFALL	RINGHALS 2 CYCLE 10	870617
	ASSEMBLY POWER	PKNF

MBS CALC
INC MEAS (SYMMETRIZED)
MBS CALC - INC MEAS

E = 6920 MWd/tU
MAP 238
ROOT MEAN SQ. = .00616
P = 80 %

	H	G	F	E	D	C	B	A
8	1150/1135 (15)	892/885 (7)	896/894 (2)	1059/1054 (5)	955/958 (-3)	1079/1085 (-6)	994/991 (3)	798/791 (8)
9		1039/1027 (12)	1258/1241 (16)	1162/1159 (3)	1237/1241 (-4)	983/985 (-2)	1196/1191 (5)	436/430 (5)
10		1257/1254 (3)	971/972 (-1)	1226/1231 (-6)	1081/1090 (-8)	964/964 (1)	1057/1049 (8)	
11		1161/1167 (-6)	1225/1235 (-10)	934/943 (-8)	1007/1013 (-6)	1185/1180 (5)	813/812 (2)	
12		1234/1242 (-8)	1081/1091 (-10)	1007/1013 (-6)	1154/1156 (-2)	494/495 (-1)		
13		982/990 (-8)	965/968 (-4)	1185/1184 (1)	494/494 (0)			
14		1195/1195 (1)	1057/1055 (2)	813/811 (3)				
15		435/433 (2)						

Fig. 5   Assembly Power Distribution RINGHALS 2, cycle 10

Fig. 6 — RINGHALS 2 CYCLE 11

VATTENFALL	RINGHALS 2 CYCLE 11	870618
	ASSEMBLY POWER	PKNF

MBS CALC
INC MEAS (SYMMETRIZED)
MBS CALC - INC MEAS

E = 3745 MWd/tU
MAP 248
ROOT MEAN SQ. = .00743
P = 80 %

	H	G	F	E	D	C	B	A
8	1077/1061 (16)	965/949 (16)	1008/993 (15)	1272/1264 (8)	858/858 (0)	1270/1262 (8)	960/956 (4)	1070/1071 (0)
9		969/955 (14)	1273/1268 (6)	1002/1005 (-3)	1072/1073 (-1)	894/893 (1)	1156/1154 (1)	881/883 (-3)
10		1276/1260 (17)	1021/1019 (2)	1269/1274 (-4)	1144/1152 (-8)	1045/1049 (-4)	1057/1066 (-9)	
11		1010/998 (13)	1272/1269 (3)	997/1002 (-5)	972/980 (-8)	1056/1063 (-7)	428/431 (-3)	
12		1073/1075 (-1)	1146/1152 (-6)	973/980 (-7)	1126/1136 (-10)	476/480 (-4)		
13		893/894 (-1)	1046/1056 (-11)	1057/1065 (-8)	476/479 (-3)			
14		1153/1153 (1)	1057/1064 (-7)	429/431 (-2)				
15		878/876 (3)						

Fig. 6   Assembly Power Distribution RINGHALS 2, cycle 11

Fig. 7 Assembly Power Distribution — RINGHALS 4, cycle 3

VATTENFALL	RINGHALS 4 CYCLE 3 ASSEMBLY POWER	870622 PKNF

MBS CALC E = 4840 MWD/TU
INC MEAS (SYMMETRIZED) MAP 073
MBS CALC - INC MEAS ROOT MEAN SQ. = .00561

Each cell: MBS CALC / INC MEAS / (MBS CALC − INC MEAS)

Row	H	G	F	E	D	C	B	A
8	1333/1339/−6	891/891/0	1161/1157/4	1025/1021/4	1232/1224/8	1212/1205/7	878/878/0	922/930/−7
9		1140/1136/4	942/943/−1	1198/1203/−5	989/987/2	967/955/12	1208/1206/2	756/762/−7
10		946/940/6	1035/1034/1	1235/1243/−8	1228/1232/−4	922/923/−1	973/976/−2	
11		1207/1194/13	1244/1234/10	1034/1035/−1	1185/1186/−8	760/769/−8	669/674/−4	
12		992/987/5	1234/1230/4	963/959/4	778/775/3	775/780/−5		
13		962/968/−5	923/927/−4	766/763/3				
14		1205/1211/−6	974/980/−7	671/673/−2				
15		754/760/−5						

Fig. 8 Assembly Power Distribution — RINGHALS 4, cycle 4

VATTENFALL	RINGHALS 4 CYCLE 4 ASSEMBLY POWER	870622 PKNF

WITH ADF - CORRECTION

MBS CALC E = 5810 MWD/TU
INC MEAS (SYMMETRIZED) MAP 088
MBS CALC - INC MEAS ROOT MEAN SQ. = .00874

Each cell: MBS CALC / INC MEAS / (MBS CALC − INC MEAS)

Row	H	G	F	E	D	C	B	A
8	1203/1188/15	1092/1073/19	1064/1051/13	1018/1010/8	1261/1257/4	1012/1007/5	998/995/2	884/881/3
9		1100/1080/20	1254/1234/21	1067/1057/10	864/864/0	984/986/−2	1185/1185/1	637/641/−3
10		1253/1245/8	1225/1218/8	916/914/2	1163/1168/−5	857/862/−5	1041/1047/−5	
11		1064/1068/−4	915/922/−7	1290/1293/−3	887/895/−8	1176/1186/−10	692/697/−5	
12		864/866/−2	1163/1173/−10	886/896/−10	1194/1209/−15	680/690/−9		
13		985/983/2	858/865/−7	1176/1192/−15	681/688/−8			
14		1187/1183/4	1043/1043/−1	693/691/1				
15		638/638/0						

3-D Full Core Calculations for the Long-Term Behaviour of PWR's

H.-J. WINTER, K. KOEBKE, M. R. WAGNER

Kraftwerk Union AG
D-8520 Erlangen, F.R. Germany

Summary

Presently, the most realistic simulation of a pressurized water reactor (PWR) core is by means of three-dimensional (3-D) full core calculations. Only by such 3-D representations can the large scope of axial effects be treated in an accurate and direct way, without the need to perform various auxiliary calculations. Although the computationally efficient burnup-corrected nodal expansion method (NEM-BC) is used, the computing effort for 3-D reactor calculations becomes rather high, e.g. a storage of about 320000 words is required to describe a 1300 MWe PWR. NEM-BC was introduced (1979) into KWU's package of PWR design codes because of its high accuracy and the great reduction of computing time and storage requirements in comparison to other methods.

The application of NEM-BC to 3-dimensional PWR design is strongly correlated with the progress achieved in the solution of the homogenization and dehomogenization problem. By means of suitable methods (equivalence theory) the transport-theoretical information of the pinwise power and burnup distribution for the heterogeneous fuel assemblies is transferred in a consistent manner to the full core reactor solution. The new methods and the corresponding code system are explained in some detail.

1. Introduction

For the evolution of PWR design during the past decade advanced nodal coarse mesh reactor analysis methods, such as the Nodal Expansion Method (NEM), became indispensable for solving the multigroup diffusion equations, due to their high efficiency in comparison to fine mesh PDQ-type calculations. The basic problem is the 3-D description of the behavior of about 50000 fuel rods in a large PWR as a function of time. The success of NEM as a coarse mesh method relies essentially on the development of complementary methods, which allow the accurate description of pinwise power and burnup distributions over the full reactor life. These results are the basis for other numerical models for full-core analysis of DNB-ratios and for the generation of fuel rod statistics.

A decisive step towards obtaining a more realistic simulation of the stationary reactor is the production use of a 2 group 3-dimensional f u l l - c o r e representation of a PWR. This model will substitute the combination of full-core 2-D and part-core 3-D reactor calculations. Of course the computational expenses for 3-D full core increase: Typical problems consist of about 4000 nodes with 14 unknowns per node and require about a factor 5 to 7 more CP-time per flux calculation than a corresponding 2-D calculation. On the other hand the manpower is considerably diminished because the number of depletion calculations is significantly reduced by eliminating the time-consuming and approximate combination of part-core 3-D and full-core 2-D calculations.

The nodal method is an essential part of the KWU code package, called Standard Design Procedure (SAV-System, Figure 1). A key property of this widely used system (e.g. in Brazil, Korea, several European utilities) is the fact that reactor- or cycle-dependent normalizations are not necessary. Therefore, reliable predictions from one cycle to the next or for the transition from conventional to modern reload schemes pose no problems.

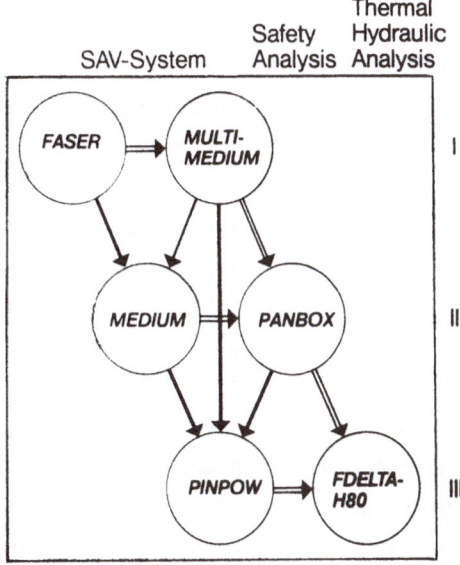

Figure 1. Schematic of KWUs Nuclear Code System
        I: Spectrum Calculation, II: Nodal Reactor
        Calculation, III: Pinwise Reactor Analysis

In the following the codes of the SAV-System are briefly characterized (see Figure 1)

- FASER-3: 1-D 85-group spectrum depletion code of MUFT/THERMOS type for computing pin or assembly average few group (2 to 10) rate conserving cross sections as a function of local burnup and several feed-back variables.
- MULTIMEDIUM: 2-D 10-group assembly transport depletion program for computing equivalent diffusion theory parameters and heterogeneous form functions per assembly [24,25]
- MEDIUM: 3-dimensional 2-group nodal reactor burnup code based on the feedback corrected version of NEM [2]
- PINPOW: Assembly dehomogenization code [5] for reconstructing pin power distributions from nodal reactor solutions

The SAV-System is coupled via data files to the thermal-hydraulics code FDELTAH-80 [22] for pinwise determination of DNB ratios and to the 3-D transient code system PANBOX for load follow and safety analysis. Only stationary reactor design with the SAV-System will be considered in this paper.

To give more insight into the principles of the nodal reactor analysis methods and KWU's SAV-System, the performance of global reactor calculations with NEM-BC is exemplified by a 2-D PWR burnup benchmark problem. In the case of large nodes the accuracy of the original NEM [3] is only sufficient for fresh initial cores . Therefore, a burnup- and feedback-corrected version of NEM [2] was developed, which takes into account the spatial variation of the macroscopic cross sections within the nodes (depletion-, temperature- or xenon-induced).

Another critical point is the generation of cross sections, because the use of nodal methods implies that the homogenization-areas are entire fuel assemblies. The derivation of the respective diffusion theory parameters (effective cross sections as a function of burnup and feedback parameters) is a crucial point for the quality and accuracy of reactor calculations; therefore the applied models for fuel assembly homogenization are discussed in some detail. Also, homogenized equivalent diffusion theory parameters for the radial reflector are computed such that the few group neutronic interaction at the core reflector interface is conserved (in comparison to a 1-D multi-group transport calculation).

The production use of 3-dimensional full core calculations at KWU for in-core fuel management and design of large 1300 MWe PWR will be a qualitative step towards a more realistic, accurate and direct simulation of the stationary reactor core, thus allowing a more economic use of existing margins. Currently, we use a combination of full-core 2-D and part-core 3-D (e.g. octant) reactor representations for computer storage reasons on our CDC 176. Thus our final reactor design is still based on 2-D full core calculations with 4 nodes per fuel assembly (FA) with precalculated 3-D correction terms [1]. These terms are derived from the above mentioned 3-D calculations. Such 3-D calculations are also needed for the determination of control rod worths, reactivity coefficients and set-up of preliminary loading patterns.

In the future we intend to use for final reactor design only 3-D full core calculations. Clearly, this type of calculation avoids the necessity of several coupled (approximately consistent) depletion calculations and the correlated extensive file handling. On the other side precaution must be taken to limit the computing costs of 3-D cases by choosing the node sizes as large as possible. The use of 2-D calculations will then be limited to to scoping and optimization studies.

A compromise with regard to accuracy and spatial resolution is to use 1 node per FA in the radial plane and about 15 axial layers in axial direction . The resulting large nodes can be treated by the burnup-corrected nodal expansion method [2] which takes into account the spatial variation of the macroscopic cross sections within the nodes due to burnup gradients. A quantitative measure for the remaining node size errors can be derived from reactor calculations with 1, 4, 9,... nodes per FA, which converge towards the exact diffusion theoretical solution. This property of the nodal expansion method NEM can be shown numerically by comparison with finite difference PDQ calculations [3,4,5,6] as well as by theoretical considerations [7]. In Section 2 of this paper the practically obtainable accuracy and the basic model assumptions of nodal reactor analysis are discussed for a 2-D PWR burnup benchmark problem [8,9,10]; moreover, the treatment of the radial shroud / water reflector and - within the framework of the simplified equivalence theory SET - the homogenization and dehomogenization techniques respectively are discussed [11,12,13]. Section 3 deals with the performance of 3-D calculations including the determination of external waterside cladding corrosion [14].

2. Nodal Reactor Analysis Methods

2.1 The Nodal Expansion Method NEM

Modern transverse nodal methods [3,4,15,16,17] save 2 to 3 orders of magnitude in computing time compared to a finite difference calculation of the same accuracy when solving the 3-dimensional two-group diffusion equations for the reactor core [4]. Mainly because of this great reduction of computing expense full-core 3-D calculations have the potential to become a standard PWR design practice. In reactor depletion calculations the computing time of NEM is nearly proportional to the number of nodes. For example, a 3-D reactor calculation with 4000 nodes needs 5 to 7 times more CPA-time than a 2-D calculation with about 1000 nodes; the storage requirements are 130000 words in small core memory (SCM) and 200000 words in large core memory (LCM). It is evident that such a realistic reactor simulation avoids the manpower consuming task to perform different approximately consistent depletion calculations.

Systematic derivations of the transverse nodal equations have been published earlier [2,3,16,17,18]. Therefore, only some basic features of NEM will be sketched: The basic problem is to solve the multigroup diffusion equations with spatially constant coefficients for a rectangular node. This solution is expressed in terms of average group fluxes ϕ and average partial in-currents J^{in} and out-currents J^{out}. A formal transverse integration of the diffusion equations yields ordinary inhomogeneous differential equations of second order for each spatial direction. Thereby, the spatial dependence of the inhomogeneous transverse leakage terms is usually approximated by a quadratic polynomial, [26]. The neutron balance equations follow from an integration over the volume of the node. For given average partial in-currents and transverse leakages the ordinary differential equations are either solved analytically [16,17] or by a weighted residual procedure [3]. By means of the matrices A, C, D the elements of which depend on the effective group cross sections and the vectors $\underline{b}$, $\underline{e}$, where the components depend additionally on the transverse leakage, the NEM equations can be formally written as

$$\underline{\phi} = A * \underline{J}^{in} + \underline{b}$$

$$\underline{J}^{out} = C * \underline{J}^{in} + D * \underline{\phi} + \underline{e}$$

(1)

In the iteration process based on equations (1) the incoming currents are given as functions of the outgoing currents of adjacent nodes. To accelerate the convergence of the iterative solution a combination of coarse mesh rebalancing and asymptotic source extrapolation is used. The high accuracy and efficiency of NEM has been shown for the well known 2-D and 3-D IAEA benchmark problems [19] which describes a fresh initial core: Maximum relative power deviations of about 1% from the reference in the case of nodes of assembly size resulted, cf. [4].

2.2 The Burnup-Corrected Nodal Expansion Method NEM-BC

At first sight a basic drawback of NEM seems to be the fact that the high accuracy for very coarse nodes can only be claimed for fresh initial cores but not for depleted reactors, for which the assumption of spatially constant cross sections within the nodes is violated. Thus the variation of the burnup in FAs adjacent to the radial reflector induces spatial variations of the macroscopic cross sections of up to 5 % [20]. In reference [2] this problem of spatially varying cross sections within the nodes is solved by a non-linear extension of NEM called NEM-BC. The NEM and the NEM-BC equations are completely analogous, except that effective flux weighted cross sections are used instead of the nodal volume average cross sections and that an additional polynomial source is added to the leakage term (the coefficients are flux-functionals, [2]). Moreover NEM-BC requires for each node the calculation and storage of the spatial burnup distributions inside the nodes:

$$B(x,y,z) = c_0 + c_1 * x + c_2 * y + c_3 * z +$$

(2)

$$+ c_4 * x * x + c_5 * y * y + c_6 * z * z$$

The CP-time of NEM-BC increases only by about 15 % [2] because suitable approximations are made for efficiently calculating the local cross section variation.

2.3 Two-Dimensional PWR Burnup Bechmark Problem

To demonstrate the accuracy of NEM-BC a 2-D burnup benchmark problem was defined [8] with the following main characteristics and assumptions:

- Simplified 2-D model of a large PWR
- Quarter core with 45° symmetry line (octant)
- 2 reactor cycles
- Typical loading and refueling patterns
- Homogenized fuel assemblies (unrodded and rodded)
- Homogeneous reflector
- Prescribed nuclide depletion chains
- Prescribed 2-group microscopic cross sections for fuel and burnable poison nuclides, prescribed macroscopic cross sections for structural materials
- The reactor is maintained critical by adjusting the amount of soluble boron
- During burnup the reactor is operated with rods fully withdrawn and with xenon in equilibrium
- At fixed time points rods are inserted and the boron equivalent (BOC1, BOC2) or shut-down reactivity (EOC1) is determined
- End of cycle is the time point for which the critical boron concentration is zero (with xenon in equilibrium)

Figure 2 shows the loading pattern of the first cycle, Figure 3 shows the refueling pattern of the second cycle. Figure 4 shows the control rod configuration and, finally, Figure 5 shows the nuclide depletion chain. In the following Table 1 the reference solutions by NEM-BC and the fine-mesh FDM code VENTURE are practical identical (the determination of "pin" peak power densities is explained below):

A coarse mesh (CM) NEM-BC calculation with only a single node per FA deviates from the foregoing reference solution only within a very small margin, see Table 2. This high accuracy of CM calculations also follows from Figure 6 which shows for the beginning of cycle 1 and 2 the maximum

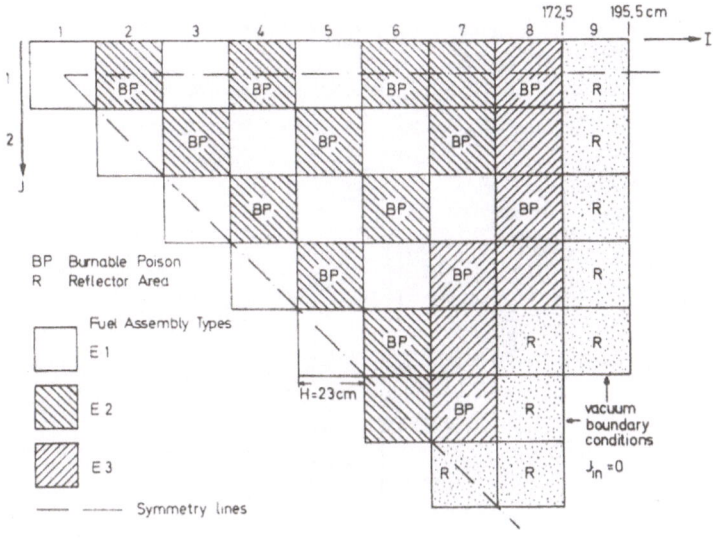

Figure 2. Loading Pattern of First Cycle

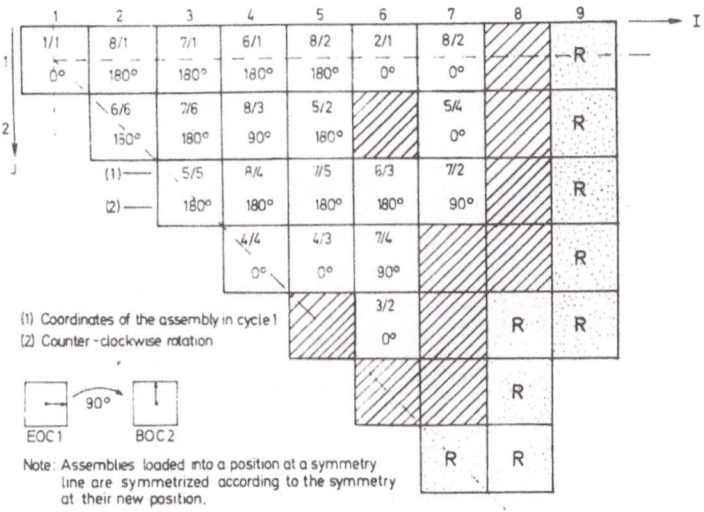

Figure 3. Refueling Pattern at BOC 2

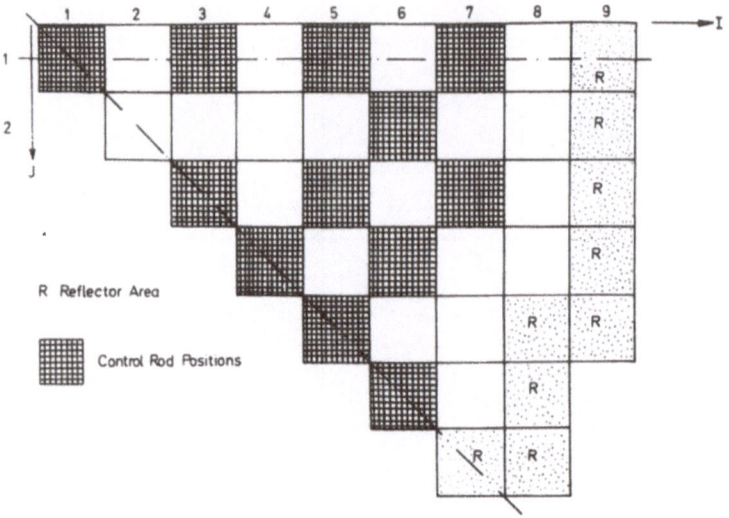

Figure 4. Control Rod Configuration

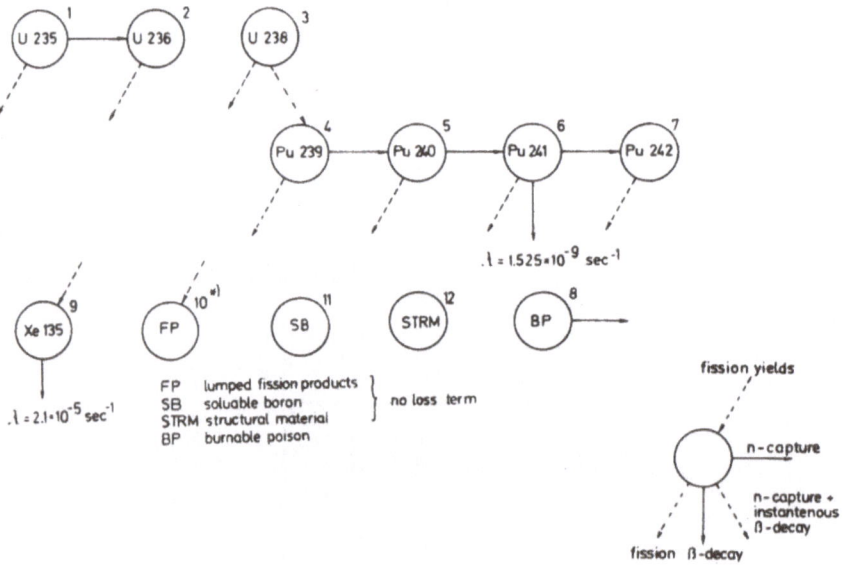

Figure 5. Nuclide Depletion Chains

ORNL-Ref: VENTURE, FDM , 256 points/FA (18496 points), 14 time points, 23 flux calc.
KWU -Ref: MEDIUM , NEM-BC, 25 nodes /FA (841 nodes), 54 time points, 62 flux calc.

	Time Point	Unit	ORNL-Ref	KWU-Ref
Cycle Length	EOC1	day	381.65	381.49
	EOC2		263.02	262.04
Boron Concentration	BOC1, no Xe	ppm	1278.6	1277.8
	Xe-equ.		1039.4	1038.6
	BOC2, no Xe		1262.9	1260.6
	Xe-equ.		977.6	979.5
"Pin" Peak Power Density	BOC1	W/cm^3	126.11	126.85
	EOC1		116.19	116.44
	BOC2		144.72	143.31
	EOC2		124.36	124.53
Control Rod Worth	BOC1	Δppm	746.4	747.7
	EOC1	$\Delta\varrho$(%)	8.0369	8.0778
	BOC2	Δppm	729.3	734.2
Computer Time		min	99.1	27.5

Table 1. Comparison of Two Reference Solutions

percentage errors of the average assembly powers as a function of the node
size. A coarse mesh (CM) NEM-BC calculation in 1 node per FA approximation
deviates from the foregoing reference solution only within a very small
margin, see the last column of Table 2. This high accuracy of CM
calculations is also seen in Figure 6 which shows the maximum percentage
errors of the average assembly powers as a function of the node size for
the beginning of cycles 1 and 2 . Obviously, the failure to account for the
spatial variation of exposure effects leads to large errors of NEM after
partial refueling at BOC2. With an assembly size mesh (H = 23 cm) the
maximum error of the average assembly power is reduced from   unacceptable
large 15 % (NEM) to 1.8 % (NEM-BC).

2.4 The Determination of Local Flux Values

A principal problem of nodal methods is the loss of detailed information
about the local flux and power distributions within the nodes. It is,
however, possible [5] to establish a local high order flux interpolation,
where per energy group 21 coefficients of a 2-dimensional polynomial are

Ref: MEDIUM, NEM-BC, 25 node/FA (341 nodes), 54 time points, 62 flux/calc., CYBER 176
CM : MEDIUM, NEM-BC, 1 node/FA (40 nodes), 54 time points, 62 flux/calc., CYBER 176

	Time Point	Unit	KWU-Ref	KWU-CM	$\frac{CM-Ref}{Ref} \cdot 100$
Cycle Length	EOC1	day	381.49	381.82	+ 0.09
	EOC2		262.04	262.05	+ 0.00
Boron Concentration	BOC1, no XE	ppm	1277.8	1278.1	+ 0.02
	Xe-equ.		1038.6	1039.8	+ 0.12
	BOC2, no XE		1260.6	1260.4	- 0.02
	Xe-equ.		979.5	980.6	+ 0.11
"Pin" Peak Power Density	BOC1	W/cm³	126.85	129.08	1.76
	EOC1		116.84	117.83	0.85
	BOC2		143.31	144.98	1.17
	EOC2		124.53	125.39	0.69
Control Rod Worth	BOC1	Δppm	747.7	747.5	- 0.03
	EOC1	$\Delta\varrho$(%)	8.0778	8.0534	- 0.30
	BOC2	Δppm	734.2	730.9	- 0.45
Computer Time		min	27.5	1.4	-94.91

Table 2. Comparison of the KWU-Reference and the Corresponding
Coarse Mesh (CM) Solution

determined from the converged nodal solution. If local continuity
conditions at the edges of the considered node and its neighbors are used
12 additional values can be determined (local fluxes and net currents in x-
and y-direction at the corners of the node) in addition to the 9 nodal
quantities (node average flux and face averaged fluxes and currents). From
this local homogeneous flux interpolation the NEM-BC power peaking values
in Tables 1 and 2 are derived. Figure 7 shows that for nodes of assembly
size a polynomial expansion yields oscillatory errors up to 3.7 % for the
local power distribution along the main axis. Therefore, an improved
interpolation scheme was developed which, in addition, exploits the typical
behavior of the neutron spectrum that the ratio of the thermal to the fast
flux in the interior of the FAs is approximately constant or a slowly
varying function, [21]. By this method of spectral interpolation the errors
can be reduced significantly, so that local and nodal (average) values are
calculated with comparable accuracy. The final conclusion is that a mesh as
coarse as one node per assembly with the presented NEM-BC is adequate for
normal PWR core design.

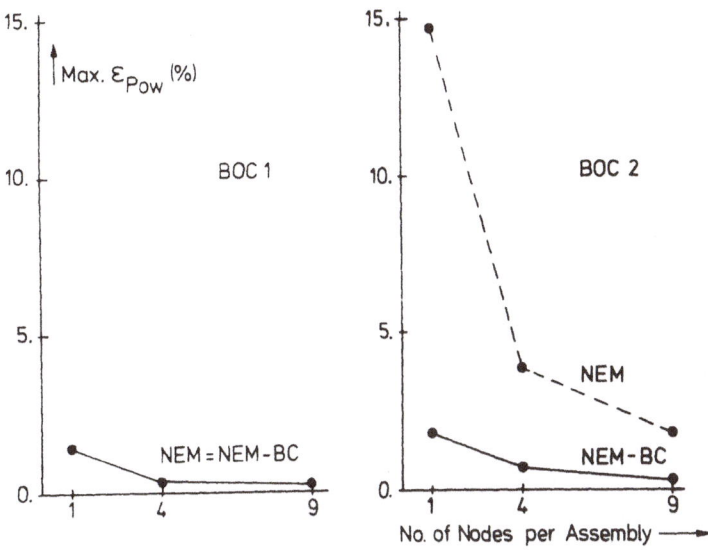

Figure 6. Maximum Percentage Error of the Average Assembly Powers

2.5 The Homogenization and Dehomogenization Problem

The burnup benchmark problem of the previous section describes an idealized reactor which neglects the heterogeneous structure of real assemblies. Solving the reactor problem by NEM-BC and by the succeeding interpolation method, it is assumed that a homogenized reactor (see e.g. Figures 2,3) behaves like a real heterogeneous reactor. Indispensable for nodal methods is therefore the existence of accurate homogenization procedures for areas including an assembly and its neutron physical environment (spectral geometry). Therefore, 2-D spectral calculations are performed in a pin by pin representation which takes into account the heterogeneities caused by the different cell types (fuel rods, guide tubes, poison rods etc.). It seems quite natural for an appropriate homogenization method to postulate that per energy group all nodal quantities such as reaction rates, average fluxes, and face average partial currents must be conserved. This problem was solved in the framework of equivalence theory [11,12] by increasing the number of degrees of freedom in the diffusion equations. Thus diffusion theory was modified by introducing directionally dependent diffusion coefficients D_x, D_y and heterogeneity factors f_x, f_y in addition to the usual rate conserving cross sections.

The solution of the homogenization problem is closely related to the so-called dehomogenization problem, that is the determination of an accurate

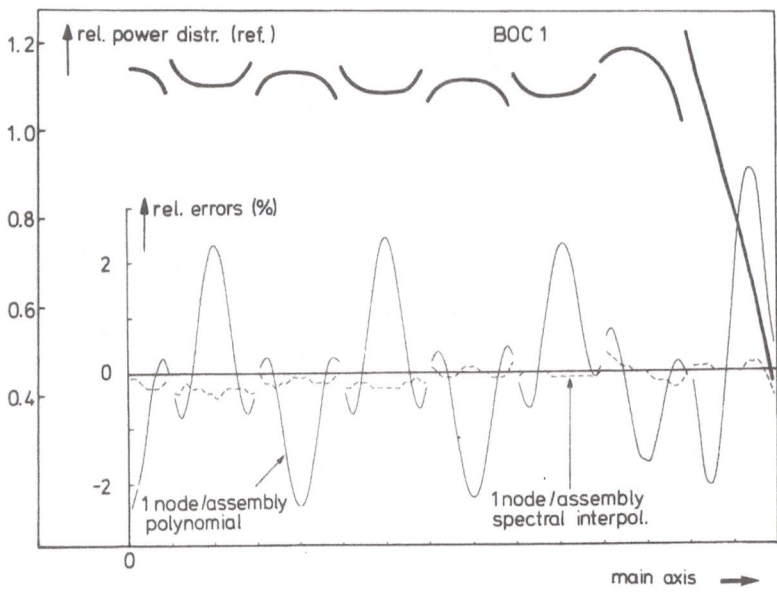

Figure 7. Accuracy of Interpolation Schemes

local heterogeneous flux and power distribution on the basis of an
equivalent reactor solution . The information about the heterogeneous flux
distribution is comprehended in the "heterogeneous form function" which is
defined as the quotient of the heterogeneous flux distribution and the
local equivalent flux distribution. Both are calculated from the 2-D
spectral solution of the different types of fuel assemblies. Note, that
the same nodal method is used to determine equivalent parameters and to
solve the global reactor problem. The generation of a heterogeneous reactor
solution then proceeds in two steps: The first step is to construct a local
homogeneous distribution (power, burnup or fluxes) by means of the   high
order interpolation method for all fuel assemblies. In the second step the
smooth nodal distributions obtained by the first step are multiplied by
precomputed heterogeneous form functions. An example for the reconstruction
of the normalized pinwise power distribution is shown in Figure 8 for a
full core representation.

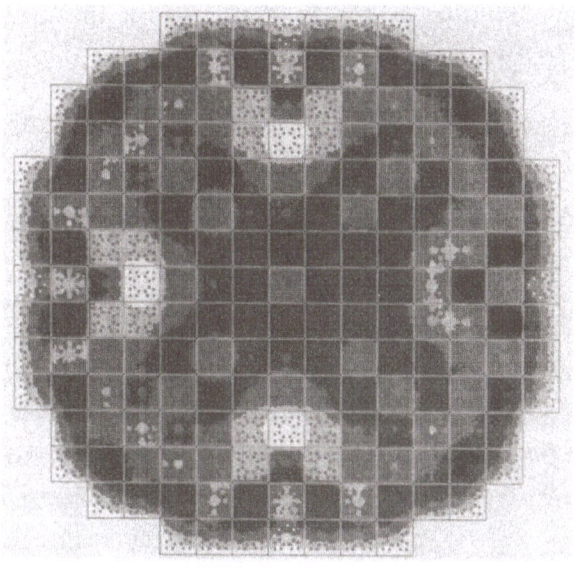

0.6 < ▦ < 0.9 < ▦ < 1.1 < ▦ < 1.4 < ■

Figure 8.   Normalized Pinwise Power Distribution of a 1300 MWe
            PWR at BOC; Three rods of a Doppler-Bank are fully
            inserted.

2.6 The Simplified Equivalence Theory SET

The main disadvantage of equivalence theory is the increased number of
equivalent parameters in comparison to conventional diffusion theory.   It
is   therefore   of great practical importance that an   approximation
(Simplified Equivalence Theory SET, [12]) of the exact equivalence theory
could be derived so that conventional diffusion theory remains applicable.
SET assumes direction independent neutronic interaction of the FAs, i.e.

$$D = D_x = D_y \text{ and } f = f_x = f_y \tag{3}$$

Then the heterogeneity factor f does no longer appear as an explicit
parameter, because all cross sections and the equivalent diffusion constant
D can be divided by f. Moreover, SET-fluxes and -net currents at the
interfaces are continuous so that the usual NEM-BC formalism can be applied
without change.

An example for SET is the homogenization of the radial shroud/water
reflector of PWRs, Figure 9. The problem is to describe the shielding
effect of the steel for the moderated fast and partially backscattered

thermal neutrons. Assuming downscattering within the reflector range the response R is given by:

$$\phi_1 = R_{11} * J_1$$

$$\phi_2 = R_{21} * J_1 + R_{22} * J_2$$

(4)

(ϕ_G and J_G are the surface fluxes and currents at the fuel/shroud interface, G = 1,2). Two different solutions for the spectral geometry of Figure 9a (with different albedos on the left side) are sufficient to calculate all response matrix elements from a linear equation system. On the other hand the coefficients of the SET-equations can be determined such that the resulting analytical solution of the homogenized reflector representation, Figure 9b, conserves the reflector response. This reflector model deduced from 1-D spectral geometries can also be applied for multidimensional reflector representations.

For a representative spectrum of reactor states, including different moderator densities, soluble boron concentrations and burnup states, rodded cores and mixed oxide FAs (plutonium) accurate results [13] have been obtained. Moreover, this reflector model improved the prediction of the local powers in the three outermost pin rows. Besides, the necessity for determining reflector albedos from independent fine mesh calculations is eliminated by this equivalent reflector model.

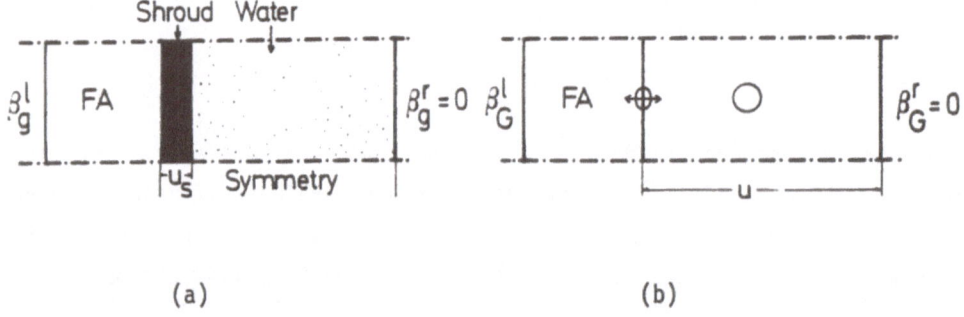

(a)   (b)

Figure 9. Spectral Geometry of the Shroud/Water Reflector of a PWR in (a) 1-D heterogeneous and (b) 1-D homogeneous representation. ß = albedos; g = microgroup index; G = macrogroup index

3. Performance of 3-D Reactor Calculations

The current trend in PWR design is to simulate the reactor core in all aspects and as realistically as possible. By an explicit 3-D representation of the full core 3-D effects can be handled in an accurate and direct way which otherwise would require many auxiliary calculations, comprehensive data file management and quality assurance work. By 3-D calculations the features can be treated directly:

- DNB ratio and maximum power peaking factor F_q
- Mutual influence of adjacent fuel assemblies with different axial burnup shapes
- Axial leakage
- Axially dependent burnout of burnable poisons
- Differential control rod worths and reactivity values of partially inserted control rod banks
- Axial blanket
- Stretchout states and load changes
- Feedback of moderator temperature, fuel temperature and xenon distribution
- Precalculation of the axial offset ratio as a function of time and control rod movements
- 3-D full core burnup calculations are the presupposition for accurate transient calculations [18]

By the transition from a 4 node/FA (present SAV-System) to a 1 node/FA (future SAV-System) representation the computing times of a 3-D PWR design are kept in reasonable limits (see chapter 2.3). This is also true for 3-D flux solutions at the beginning of higher cycles, which are most difficult to describe. Nevertheless, 2-D solutions will always be used for long term fuel management or scoping calculations. But also in that case accurate 2-D calculations can only be performed if appropriate 3-D correction terms are available [1] from previous representative 3-D solutions. Besides the pin power peaking value and the maximum pellet burnup as the leading design limits, the determination of the pinwise DNB-ratios [22] and of the fuel rod waterside corrosion [14] thickness became increasingly important for core refueling optimization procedures [23]. Thus both quantities have also to be calculated on the basis of the 3-D nodal code system. The DNB-ratio is determined by the program FDELTA-H80 (see Fig.1), while the calculation of corrosion thickness is incorporated directly in the nodal codes.

References

1. Winter H.-J., Koebke K., "Treatment of 3D-effects in Reactor Calculations", Trans. Am. Nucl. Soc., 31, p.246, (1979)

2. Wagner M.R., Koebke K., Winter H.-J., "A Nonlinear Extension of the Nodal Expansion Method", Proc. ANS/ENS Intl. Topical Mtg., Munich, FRG, 2, p.43, (April 1981)

3. Finnemann H., Bennewitz F., Wagner M.R., "The Nodal Expansion Method - A Consistent Approximation to the Multidimensional Neutron Diffusion Equation", Atomkernenergie, 30, p.123, (1977)

4. Wagner M.R., Finnemann H., Koebke K., Winter H.-J., "Validation of the Nodal Expansion Method and the Depletion Program MEDIUM-2 by Benchmark Calculations and direct Comparison with Experiment", Atomkernenergie, 30, p.129, (1977)

5. Koebke K., Wagner M.R.: "The Determination of the Pin Power Distribution in a Reactor Core on the Basis of Nodal Coarse Mesh Calculations", Atomkernenergie, 30, p.136, (1977)

6. Cadwell W.R., "PDQ-7 reference manual, WAPD-TM-678, Bettis Atomic Power Laboratory, (1967)

7. Huang Aixiang, Zhang Bo, "The Convergence for Nodal Expansion Method", IMA Preprint Series # 281, Institute for Mathematics and its Applications, University of Minnesota, Minneapolis, (November 1986)

8. Koebke K., Wagner M.R., Winter H.-J., Wörner A., "PWR Burnup Benchmark Problem", ANL-7416, Supplement 3 (DE86012678), p. 880-924 (accepted October 1985)

9. Winter H.-J., Koebke K., and Wagner M.R., "Solution of the two-dimensional PWR burnup benchmark problem by coarse mesh methods", Proc. ANS/ENS Intl. Topical Mtg., Knoxville, 1, (April 1985)

10. Vondy D.R., "Solution of a Two-Dimensional PWR Benchmark Problem Modeling Two Cycles with Refueling and Fuel Assembly Repositioning by Finite-Difference Method", ANS April 1985 Knoxville Topical Meeting, Poster Session, p.159

11. Koebke K., "A New Approach to Homogenization and Group Condensation", IAEA-TECDOC 231, p. 303, IAEA Technical Comm. Mtg., Lugano, Switzerland (Nov. 1978)

12. Koebke K., "Advances in Homogenization and Dehomogenization" Proc. ANS/ENS Intl. Topical Mtg., Munich, FRG, 2, p. 59, (April 1981)

13. Koebke K., Haase H., Hetzelt L., Winter H.-J. "Application and Verification of the simplified Equivalence Theory for Burnup States", Proc. ANS/ENS Intl. Topical Mtg., Knoxville, 2, p. 607, (April 1985)

14. Zimmermann R., Eberle R., Fuchs H.-P., Groß H., Wunderlich F., "On the Influence of Thermal-Hydraulic Plant Design Parameters and Power History on PWR Cladding Corrosion", IAEA Technical Committee Meeting on External Cladding Corrosion in Water Power Reactors, CEN-Cadarache-France, (October 1985)

15. Dorning J.J., "Modern Coarse-Mesh Methods - A Development of the '70s", Proc. ANS Topical Mtg., Williamsburg, Virginia, $\underline{1}$, p. 3-1, (April 1979)

16. Fischer H.D., Finnemann H., "The Nodal Integration Method - A Diverse Solver for Neutron Diffusion Problems," Atomkernenergie / Kerntechnik, $\underline{39}$, p. 229, (1981)

17. Greenmann G., Smith K., Henry A.F., "Recent Advances in an Analytic Nodal Method for Static and Transient Reactor Analysis", Proc. ANS Topical Mtg. Williamsburg, Virginia, $\underline{1}$, p. 3-49 (April 1979)

18. Finnemann H., Gundlach W., "Space-time kinetics code IQSBOX for PWR and BWR", Atomkernenergie/Kerntechnik, $\underline{37}$, p. 176, (1981)

19. Misfeldt I.,"2-D IAEA Benchmark Problem, "DAEC RP-3-75/ NEACRP-L-138, see also ANL 7416, Benchmark Problem Book, Supplement 2, p. 277 - 472 (1977)

20. Koebke K., Hetzelt L., Wagner M.R., Winter H.-J.: "Principles and Application of Advanced Nodal Reactor Analysis", ANS Topical Mtg., Chicago, (Sept. 17-19, 1984)

21. Koebke K., Hetzelt L.: "On the Reconstruction of Local Neutron Flux and Current Distributions from Nodal Schemes", Nucl. Sci. Eng., $\underline{91}$, p.123, (1985)

22. Suchy P., Ulrych G.: "Höhere Betriebsflexibilität durch verbesserte thermohydraulische Ganzcorerechnungen in KWU- Druckwasserreaktoren und Übergang von FΔH auf DNBR", Proc. KTG Conference, Munich, (May 1985)

23. Böhm W., Kiehlmann H.-D., Neufert A., Ulrych G.: "Advanced Incore Fuel Management for KWU and other Vendor PWRs",Proc. ENC '86, Geneve (1986)

24. Wagner M.R.: "A Nodal Discrete Ordinates Method for the Numerical Solution of the Multidimensional Transport Equation", Proc. ANS Topical Mtg., Williamsburg, $\underline{2}$, page 4 (April 1979)

25. Wagner M.R., Müller B.: "The Nodal Discrete Ordinates Method and its Application to LWR Lattice Problems", Proc. ANS Topical Mtg., Chicago, (Sept.17-19, 1984)

26. Bennewitz F., Finnemann H., Wagner M.R.: "Higher Order Corrections in Nodal Reactor Calculations", Trans. Am. Nucl. Soc., 22, page 250 (1975)

Numerical Methods for Advanced LWR Core Simulators

R. Böer, H. Finnemann

Kraftwerk Union AG
D-8520 Erlangen, F.R. Germany

E. Michel

Department of Computer Science (IMMD)
University of Erlangen-Nuremberg

Abstract

The subject of this paper is the solution of partial differential equations for the simulation of the reactor core on high-
performance computers. The multi-level methods used for the
calculation of core power distributions are well-suited for the
considered multiprocessor systems. The implementation of a
multigrid nodal diffusion method on a special array configuration of the memory-coupled multiprocessor DIRMU is outlined. The
problem of time integration is discussed from the viewpoint of
advanced multi-level techniques and higher-order integration
methods. For the reconstruction of local power distributions by
means of coarse-mesh solutions the variable-order ECD-method
is introduced which yields accurate flux expansion coefficients
in the center of a calculational node.

1. Introduction

The numerical solution of partial differential equations (PDE)
describing the physical behaviour of the reactor core plays an
important role in the simulation of nuclear power plants. In this
paper efficient numerical algorithms for LWR core simulators as
well as their implementation on advanced computer architectures
will be discussed. The investigations are still in progress and
part of KWU's contribution to the German Supercomputer Project
SUPRENUM /1/. The performance of computers can be increased by
technological progress and parallel architecture. Since technological improvements are limited by physical laws, the desired
performance can only be achieved by massive parallelism. In the
last few years the parallel principle of pipelining has been

successfully used. But now the limits of such vector computers have nearly been reached. Therefore a further great leap forward can only be made by additionally exploiting the possibilities of concurrency. Such multiprocessor systems can in principle be built up to unlimited size. Multiprocessors are determined by the architecture of their nodes (computing elements) and the structure of their interconnection network. The nodes of the available or announced multiprocessors are - for cost reasons - very similar: a multiprocessor, private memory and a coprocessor (mostly a custom-designed vector-unit).

With regard to the interconnection structure there are bigger differences. There exist two main types: bus-coupled and memory-coupled systems. From the programmer's point of view, in the first case communication must be carried out via messages, whereas in the second case it is additionally possible to use common data.

SUPRENUM can be characterized as a multiprocessor with a hierarchical bus system. One of the goals of the SUPRENUM project is to provide the potential user of the SUPRENUM-computer with attractive applications software. KWU is developing a space-time kinetics code system for the analysis of transients in light water reactors with emphasis on the behaviour of the reactor core. For the neutronic part of the problem nodal multigrid (MG) methods are presently being investigated. The multi-level principle has proven to accelerate classical grid methods by orders of magnitude /2/. Section 2 discusses some extensions of standard MG methods to nodal diffusion equations. Higher-order time integration methods applicable to neutron diffusion equations are introduced in section 3. The reconstruction of the fine-mesh solution from a coarse-mesh solution is treated in some detail in section 4. The implementation of an MG algorithm on the memory-coupled multiprocessor system DIRMU (DIstributed Reconfigurable MUltiprocessor kit) /3/ is discussed in section 5. A 25-processor-system is running at the University of Erlangen-Nuremberg since 1985 and available as a testbed for SUPRENUM partners. Finally, section 6 contains some concluding remarks and recommendations for future investigations.

2. Nodal Multigrid Techniques

2.1 Semi-discrete Transport Equations

The semi-discrete neutronics equations resulting from spatial
discretization of the transport equation by means of the nodal
expansion method (NEM) /4/ can be written in the following form

$$\frac{1}{V_g}\frac{d\phi_g^m}{dt} + \left(\Sigma_{ag}^m + \sum_{g'\neq g}^{G}\Sigma_{g'g}^m + \sum_{u=x,y,z}\frac{2\,C_{1gu}^m}{a^m}\right)\phi_g^m$$

$$= \sum_{g'\neq g}^{G}\Sigma_{gg'}^m\phi_{g'}^m + \frac{1}{\lambda}\sum_{g'=1}^{G}\sum_{j=1}^{J}(1-\beta^j)\,\chi_{pg}^j\,\nu\Sigma_{fg'}^{jm}\phi_{g'}^m \tag{1}$$

$$+ \sum_{u=x,y,z}\frac{1}{a^m}\left[(1 - C_{2gu}^m - C_{3gu}^m)(j_{gul}^{+m} + j_{gur}^{-m}) - 2\,C_{1gu}\,a_{4gu}^m\right]$$

$$+ \sum_{i=1}^{I}\chi_{dg}^i\lambda_i\,C_i^m + \chi_{eg}\,S_{ext}^m.$$

$$\frac{dC_i^m}{dt} + \lambda_i C_i^m = \frac{1}{\lambda}\sum_{g'=1}^{G}\sum_{j=1}^{J}\beta_i^j\,\nu\Sigma_{fg'}^{jm}\phi_{g'}^m \tag{2}$$

$$j_{gul}^{-m} = C_{1gu}^m(\phi_g^m + a_{4gu}) + C_{2gu}^m j_{gul}^{+m} + C_{3gu}^m j_{gur}^{-m} - C_{4gu}^m a_{3gu}$$

$$j_{gur}^{+m} = C_{1gu}^m(\phi_g^m + a_{4gu}) + C_{3gu}^m j_{gul}^{+m} + C_{2gu}^m j_{gur}^{-m} + C_{4gu}^m a_{3gu} \tag{3}$$

The notation is fairly standard and in accordance with previous
usage /4/. The equations are to be solved subject to the appropriate boundary conditions.
The exponential transformation

$$\emptyset\,(t) = \exp\left[\omega\,(t-t_o)\right]\varphi(t)$$

leads to an entirely implicit method of solution if $t = t_1$ is
set on the right side of the balance equations. For known ω
the system of balance equations may be solved iteratively for
fluxes $\emptyset\,(t_1)$ and outgoing currents $j\,(t_1)$. The method has a
local error proportional to the square of the time step length.
By choosing frequencies skillfully, the proportionality factor

may become small and vanish in the ideal case when the solution assumes asymptotic behaviour. Methods are available permitting to estimate frequencies on the basis of the behavior of the solution of the current time step. In addition, a new method has been introduced which, while being only slightly more complicated compared to the Euler method, has the advantage that the local error is proportional to the fourth power of the time step duration. The method is strongly absolutely stable and well suited for the solution of the time-dependent nodal transport equation. Further details are presented in section 3.

Different thermal-hydraulic models can be coupled with the neutron transport equation. The homogeneous model for a steam water mixture assumes that both phases are in thermodynamic equilibrium. Though overly simplistic, this model yields sufficiently accurate results for low-quality two-phase flow and therefore will be utilized in on-line simulators. For severe transients and detailed analysis more sophisticated models are also available /5/.

Because of space limitations we only discuss the application of MG methods to nodal transport equations in this work.

2.2 Parallel Multigrid Techniques

Coarse-grid acceleration techniques were recommended by many authors. Techniques of multiplicative coarse-grid corrections were developed by Wachspress /6/. This work motivated several studies and was successfully applied in nuclear reactor design calculations.

A. Brandt recognized the actual efficiency of MG methods /7/. Since then they have gained wide acceptance. Standard sequential MG algorithms are known to be optimal in the sense that the number of arithmetic operations is proportional to the number of discrete unknowns. The basis of MG methods are local smoothing techniques applied on a hierarchical system of grids. On each grid relaxation steps can be performed in parallel by using the obvious approach of domain splitting. Thus the original sequential algorithm can be adapted to the multiprocessor architecture by breaking up the problem into parallel subtasks. Data have to be exchanged with nearest neighbors only, i.e.

the necessary communication is local with respect to the corresponding grid. To avoid that the number of processors exceeds the number of grid points the coarsest grid could be chosen finer than on a sequential computer. A natural subdivision of the reactor core is obtained by allocating a (cluster of) processor(s) to each fuel assembly.

2.3 Multigrid Operators

MG methods are normally formulated with additive corrections. This procedure is more general than a multiplicative technique /6/. We have chosen a few candidates for the coarse-mesh operators, the simplest of which is the coupling coefficient model

$$
\frac{1}{V_g} \frac{d\phi_g^m}{dt} + \sum_{u,m_s(u)} \frac{1}{a_u^m} \left[(\gamma_{gul}^m + \gamma_{gur}^m) \phi_{g}^m - \gamma_{gur}^m \phi_g^{m_l} - \gamma_{gul}^m \phi_g^{m_r} \right]
$$

$$
+ (\Sigma_{ag}^m + \sum_{g=1}^G \Sigma_{g'g}^m) \phi_g^m \tag{4}
$$

$$
= \sum_{g=1}^G (\Sigma_{gg'}^m + \frac{1}{\lambda} \sum_{j=1}^J (1 - \beta^j) \chi_{pg}^j v\Sigma_{fg'}^{jm}) \phi_{g'}^m
$$

$$
+ \sum_{i=1}^I \chi_{dg}^i \lambda_i C_i^m + \chi_{eg} S_{ext}^m
$$

The multi-level approach implemented on DIRMU (s. section 5) is based on this equation. The full NEM equations (1-3) are used on the finest mesh only. A coarse grid can also be defined in energy space. In two- to four-group calculations we consider a condensation to one or two broader energy groups. Because of the dependence of the coefficients on the solution, the problem is nonlinear.

The MG algorithms based on (4) are very efficient because the amount of computation on all but the finest grid is very small. Nevertheless MG formulations based on the complete model (1-3) are of theoretical and practical interest. These higher-order coarse-mesh operators are presently being investigated. The Full Approximation Scheme (FAS) is a version of multigrid mainly being used in solving nonlinear problems /7/. The method is illustrated here for the reactor eigenvalue problem which may be written formally in the following form

$$L(\varphi, \lambda) = 0$$

where the eigenvalue is denoted by λ. If the differential operator L is approximated on two consecutive grids m and M by L^m and L^M, respectively, then the coarse grid equations can be written

$$L^M(\varphi^M, \lambda) = \tau^M_m$$

$$= L^M(\hat{I}^M_m \tilde{\varphi}^m, \tilde{\lambda}) - I^M_m L^m(\tilde{\varphi}^m, \tilde{\lambda}) \qquad (5)$$

$$\varphi^m_{NEW} = \tilde{\varphi}^m + I^m_M(\tilde{\varphi}^M - \hat{I}^M_m \tilde{\varphi}^m) \qquad (6)$$

where I^M_m and I^m_M are restriction and prolongation operators, respectively, $\tilde{\varphi}^m, \tilde{\varphi}^M, \tilde{\lambda}$ approximations to the exact solution $\varphi^m, \varphi^M, \lambda$ on grids m and M, respectively. If $\tilde{\varphi}^m = \varphi^m$, then the second term on the RHS of (5) vanishes and the level M solution is given by $\hat{I}^M_m \varphi^m$. Thus τ^M_m is the quantity needed to obtain an approximation of the fine-mesh solution by solving the coarse-mesh equation. For example, if $\hat{I}^M_m$ is the operator which performs spatial averages, then the converged coarse-mesh solution reproduces averages of the fine-mesh solution. The fine-mesh equations can therefore be considered as a device for caculating corrections τ for the coarse-mesh equations. This dual point of view - as Brandt /7/ calls it - of the multigrid process is the key to many algorithmic possibilities which will also be exploited in the development of advanced core simulators at KWU. Of particular importance is the possibility of applying MG methods directly to the parabolic problem (1-3). Discretizing (1-3) by implicit time integration methods presented in the next section, FAS-MG cannot only be used to accelerate the solution process at each time step but also in the time evolution process itself (frozen τ-techniques /7/). With the advent of high-performance computers the "on-line" reconstruction of pin-by-pin solutions will also become feasible using these advanced MG-techniques. At present sophisticated a posteriori interpolation techniques are used for this purpose. These methods are discussed in section 4 in some detail.

3. Higher Order Time Integration Methods

For the solution of the time dependent diffusion equation the implicit Euler method modulated by an exponential expression has been used up to now /4/. This method has a local error proportional to the square of the time step length and is characterized by the fact that a linear expression

$$y(t) = y_0 + a (t - t_0) \qquad t_0 \leq t \leq t_1 \qquad \Delta t = t_1 - t_0$$

is exactly satisfying the system of differential equations at the final point t_1 of the time interval. Thus a quadratic expression for $y(t)$ can serve to obtain a higher order method. The special quadratic expression

$$y(t) = \bar{y} + \frac{y_1 - y_0}{2} (2\tau - 1) + (\bar{y} - \frac{y_1 + y_0}{2}) (6\tau(1-\tau) - 1) \qquad (7)$$

where

$$\tau = (t - t_0) / \Delta t \qquad \text{and} \qquad \bar{y} = \frac{1}{\Delta t} \int_{t_0}^{t_1} y(t) \, dt$$

is the time average in the intervall $\llcorner t_0, t_1 \lrcorner$, yields a 3rd order method being strongly absolutely stable if one requires the test equation $\dot{y} = \lambda y$ to be satisfied not only at point $t = t_1$ but also integrally. The formula

$$y_1 = \frac{1 + \frac{1}{2} \lambda \Delta t}{1 - \frac{2}{3} \lambda \Delta t + \frac{1}{6} (\lambda \Delta t)^2} \, y_0$$

results which is identical with the well known Padé approximation $(2,1)$ of the exponential function. The above mentioned characteristics of the method follow from this fact. It should be noticed that even a 4th order method can be obtained by a quadratic expression $y(t) = a_0 + a_1 \tau + a_2 \tau^2$ if the assumption is made that $y(t)$ fullfills the differential equation in the subdomains $\llcorner 0, \frac{1}{2} \lrcorner$ and $\llcorner \frac{1}{2}, 1 \lrcorner$. In this case the Padé approximation $(2,2)$ results. The method is therefore not strongly absolutely stable and still method (7) might be preferable. To solve the neutron balance equations, the corresponding expression has to be formulated for each energy group:

$$\emptyset_g(t) = \overline{\emptyset}_g + \frac{\emptyset_{g1} - \emptyset_{g0}}{2}(2\tau - 1) + (\overline{\emptyset}_g - \frac{\emptyset_{g1} + \emptyset_{g0}}{2})(6\tau(1-\tau) - 1) \qquad (8)$$

This expression is formally identical with the parabolic expression used in NEM for spatial integration. To realize the advantages of the method, an algorithm has to be developed for calculating the time-averaged value of the neutron flux in (1). A system for determining $\emptyset_g(t_1)$ and the time-averaged values $\overline{\emptyset}_g$, representing auxiliary parameters only, is obtained if (8) is substituded into (1), (2), and (3) and the condition is made that the residue vanishes integrally and at time $t = t_1$. Thus, the time derivative of (1) can be approximated by

$$\frac{-6\overline{\emptyset}_g + 4\emptyset_{g1} + 2\emptyset_{g0}}{v_g \Delta t} \qquad \text{and} \qquad \frac{\emptyset_{g1} - \emptyset_{g0}}{v_g \Delta t}$$

respectively. The precursor concentrations occurring in (1) are completely replaced by the integral formed using (8). Except for a few obvious approximations, the additional work compared to the Euler method consists in having to solve not one but two coupled balance equations per energy group. Additional memory requirements do not exist. The solution procedure in the time range itself is similar to the iteration procedure for solving a stationary problem. Once the neutron fluxes $\emptyset_{g0}$ for timepoint t_0 are known, the calculation of the new fluxes $\emptyset_{g1}$ takes place in the same way as solving an external source problem in the stationary case. Therefore all considerations made for parallelization techniques to run the stationary problem on a multiprocessor configuration can be adopted for the time dependent problem.

Finally, it has to be noticed that the neutronic equations are representing only a part of the coupled neutron/thermal-hydraulic system. Therefore, one cannot generally count on utilizing the entire convergence order of the introduced method.

4. Nodal Interpolation Schemes

In reactor calculations based on a mesh size comparable with a fuel assembly size in addition to the solution of the nodal flux distribution local flux values are required for determining maximum power peaking factors in order to calculate e.g. DNB-ratios and maximum fuel temperatures. Thus nodal interpolation schemes are to be provided to reconstruct the local fluxes from the nodal flux solution by solving the following problem:

If for a calculational node the nodal flux functions

$$r(x) = \frac{1}{a} \int_{-\frac{a}{2}}^{\frac{a}{2}} \varphi(x,y) \, dy \qquad s(y) = \frac{1}{a} \int_{-\frac{a}{2}}^{\frac{a}{2}} \varphi(x,y) \, dx \qquad (9)$$

are given, find $\tilde{\varphi}(x,y)$ which approximates $\varphi(x,y)$ with sufficient accuracy. (The axial problem is considered as separated.)

The functions $r(x)$ and $s(y)$ are given in the NEM-theory by polynomials of up to order 4 or in the NIM-theory by a linear combination of trigonometric and hyperbolic functions /4/, /8/.

In recent years at KWU the polynomial based interpolation schemes MSS /8/ and PINPOW /9/ as well as an exponential based method /9/ have been developed. MSS and PINPOW are characterized by constructing an interpolation function using values and derivatives at corners calculated from the nodal flux solution.

In section 4.1 a new method, the ECD-method (Expansion Coefficients by means of the Differential Equation) is introduced with the goal of calculating expansion coefficients of any even/even order and mixed even/odd order in the midpoint of fuel assemblies or reflector nodes. Thus ECD provides a powerfull tool to construct an interpolation function which is very close to the reference solution in the interior of the considered node. An example for such an interpolation scheme is given in section 4.2. Furthermore in section 4.3 the calculation of transverse leakage functions by means of ECD is discussed.

Examples are given for the 2D-IAEA benchmark problem which contains homogeneous regions, rodded fuel assemblies and reflector nodes and is therefore well suited to qualify the presented theory. To eliminate the effect of nodal coarse mesh errors a reference solution with a meshsize of 1.6667 cm was generated and the nodal flux functions with respect to the fuel assembly size of a = 20 cm were obtained by integration of the reference solution. All flux values refer to the thermal power of 2.124 E-4 W and $\varkappa$ = .3044 E-10 Ws/fission.

4.1 The ECD-Method

Starting from a Taylor expansion in the midpoint of a node

$$\varphi(x,y) = \sum_{i,j} c_{ij} x^i y^j \qquad -\frac{a}{2} \leq x,y \leq \frac{a}{2} \qquad (10)$$

and defining (see formula (9))

$$\tilde{r}(x) = \frac{1}{2}(r(x) + r(-x))$$
$$\tilde{s}(y) = \frac{1}{2}(s(y) + s(-y))$$
$$\bar{\varphi} = \frac{1}{a} \int_{-\frac{a}{2}}^{\frac{a}{2}} r(x)\ dx \qquad (11)$$

it turns out that $\tilde{r}(x)$, $\tilde{s}(y)$ and $\bar{\varphi}$ contain only coefficients of even/even order in x/y.

The same coefficients appear in the balance equation and its higher order derivatives evaluated at the midpoint (0,0), if homogeneous material constants are assumed in the considered node as it is usual in nodal reactor calculations. In this case the balance equation in energy group g (B^g) can be written as:

$$B^g: \quad -D^g\left(\frac{\partial^2}{\partial x^2}\varphi^g + \frac{\partial^2}{\partial y^2}\varphi^g\right) + \Sigma_1^g \varphi^1 + \Sigma_2^g \varphi^2 = 0$$

Thus the following quantities serve to establish an energy group coupled system of linear equations of m unknowns to calculate all coefficients of even/even order up to order n:

n = 4, m = 12

$\widetilde{r}^g(0)$, $\widetilde{s}^g(0)$, $\overline{\varphi}^g$

B^g , $\dfrac{\partial^2}{\partial x^2} B^g$, $\dfrac{\partial^2}{\partial y^2} B^g$

n = 6, m = 20

$\widetilde{r}^g(0)$, $\widetilde{s}^g(0)$, $\overline{\varphi}^g$, $\widetilde{r}^g(z_1)$, $\widetilde{s}^g(z_1)$ $\qquad$ $0 < z_1 < \frac{a}{2}$

B^g , $\dfrac{\partial^2}{\partial x^2} B^g$, $\dfrac{\partial^2}{\partial y^2} B^g$, $\dfrac{\partial^4}{\partial x^4} B^g$, $\dfrac{\partial^4}{\partial x^2 \partial y^2} B^g$, $\dfrac{\partial^4}{\partial y^4} B^g$

The extensions to orders n ≥ 8 are obvious and need not explicitely be noted here.

Table 1 shows results for the ECD method of order up to 6 for the flux values in the midpoints of selected fuel assemblies and reflector nodes. It can clearly be seen that the flux values for both groups converge to their reference values if the order n is increasing. The same result is obtained for a l l o t h e r fuel assemblies and reflector nodes. Furthermore it turns out that also the coefficients c_{20} and c_{22} converge to their reference values as Table 2 shows for FA (1,1).

The ECD-method can be extended to calculate mixed even/odd or odd/even coefficients in x/y. If $\widetilde{r}(x)$ and $\widetilde{s}(y)$ are replaced by

$\widetilde{dr}(x) = \frac{1}{2}(r(x) - r(-x))$ or $\widetilde{dr}(x) = \frac{1}{2}(r'(x) + r'(-x))$

$\widetilde{ds}(y) = \frac{1}{2}(s(y) - s(-y))$ or $\widetilde{ds}(y) = \frac{1}{2}(s'(y) + s'(-y))$

the following quantities establish two 6 x 6 systems of linear equations to calculate e.g. the coefficients c_{10}, c_{01}, c_{21}, c_{12}, c_{30} and c_{03}:

$\widetilde{dr}^g(z_1)$, $\widetilde{dr}^g(z_2)$, $\dfrac{\partial}{\partial x} B^g$, $\widetilde{ds}^g(z_1)$, $\widetilde{ds}^g(z_2)$, $\dfrac{\partial}{\partial y} B^g$

$0 < z_1, z_2 < \frac{a}{2}$

The elimination of higher order coefficients ($n = 5,7$) again is obvious. Table 1 shows examples for the good approximations of c_{10} in the cases $n = 3$ and $n = 5$.

At this very point it should be noticed that all coefficients calculated up to now are based o n l y on the nodal flux functions of the considered node and n o   i n f o r m a t i o n s of neighbouring nodes have been used.

The coefficients of odd/odd order in x/y such as c_{11}, c_{31}, c_{13}, etc. are not contained in the expansions of $r(x)$ and $s(y)$. Therefore they cannot be calculated by means of ECD. But clearly they can be determined in conventional ways e.g. by precalculated corner values or flux derivatives at the corners of a node. Surely additionally the exact equations given by $\frac{\partial^2}{\partial x \partial y} B^g$ etc. can be used.

4.2   An Interpolation Scheme Based on the ECD-Method

Apart from the problem of determining the coefficients of odd/odd order, the coefficients calculated by means of ECD can be used to construct a simple interpolation scheme. The idea is to correct the basic interpolation function

$$\tilde{\varphi}^0(x,y) = r(x) + s(y) - \bar{\varphi}$$

which is identical with the zeroth order MSS-interpolation function /8/. Of course the expansion of $\tilde{\varphi}^0(x,y)$ in terms of c_{ij} does not agree with (10). Therefore the corrections are defined as polynomials constructed from ECD-coefficients in such a way that expansion (10) results. The interpolation function $\tilde{\varphi}_4^{ECD}$ e.g. considering all coefficients up to order 4 is given by

$$\tilde{\varphi}_4^{ECD}(x,y) = \tilde{\varphi}^0(x,y) + c_{11} xy + c_{31} x^3 y + c_{13} xy^3$$
$$+ c_{21} (x^2 - \frac{a^2}{12}) y + c_{12} x(y^2 - \frac{a^2}{12})$$
$$+ c_{22} (x^2 y^2 - x^2 \frac{a^2}{12} - y^2 \frac{a^2}{12} + \frac{a^4}{144})$$

Similarly the extensions to $\tilde{\varphi}_n^{ECD}$ with order $n > 4$ are obtained.

Table 3 shows average errors of fuel assemblies lying at the
symmetry line y = 0, where the coefficients c_{11}, c_{31}, c_{13} etc.
are zero due to symmetry conditions. The errors are calculated
for the 13 x 13 = 169 points belonging to the 1.6667 cm reference solution and are separated into two parts: e_i includes
only the 11 x 11 = 121 interior points whereas e_t includes the
total amount of 169 points. The result is that even in the
rodded fuel assemblies (1,1) and (5,1) the errors e_i of the
thermal flux, which is strongly varying within the fuel assemblies, decrease to about 1 %o if the order is increased to
n = 6. A highly accurate solution which reproduces almost the
reference solution is obtained in fuel assembly (3,1) which lies
within a homogeneous region. Furthermore, in Table 2 the errors
within fuel assembly (1,1) are analyzed in more detail up to
order 12. Here the coefficients c_{10}, c_{12} etc. are zero too and
therefore only even coefficients are to be calculated. The
errors in thermal flux in the interior decrease to about .1 0/00
and show that even in cases where the thermal flux locally is
strongly depressed a highly accurate approximation can be obtained if the order of approximation is increased sufficiently.

In addition the corner values $\tilde{\varphi}_n^{ECD}(c)$ show that in the fast
group a better value than by MSS is given already for n = 4.
In the thermal group the corner values decrease monotonously
to the reference value even if the accuracy of the MSS value is
still not attained for n = 12. But it has to be noticed that
the values in thermal flux for the point c + Δ which is diago-
nally adjacent to corner c are very accurate compared with the
reference value already for n $\geq$ 8.

4.3   The Calculation of Transverse Leakage Using ECD

The calculation of transverse leakage functions $-DL_u(u)$ /4/
influences the accuracy of NEM or NIM. Up to now a quadratic
approximation built up by leakage information from adjacent
nodes has been used /4/.
It turns out that in terms of the expansion coefficients (10)
all coefficients of $-DL_u(u)$ can be approximated by coefficients
calculated with ECD. Therefore it might be possible to expand

-DL$_u$(u) into a higher order polynomial to improve the accuracy of NEM or NIM. The determination of -DL$_u$(u) is to be performed during inner flux iterations and therefore a reduction of computational effort in solving the ECD-equations is desirable. This problem can be solved e.g. for the even order coefficients approximating -DL$_u$(0) by the following considerations (e.g. for n = 4): If $\tilde{r}(x)$ and $\tilde{s}(y)$ are given, the quantities

$$\tilde{r}''(0) = 2 \, c_{20} + c_{22} \, \frac{a^2}{6} \qquad \tilde{r}^{(4)}(0) = 24 \, c_{40}$$

$$\tilde{s}''(0) = 2 \, c_{02} + c_{22} \, \frac{a^2}{6} \qquad \tilde{s}^{(4)}(0) = 24 \, c_{04}$$

can easily be constructed and with their aid the unknowns c_{20}^g, c_{02}^g, c_{40}^g and c_{04}^g eliminated in $\frac{\partial^2}{\partial x^2} B^g$, $\frac{\partial^2}{\partial y^2} B^g$ yielding a 2 x 2 system for c_{22}^g, g = 1,2. Thus the original effort in solving a 12 x 12 system is remarkably reduced.

Table 1: ECD Results for Selected Fuel Assemblies and Reflector Nodes of the 2D-IAEA-Problem

FA	order of ECD	group 1 c_{00}	group 2	order of ECD	group 1 c_{10}	group 2
(1,1)	4	21.316	3.363	3		
	6	21.320	3.358	5	= 0.	
	ref	21.319	3.355	ref		
(5,1)	4	17.440	2.784	3	-.1385	-.2212-1
	6	17.443	2.781	5	-.1392	-.2215-1
	ref	17.443	2.778	ref	-.1394	-.2212-1
(5,5)	4	13.394	2.114	3	-.1493	-.2370-1
	6	13.396	2.110	5	-.1500	-.2371-1
	ref	13.396	2.109	ref	-.1501	-.2368-1
(3,1)	4	33.485	7.860	3	-.6373-1	-.1496-1
	6	33.485	7.860	5	-.6368-1	-.1495-1
	ref	33.485	7.860	ref	-.6367-1	-.1495-1
(9,1)	4	1.785	6.552	3	-.2629	-.6098
	6	1.783	6.577	5	-.2623	-.6131
	ref	1.783	6.578	ref	-.2621	-.6172
(7,7)	4	.333	1.709	3	-.4017-1	-.1673
	6	.332	1.716	5	-.3814-1	-.1724
	ref	.332	1.714	ref	-.3747-1	-.1734

Table 2: ECD Results for FA (1,1) of the 2D-IAEA-Problem

	order of ECD	c_{20}	c_{22}	$\tilde{\varphi}_n^{ECD}(c)$	$\tilde{\varphi}_n^{ECD}(c+\Delta)$
group 1	4	.3351-1	-.1017-3	27.310	25.698
	6	.3365-1	-.8769-4	27.277	25.698
	8	.3368-1	-.9687-4	27.270	25.695
	10	.3369-1	-.9598-4	27.276	25.696
	12	.3369-1	-.9669-4	27.279	25.695
	ref	.3370-1	-.9668-4	27.284	25.695
	MSS			27.426	
group 2	4	.6878-2	-.1421-4	6.075	4.708
	6	.6373-2	-.1927-4	6.052	4.702
	8	.6252-2	-.1641-4	5.985	4.694
	10	.6220-2	-.1545-4	5.918	4.692
	12	.6212-2	-.1600-4	5.898	4.690
	ref	.6201-2	-.1594-4	5.827	4.692
	MSS			5.859	

Table 3: Average Errors of ECD Interpolation Scheme for Selected Fuel Assemblies on the Symmetry Axis y = 0 of the 2D-IAEA-Problem

FA	order of ECD	group 1		group 2	
		e_t 0/00	e_i 0/00	e_t 0/00	e_i 0/00
(1,1)	4	.113	.054	3.91	1.66
	6	.080	.046	3.16	1.19
	8	.042	.013	1.71	.496
	10	.025	.010	.781	.177
	12	.012	.005	.567	.093
(3,1)	4	.019	.011	.018	.011
	6	.003	.002	.007	.004
(5,1)	4	.400	.245	3.95	1.71
	6	.123	.066	3.16	1.19

5. Implementation of NEM-Algorithm on DIRMU

In this section the implementation of the parallelization of
the stationary NEM-algorithm on the memory-coupled multiprocessor system DIRMU /3/ is outlined /10/. The implementation
was performed for a sample problem consisting of 24 fuel assemblies (FA) (Fig. 1). Each FA consists of 15x15 pins each of
which represents a calculational node on the finest grid (Fig.2).

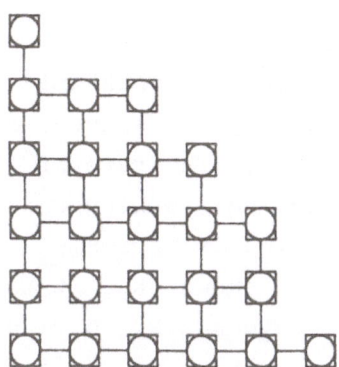

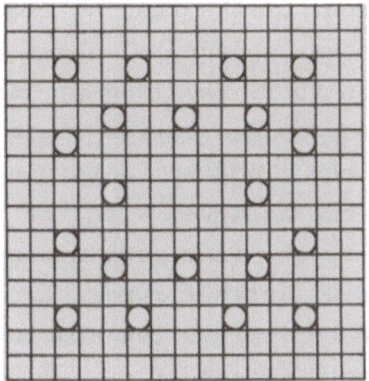

Fig. 1 Core and
        processor
        configuration

Fig. 2   FA-geometry

☐ fuel rod   ◘ guide tube

The task to relax on all gridpoints of a FA was assigned to one
of the processors of the multiprocessor array as indicated by
Fig. 1. In this way a nearly perfect load balance could be
achieved.

One of the processors is the master which determines the global
data such as eigenvalue, maximum error and so on. The coarse
mesh operator (4) condensed to one energy group has been used
to accelerate the fine mesh iteration performed by the red-black
Gauss-Seidel method.

Coarse mesh iteration was performed within a multigrid V-cycle
of up to 5 levels. The eigenvalue has been updated only on the
finest grid and therefore a correction scheme (CS) could be
used. The algorithm was programmed in Modula-2. The data of
the grid points at the boundary of the FA were allocated to the
multiport memory of the DIRMU modules to allow access of

neighbouring processors to these data. The programming language
Modula-2 offers a simple solution to this problem by type and
pointer declarations. Consequently, there is no difference in
the programming technique for nodes allocated to the multiport
or private memory.

For the computation of the eigenvalue and the status of convergence the processors were used in an open ring configuration.
This configuration allows data to be easily passed from one
processor to the other.

Measurement of the running time for the 24 processor configuration shows a speedup of about 22 and thus an efficiency of
more than 90 %. The loss of efficiency is caused by synchronization and communication times and, of course, by the sequential
part of the algorithm determining the eigenvalue. The rate of
convergence achieved with the multigrid method per working unit
defined by one fine mesh Gauss-Seidel relaxation is approximately 0.8. Further numerical experiments have shown that it is
possible to improve the absolute rate of convergence by solving
the reduced NEM-equations without group condensation or, additionally, by using the full NEM equations also on the coarser
grids (see section 2).

At the moment the DIRMU kit contains only 25 elements. Of
course, a much larger number of modules would be necessary to
solve realistic reactor problems. Also, the coprocessor 8087
used in the modules is too weak. By using more powerful copro-
cessors substantial improvements of the results are possible
/11/.

6. Conclusions

Efficient numerical algorithms are available for the solution
of partial differential equations describing the physical behavior of the reactor core. Much work is still required to adapt
the numerical methods to parallel computers. Recent trends towards increased parallelization in supercomputer architecture
have motivated intensive study of parallel algorithms. The fundamental question how to use multiprocessors efficiently will
remain a challenging problem for numerical analysts and computer
scientists. On the other hand, as shown in this paper, even

simple techniques can work very well in many cases. Multi-level
methods seem to be similarly effective in parallel and serial
environments. For large problems the speedup is roughly proportional to the number of processors. These results encourage
additional research activities in this area.

References

1. Trottenberg U.: On the SUPRENUM Conception. International
    Topical Meeting on Advances in Reactor Physics, Mathema-
    tics and Computation, Paris 1987, Vol. 1, 447-455,

2. Hackbusch W.; Trottenberg U., eds.: Multigrid Methods.
    Proceedings of the Conf. Köln-Porz, Nov. 23-27, 1981,
    Lecture Notes in Mathematics. Springer Berlin 1982

3. Händler W.; Maehle E.; Wirl K.: The DIRMU Testbed for
    High Performance Multiprocessor Configurations. Proc.
    SCS '85, First International Conference on Supercomputing
    Systems, St. Petersburg, Florida 1985

4. Finnemann H.; Gundlach W.: Space Time Kinetics Code IQSBOX
    for PWR and BWR, Part I: Description of Physical and Thermo-
    hydraulic Models, Atomkernenergie 37 (1981), 176-182

5. Reed W.H.; Stewart H.B.; Wolf L.: Applications of the
    THERMIT Code to 3D Thermal Hydraulic Analysis of LWR Cores.
    Proc. Top. Meeting on Computational Methods in Nuclear En-
    gineering, Williamsburg, Virginia, April 23-25, 1979

6. Wachspress E.L.: Iterative Solution of Elliptic Systems
    and Applications to the Neutron Diffusion Equations of
    Reactor Physics, Prentice Hall, Inc. 1966

7. Brandt A.: Multigrid Techniques: 1984 Guide with Appli-
    cations to Fluid Dynamics, GMD-Studie Nr. 85, May 1984

8. Fischer H.D.; Finnemann H.: The nodal integration method
    - A diverse solver for neutron diffusion problems. Atom-
    kernenergie 39 (1981) 229-236

9. Koebke K.; Hetzelt L.: On the Reconstruction of Local
    Homogeneous Neutron Flux and Current Distributions of
    Light Water Reactors from Nodal Schemes. Nuclear Science
    and Engineering 91 (1985) 123-131

10. Michel E.: Parallele Lösung der Neutronen-Diffusionsglei-
    chung mit Hilfe eines Mehrgitterverfahrens auf verschie-
    denen DIRMU-Konfigurationen
    Diplomarbeit, Department of Computer Science, IMMD 3,
    University of Erlangen-Nuremberg 1987

11. Finemann H.; Volkert J.: Parallel Multigrid Algorithms Im-
    plementation on Memory-Coupled Multiprocessors, Int. Top.
    Meeting on Advances in Reactor Physics Math. and Comp.,
    Paris, France, April 27-30, 1987

Evaluation of Systematically Derived Neutron Kinetics Models

ANTONIO F. V. DIAS, ALLAN F. HENRY

Massachusetts Institute of Technology
Cambridge, Massachusetts, U.S.A.

Summary

A variational principle made stationary by the solution of the
two-group, three-dimensional, time-dependent matrix equations and
adjoint matrix equations embodied in the nodal code QUANDRY is
used to derive a number of more approximate neutron kinetics
models varying from point kinetics to a group-dependent,
one-dimensional scheme. Numerical tests of the accuracy of these
models substantiate the value of using adjoint weighted kinetics
parameters. They also suggest that using unity weighted
parameters (while not acceptable for all cases) is usually
preferable to Galerkin weighting, and that certain one-dimensional
models are actually inferior to the adjoint-weighted, point
kinetics scheme.

Introduction

It is extremely difficult to evaluate, by comparison with
experiments, numerical methods for the prediction of transient
neutron behavior in a power reactor. The measurements themselves
are very challenging, and disentangling errors due to
thermal-hydraulic feedback modeling and cross section libraries
from those due to the nuclear model is virtually impossible.
Thus, if theory and experiment disagree, it is difficult to know
why, and if they agree, it may be because of cancellation of
errors.

Testing a simplified nuclear model by comparison with a more exact
one involves a complementary set of advantages and disadvantages:
If the same thermal-hydraulic model and cross section information
is used for both the reference and the approximate model, it is
possible to make a quantitative assessment of the accuracy of the
latter. However, the reference model itself may be inaccurate

either because it too involves approximations or because of errors in the nuclear data it employs.

We conclude that a complete evaluation of the accuracy of predictions of neutron kinetic behavior requires comparisons with both measurements and with the predictions of more accurate numerical models.

The present paper deals with the latter kind of testing. Specifically, we shall compare predictions of a number of kinetics models, ranging from point-kinetics to two-group, one-dimensional, with those of a two-group, time-dependent, three-dimensional, nodal model. To make the evaluation as definitive as possible, the approximate models will all be derived from a variational principle made stationary by solutions of the 3D, time-dependent, reference nodal equations and corresponding adjoint equations. Also, the same thermal-hydraulics model will be used to predict feedback effects for all cases.

Results will suggest that both point kinetics and one-dimensional models should be used with great caution.

Theory

As a numerical model, to serve both as a reference against which to compare more approximate models and as a starting point from which to derive those approximate models, we have used the three-dimensional, two-group, time-dependent nodal code, QUANDRY; [G1], [S1]. There is considerable evidence [K1], [K3], [S3] for static cases that QUANDRY and similar advanced nodal codes (with properly homogenized parameters and heterogeneity or discontinuity factors [K2], [S2]) predict the neutronic properties of light water reactors very well. Also, their predictions of transient behavior agree with each other [S1]. However, here, we know of no comparison with measurements.

The unknowns of the QUANDRY equations are the node-averaged group-fluxes, net group-leakage rates and precursor concentrations for each node in the reactor. These quantities are specified by the following matrix equation, the derivation of which is sketched in Reference [G1] and presented in detail in Reference [S1].

$$
\begin{bmatrix}
[V^{-1}] & [0] & [0] & [0] & [0] & \cdots & [0] \\
[0] & [0] & [0] & [0] & [0] & \cdots & [0] \\
[0] & [0] & [0] & [0] & [0] & \cdots & [0] \\
[0] & [0] & [0] & [0] & [0] & \cdots & [0] \\
[0] & [0] & [0] & [0] & [I] & \cdots & [0] \\
\cdot & \cdot & \cdot & \cdot & \cdot & & \cdot \\
\cdot & \cdot & \cdot & \cdot & \cdot & \cdot & \cdot \\
\cdot & \cdot & \cdot & \cdot & \cdot & & \cdot \\
[0] & [0] & [0] & [0] & [0] & \cdots & [I]
\end{bmatrix}
\frac{d}{dt}
\begin{bmatrix}
[\bar{\phi}(t)] \\
[\bar{L}_x(t)] \\
[\bar{L}_y(t)] \\
[\bar{L}_z(t)] \\
[\bar{C}_1(t)] \\
\cdot \\
\cdot \\
\cdot \\
[\bar{C}_D(t)]
\end{bmatrix}
=
$$

$$
=
\begin{bmatrix}
[M_p - \Sigma_T] & -h_y^j h_z^k[I] & -h_x^i h_z^k[I] & -h_x^i h_y^j[I] & \lambda_1[I] & \cdots & \lambda_D[I] \\
[F_x] & -[I] & \frac{1}{h_y^j}[G_x] & \frac{1}{h_z^k}[G_x] & [0] & \cdots & [0] \\
[F_y] & \frac{1}{h_x^i}[G_y] & -[I] & \frac{1}{h_z^k}[G_y] & [0] & \cdots & [0] \\
[F_z] & \frac{1}{h_x^i}[G_z] & \frac{1}{h_y^j}[G_z] & -[I] & [0] & \cdots & [0] \\
[M_1] & [0] & [0] & [0] & -\lambda_1[I] & \cdots & [0] \\
\cdot & \cdot & \cdot & \cdot & \cdot & \cdot & \cdot \\
\cdot & \cdot & \cdot & \cdot & \cdot & \cdot & \cdot \\
\cdot & \cdot & \cdot & \cdot & \cdot & \cdot & \cdot \\
[M_D] & [0] & [0] & [0] & [0] & \cdots & -\lambda_D[I]
\end{bmatrix}
x
$$

$$
x \, \text{Col} \, \{ \, [\bar{\phi}], \, [\bar{L}_x], \, [\bar{L}_y], \, [\bar{L}_z], \, [\bar{C}_1], \, \cdots \, , \, [\bar{C}_D] \, \} \qquad (1)
$$

where: $[\bar{\phi}(t)] \equiv$ a column vector of length $G \cdot I \cdot J \cdot K \, (\equiv N)$ containing the node averaged fluxes (ordered first by group, then x-direction, then y-direction, and finally z-direction)

$[\bar{L}_u(t)] \equiv$ a column vector of length N containing the u-directed net leakages for each node ; $u =$ x, y, or z

$[\bar{C}_d(t)] \equiv$ a column vector of length N containing, for the dth precursor family, the elements of $V_{ijk}[\chi_d]\bar{C}_{d_{ijk}}(t)$

$[V^{-1}] \equiv$ a block-diagonal matrix of order N x N containing the elements of $V_{ijk}[v]^{-1}$

$[M_p(t)] \equiv$ a block-diagonal matrix of order N x N containing the elements of $(1-\beta)V_{ijk}[\chi_p][v\Sigma_{f_{ijk}}(t)]^T$

$[\Sigma_T(t)] \equiv$ a block-diagonal matrix of order N x N containing the elements of $V_{ijk}[\Sigma_{T_{ijk}}(t)]$ with $[\Sigma_{t_{ijk}}(t)]$ equal to the G x G matrix $\{\delta_{gg'}\Sigma_{t_{gijk}} - \Sigma_{gg'_{ijk}}\}$

$h_u^\ell \equiv$ width of the ℓth nodes in the u-direction

$[F_u(t)] \equiv$ a block-tridiagonal matrix of order N x N containing the elements of $[F_{u_{\ell mn}}^\ell(t)]$ specifying leakage in the u-direction

$[G_u(t)] \equiv$ a block-pentadiagonal matrix of order N x N containing the elements of $[G_{u_{\ell mn}}^\ell(t)]$ specifying leakage transverse to the u-direction

$[M_d(t)] \equiv$ a block-diagonal matrix of order N x N containing, for the dth precursor family, the elements of $\beta_d V_{ijk}[\chi_d][v\Sigma_{f_{ijk}}(t)]^T$

Detailed expressions for the elements of the vectors and matrices cited above are given in Reference [S-1].

A simple, one-dimensional, constant-pressure, thermal-hydraulic model, based on the "average" fuel rod and coolant channel in each node, is solved in tandem with Equation (1). Changes in fuel and moderator temperatures then result in changes in the cross sections comprising Σ_T, F, G and M matrices.

In order to derive systematically a hierarchy of approximate models all based on Eq. (1), we introduce a functional made stationary by solutions of (1) and the equations adjoint to (1).

The functional in question is defined for a set of functions:
$$\{[u(t)], [v_x(t)], [v_y(t)], [v_z(t)], [c_1(t)], \ \cdots \ , [c_D(t)],$$
$$[u^*(t)], [v_x^*(t)], [v_y^*(t)], [v_z^*(t)], [c_1^*(t)], \ \cdots \ , [c_D^*(t)]\},$$
continuous in time within the time interval $(t_i, \ t_f)$, during which the simulation takes place. Each of these functions is actually a column vector of length $G \cdot I \cdot J \cdot K$.

The expression for the functional is:

$$F\left[[u], [v_x], [v_y], [v_z], [c_1], \ \cdots \ , [c_D], [u^*], [v_x^*],\right.$$
$$\left.[v_y^*], [v_z^*], [c_1^*], \ \cdots, [c_D^*]\right] =$$

$$= \int_{t_i}^{t_f} dt \left\{ \left[[u^*]^T \ [v_x^*]^T \ [v_y^*]^T \ [v_z^*]^T \ [c_1^*]^T \ \cdots \ [c_D^*]^T \right] \ \times \right.$$

$$\times \begin{bmatrix}
[H-V^{-1}\frac{d}{dt}] & -h_y^j h_z^k[I] & -h_x^i h_z^k[I] & -h_x^i h_y^j[I] & \lambda_1[I] & \cdots & \lambda_D[I] \\
[F_x] & -[I] & \frac{1}{h_y^j}[G_x] & \frac{1}{h_z^k}[G_x] & [0] & \cdots & [0] \\
[F_y] & \frac{1}{h_x^i}[G_y] & -[I] & \frac{1}{h_z^k}[G_y] & [0] & \cdots & [0] \\
[F_z] & \frac{1}{h_x^i}[G_z] & \frac{1}{h_y^j}[G_z] & -[I] & [0] & \cdots & [0] \\
[M_1] & [0] & [0] & [0] & -\lambda_1[I]-\frac{d}{dt} & \cdots & [0] \\
\cdot & \cdot & \cdot & \cdot & \cdot & & \cdot \\
\cdot & \cdot & \cdot & \cdot & \cdot & & \cdot \\
\cdot & \cdot & \cdot & \cdot & \cdot & & \cdot \\
[M_D] & [0] & [0] & [0] & [0] & \cdots & -\lambda_D[I]-\frac{d}{dt}
\end{bmatrix} \times$$

$$\left. \times \ \mathrm{Col}\{ [u], [v_x], [v_y], [v_z], [c_1], \ \cdots \ , [c_D] \} \right\} \qquad (2)$$

where $[H] = [M_p - \Sigma_T]$ and $\frac{d}{dt}$ is considered as an operator. In order to keep the notation simple, time-dependence has been suppressed in Equa. (2).

The defining equations for approximate models based on Equa. (1) can now be derived by selecting various subspaces of the spaces for which (2) is defined and requiring (2) to be stationary for arbitrary variations of the variable parts of the vectors comprising those subspaces. For example, for a point kinetics model we express the $(gijk)^{th}$ elements $[u(t)]$ and $[v_u(t)]$; $u = x,y,z$, as $\bar{\phi}_{gijk}(0)\, T(t)$ and $\bar{L}_{ugijk}(0)\, T(t)$; $u = x,y,z$, where $\bar{\phi}_{gijk}(0)$ is the average, group-g flux in node-ijk and the $\bar{L}_{ugijk}(0)$ are the group-g, net leakage rates across the two faces of node-ijk perpendicular to the u-direction; $u = x,y,z$, both $\bar{\phi}_{gijk}(0)$ and the $\bar{L}_{ugijk}(0)$ having known values corresponding to some reference condition. The same (unknown) scalar "amplitude function," $T(t)$, multiplies all group-fluxes and leakage rates for all nodes. If an analogous subspace is defined for the starred vectors and the first order variation of (2) is required to vanish for an arbitrary variation in $T^*(t)$, the point kinetics equation for $T(t)$ results with the point kinetics parameters ρ, β, and Λ defined in terms of inner products involving the $\bar{\phi}_{gijk}(0)$ $\bar{L}_{ugijk}(0)$, and their starred counterparts. Notice that with Equa. (1) used as a starting point, the problem of inferring integrals of gradient terms $\int_{V_{ijk}} \nabla\phi_g^*(\underline{r})\cdot\nabla\phi_g(\underline{r})\, dV$ from a nodal solution is avoided [L1], [T1]. Such terms appear in the standard expression for reactivity derived from the differential form of the group equations. Notice also that there is no need to define a subspace for the $[C_i]$ and $[C_i^*]$ vectors. Why this is so will become clearer below.

Another example is a model which we shall call the one-dimensional, unrestrained leakage approximation. Here we define a subspace for which the $(gijk)^{th}$ elements of $[u(t)]$ and the $[v_u(t)]$ are $\bar{\phi}_{gijk}(0) T_{gk}(t)$ and $\bar{L}_{ugijk}(0) U_{gk}(t)$, $u = x,y,z$; $U = X,Y,Z$. Thus here the reference nodal fluxes and leakages for each radial plane and each group are allowed to vary independently in time.

In order to avoid deriving a separate set of kinetics equations for each approximate model to be investigated, we introduce a mathematical form sufficiently general to accommodate all the models of interest. With $[T]$, $[X]$, $[Y]$, $[Z]$ and the $[\bar{C}_d]$ (and corresponding starred quantities) representing unknown, time-dependent vectors (or scalars) and the $[\bar{\psi}]$, $[\bar{\xi}]$, $[\bar{\eta}]$ and $[\bar{\zeta}]$ (and corresponding starred quantities) representing known expansion matrices (or vectors) the general mathematical form to be used in Equa. (2) is

$$
\begin{aligned}
[u] &= [\bar{\psi}][T] & [u^*] &= [\bar{\psi}^*][T^*] \\
[v_x] &= [\bar{\xi}][X] & [v_x^*] &= [\bar{\xi}^*][X^*] \\
[v_y] &= [\bar{\eta}][Y] & [v_y^*] &= [\bar{\eta}^*][Y^*] \quad\quad (3) \\
[v_z] &= [\bar{\zeta}][Z] & [v_z^*] &= [\bar{\zeta}^*][Z^*] \\
[c_1] &= [\bar{C}_1] & [c_1^*] &= [\bar{C}_1^*] \\
&\;\;\vdots & &\;\;\vdots \\
[c_D] &= [\bar{C}_D] & [c_D^*] &= [\bar{C}_1^*]
\end{aligned}
$$

where all matrices and vectors have structures and dimensions dependent on which model they represent. In the present paper we shall investigate four models. For all of them, the elements of the matrices $[\bar{\psi}]$, $[\bar{\xi}]$, $[\bar{\eta}]$ and $[\bar{\zeta}]$ are the node averaged group fluxes $\bar{\phi}_{gijk}(0)$ and net leakages $\bar{L}_{ugijk}(0)$; $u = x,y,z$ for some reference state of the reactor (usually the initial critical state). Values for corresponding starred matrices are a matter of choice. We shall examine three possibilities:
(a) all matrix elements have unit value (unity weighting).

(b) $[\vec{\psi}^{*}] = [\overline{\psi}]$; $[\xi^{*}] = [\overline{\zeta}]$ etc. (Galerkin weighing).

(c)   Elements $[\vec{\psi}^{*}]$ are $\vec{\phi}^{*}_{gijk}$ (0) and those of $[\overline{\xi}^{*}]$, $[\vec{\eta}^{*}]$ and $[\overline{\zeta}^{*}]$ are $[L^{*}_{ugijk}$ (0)]; u = x,y,z, found from a steady state solution of the equations adjoint to (1) (adjoint weighting).

For any given model, the structure of $[\overline{\xi}]$, $[\overline{\eta}]$, and $[\overline{\zeta}]$, as well as the starred matrices (including $[\vec{\psi}^{*}]$) is identical with that of $[\overline{\psi}]$. In addition [X], [Y], and [Z] and all the analogous starred vectors have the same structure as [T]. Thus, in describing the four models investigated, we give in what follows the structure of only $[\overline{\psi}]$ and [T].

Point Kinetics (P.K.)

$$[\overline{\psi}] = Col \ \{\overline{\phi}_{gijk} \ (0)\} \tag{4}$$

a column vector having G·I·J·K scalar elements (G = no. of groups; I = no. of nodes in the X-direction ; J = no. of nodes in the Y-direction; K = no. of nodes in the Z-direction)

$$[T] = [X] = [Y] = [Z] = T(t) \tag{5}$$

where T(t) is a scalar.

Thus the reference, nodal-group flux-leakage shape is assumed to remain rigid with a change in amplitude during a transient specified by T(t).

Zero-Dimensional Kinetics (O-D)

$$[\overline{\psi}] = Col \ \{[\overline{\psi}_{ijk}]\} \tag{6}$$

an I·J·K element column vector whose elements are diagonal, G × G matrices given by

$$[\overline{\psi}_{ijk}] = Diag \ \{\overline{\phi}_{gijk} \ (0)\} \tag{7}$$

$$[T] = [X] = [Y] = [Z] = \text{Col } \{T_g(t)\} \qquad (8)$$

a G-element column vector, each $T_g(t)$ being a scalar.

Here the reference three-dimensional, flux-leakage shapes for each group are allowed to change independently in time.

One-Dimensional Kinetics (1-D)

$$[\psi] = \text{Diag } \{[\overline{\psi}_k]\} \qquad (9)$$

a diagonal K·K matrix whose elements are I·J-element column vectors

$$[\overline{\psi}_k] = \text{Col } \{[\overline{\psi}_{ijk}]\}; \quad k = 1,2 \ldots k \qquad (10)$$

the $[\overline{\psi}_{ijk}]$ being the G·G matrices defined by Equa. (7)

$$[T] = [X] = [Y] = [Z] = \text{Col } \{[T_k(t)]\} \qquad (11)$$

a column vector of K elements, each element in turn being a column vector of G elements.

$$[T_k(t)] = \text{Col } \{T_{gk}(t)\} \quad ; \quad k = 1,2 \ldots k \qquad (12)$$

With this model, the reference planar flux-leakage shapes for each group are allowed to change independently.

Unrestrained Leakage (UL)

Here $[\psi]$ and $[T]$ have the same structure as (4,5), (6,8) or (9,10). However the nodal leakages for each plane and group are allowed to behave independently in time. Thus $[X]$, $[Y]$ and $[Z]$ are independent vectors having the same structure as $[T]$ in (5) or (8) or (12). We thus have the (PK,UL), (0-D,UL) and (1-D,UL) approximations.

If the trial functions (3) are inserted into the functional (2), and the first order variation of this functional, with respect to arbitrary and independent variations in $[u^*]$, $[v_x^*]$, $[v_y^*]$, $[v_z^*]$, and $[c_d^*]$ (d = 1,2, ... , D) is set to zero, the following equations result:

$$[\bar{\psi}^*]^T[V^{-1}] \frac{d}{dt} [\bar{\psi}][T] = [\bar{\psi}^*]^T[H][\bar{\psi}][T] - h_y^j h_z^k [\bar{\psi}^*]^T[\bar{\xi}][X] -$$

$$- h_x^i h_z^k [\bar{\psi}^*]^T[\bar{\eta}][Y] - h_x^i h_y^j [\bar{\psi}^*]^T[\bar{\zeta}][Z] + \sum_{d=1}^{D} \lambda_d [\bar{\psi}^*]^T[\bar{C}_d] \quad (13)$$

$$[\bar{\xi}^*]^T[F_x][\bar{\psi}][T] - [\bar{\xi}^*]^T[\bar{\xi}][X] + \frac{1}{h_y^j} [\bar{\xi}^*]^T[G_x][\bar{\eta}][Y] +$$

$$+ \frac{1}{h_z^k} [\bar{\xi}^*]^T[G_x][\bar{\zeta}][Z] = 0 \quad (14)$$

$$[\bar{\eta}^*]^T[F_y][\bar{\psi}][T] + \frac{1}{h_x^i} [\bar{\eta}^*]^T[G_y][\bar{\xi}][X] - [\bar{\eta}^*]^T[\bar{\eta}][Y] +$$

$$+ \frac{1}{h_z^k} [\bar{\eta}^*]^T[G_y][\bar{\zeta}][Z] = 0 \quad (15)$$

$$[\bar{\zeta}^*]^T[F_z][\bar{\psi}][T] + \frac{1}{h_x^i} [\bar{\zeta}^*]^T[G_z][\bar{\xi}][X] +$$

$$+ \frac{1}{h_y^j} [\bar{\zeta}^*]^T[G_z][\bar{\eta}][Y] - [\bar{\zeta}^*]^T[\bar{\zeta}][Z] = 0 \quad (16)$$

$$\frac{d}{dt} [\bar{C}_d] = [M_d][\bar{\psi}][T] - \lambda_d [\bar{C}_d] \quad ; \quad d = 1, 2, \ldots , D \quad (17)$$

In these equations the $[\bar{C}_d]$ are I·J·K-element column vectors, each element, in turn being a G-element column vector. (See the definition of $[\bar{C}_d(t)]$ following Equa. 1.) However, all that is needed to solve Equas. (13)-(16) is the much smaller vector $[\bar{\psi}^*]^T [\bar{C}_d]$ (which, for the point model, reduces to a scalar). Accordingly, we multiply Equation (17) by $[\bar{\psi}^*]^T$ and replace it by

$$\frac{d}{dt}\{[\overrightarrow{\psi^*}]^T [C_d]\} = [\overrightarrow{\psi^*}]^T [M_d] [\overline{\psi}] [T] - \lambda_d [\psi^*]^T [C_d],$$
$$d = 1,2 \ldots 0 \tag{18}$$

For the "restricted leakage," P.K., 0-D and 1-D models, [X] = [Y] = [Z] = [T], and for this case the variational principle yields a single equation involving $\frac{d}{dt}$ [T]. This equation may be obtained by adding Equas. (13)-(16), replacing [X], [Y], and [Z] by [T].

Equations (13)-(16) and (18) are thus valid for all the models we have defined, although their degree of complexity depends on the model chosen. A much more detailed discussion of these matters along with a description of the numerical methods used to solve the equations is given in Reference [D1].

Numerical Results

With two-group, three-dimensional QUANDRY computations used to provide numerical standards for all cases and employing three different weight functions, we shall examine the accuracy of the four models of interest when applied to two significantly different problems, a homogeneous perturbation problem and a coolant temperature drop problem. The detailed geometry and input data for these problems, which are both simulations of light water reactors, are provided in Reference [D1].

The Homogeneous Reactor Problem

The first transient analyzed is a very simple one. A homogeneous cubic reactor, 50 cm on a side, is perturbed by reducing its thermal absorption cross section in a step fashion from 0.06669 cm^{-1} to 0.06630 cm^{-1}. Only one group of delayed neutrons is simulated, and thermal feedback is ignored, the initial power level being 1 watt in each quarter of the core. The original, critical cosine shapes of the fast and slow fluxes are unaltered by this perturbation. Only the fast-to-slow flux ratio changes. Results are given in Table 1 along with per cent differences from the QUANDRY reference values. For this case, since flux shapes do not change, the 1-D approximation yields results which are identical with the 0-D numbers. The unrestricted leakage options were not tested.

Table 1: Quarter Core Reactor Power Behavior (Watts) Following a
Step Decrease in Thermal Absorption in a Homogeneous
Reactor

Model :		Point Kinetics			Zero-Dimensional		
Weighting fct.:		Unity	Galerk.	Adjoint	Unity	Galerk.	Adjoint
T (s)	Refer.						
0.	1.0000	1.0000	1.0000	1.0000	1.0000	1.0000	1.0000
1.	2.7768	1.4692 (-47.%)	1.0852 (-61.%)	2.7744 (-0.1%)	2.8108 (3.4%)	2.8103 (3.4%)	2.8023 (2.6%)
2.	3.1644	1.5202 (-52.%)	1.0921 (-66.%)	3.1335 (-1.0%)	3.1767 (0.4%)	3.1761 (0.4%)	3.1702 (0.2%)
3.	3.5779	1.5739 (-56.%)	1.0970 (-69.%)	3.5353 (-1.2%)	3.5903 (0.3%)	3.5896 (0.3%)	3.5828 (0.1%)
4.	4.0475	1.6275 (-60.%)	1.1060 (-73.%)	3.9886 (-1.5%)	4.0577 (0.3%)	4.0568 (0.2%)	4.0491 (0.04%)
5.	4.5772	1.6839 (-63.%)	1.1130 (-76.%)	4.5000 (-1.7%)	4.5860 (0.2%)	4.5849 (0.2%)	4.5762 (-.02%)
6.	5.1719	1.7423 (-66.%)	1.1201 (-78.%)	5.0770 (-1.8%)	5.1831 (0.2%)	5.1817 (0.2%)	5.1718 (-.00%)

Perhaps the most striking feature of these results is the
magnitude of the errors when unity or Galerkin weighting is used
with the point kinetics model. Power is underpredicted by a
factor of ~ 2 at one second, and that factor grows to the range
3-4 at six seconds. Yet there is no perturbation in the spatial
shape of the flux; only the spectrum is perturbed.

Very often a variant of the Galerkin scheme is used wherein the
spatial shapes of the group-fluxes are taken to be those of the
regular fluxes, but the group-to-group flux ratios are altered by
various prescriptions so as to match better the adjoint,
group-to-group ratios (M1). Our results suggest that these
prescriptions need to be quite accurate.

Almost as striking in Table (1) is the accuracy of the point model
with adjoint weighting and the excellence of the zero-dimensional

models. The latter scheme is little more complicated than the
point model and, with unity or Gallerkin weighting , avoids the
need to determine the adjoint flux.

The Salem Coolant Temperature Drop Problem

Salem-1 is a Westinghouse PWR. The final transient to be analyzed
is a simulation of a 2 second, $20^{o}K$ drop in inlet coolant
temperature for this reactor initially operating at full power.
Selected Results are shown in Table (2).

Table 2: Quarter Core Reactor Power (Megawatts) for the Salem-1
Inlet Temperature Drop Transient

Model :	Pt. Kin.	1D	1D	1D, UL	
Weighting :	Adjoint	Adjoint	Galerkin	Galerkin	
Time (s)					Reference
0.0	834.67	834.67	834.67	834.67	834.50
0.4	1002.8 (0.2%)	1002.6 (0.2%)	974.61 (-2.6%)	983.16 (-1.8%)	1000.7
0.8	1371.0 (0.7%)	1368.8 (0.5%)	1296.6 (-4.8%)	1311.4 (-3.7%)	1361.7
1.2	1585.1 (1.5%)	1567.3 (0.4%)	1524.1 (-2.4%)	1515.6 (-2.9%)	1561.6
1.6	1681.4 (2.7%)	1642.0 (0.3%)	1568.6 (-4.2%)	1603.8 (-2.1%)	1637.8
2.0	1754.1 (3.9%)	1691.3 (0.2%)	1562.4 (-7.4%)	1656.9 (-1.8%)	1687.5
2.4	1500.8 (4.9%)	1428.4 (-0.2%)	1318.3 (-7.9%)	1426.6 (-0.3%)	1430.9
2.8	1291.7 (5.4%)	1223.8 (-0.2%)	1154.9 (-5.8%)	1228.6 (-0.2%)	1225.9
3.2	1229.0 (6.3%)	1156.4	1094.0 (-5.4%)	1158.4 (0.2%)	1156.2
3.6	1209.4 (6.9%)	1133.3 (0.2%)	1074.2 (-5.8%)	1132.3 (0.1%)	1131.2
4.0	1204.4 (6.9%)	1127.1 (0.1%)	1072.2 (-4.8%)	1123.7 (-0.2%)	1126.2

Feedback effects on the homogenized nodal cross sections due to
fuel temperature and coolant temperature and density changes are

accounted for. Six delayed precursor groups are simulated.
Control rods are assumed to be fully withdrawn throughout the
transient.

Errors for the 0-D, adjoint-weighted model are about 0.2% greater
than those for the P.K. case, and all the adjoint-weighted,
unrestrained leakage models yield results very close to the
corresponding restrained results. The agreement between the point
and 1-D models is, however, misleading. Table (3) compares the
reconstructed axial power traverses for the three adjoint-weighted
(restrained leakage) models for one of the high power assemblies
at the time of peak power. The individual nodal powers are
clearly predicted much more accurately by the 1-D model.

Table 3: Comparison of the Axial Nodal Power Profiles (kilowatts)
for an Assembly, at time t = 2 s, Obtained with the
Restrained Leakage Approximation for the Salem-1 Problem

Plane k	Point kin Model	0-dimen. Model	1-dimen. Model	QUANDRY
1	942.04 (-13.2%)	945.85 (-12.8%)	1064.4 (-1.9%)	1085.1
4	2343.96 (-5.6%)	2351.9 (-5.3%)	2470.31 (0.5%)	2482.3
7	1877.3 (7.0%)	1882.8 (7.3%)	1765.8 (0.7%)	1754.1
10	1054.0 (20.2%)	1056.5 (20.5%)	892.53 (1.8%)	876.63
13	503.26 (32.7%)	504.33 (33.0%)	388.90 (2.5%)	379.23
16	196.28 (+41.5%)	196.67 (41.8%)	142.61 (2.8%)	138.74
18	64.688 (43.5%)	64.846 (43.8%	46.279 (2.6%)	45.089

Whether the restrained or unrestrained leakage approximation is
used, the power behavior is very poorly predicted by both the
unity-weighted and Galerkin-weighted point models. In fact, both
models predict a monotonic decrease in reactor power throughout
the transient. The Galerkin-weighted point models are the worst:
They predict a quarter-core power of ~ 47 MW rather than 1126.2 MW

at 4 = s.

The 0-D models whether unity or Galerkin-weighted and whether the restrained or unrestrained leakage approximation is used are about as accurate as the P.K., adjoint-weighted model (column 1 of Table 2).

Errors for the 1-D unity-weighted model are very close to those of the 1-D Galerkin-weighted case shown in Table 2, while those for the unity-weighted 1-D, UL approximation are about twice as large as the ones shown for the 1-D, UL, Galerkin-weighted case.

While the unrestricted leakage approximation does nothing to improve accuracy when adjoint weighting is used, it helps the 1-D models a great deal if unity or Galerkin weighting is used. This fact is not apparent from Table 2 since only total reactor power is displayed. Comparison of individual predicted nodal powers brings out the advantages of the UL scheme. Table (4) shows such comparisons.

Table 4: Errors.in Predicted Power on a Node-by Node Basis for the Salem-1 Inlet Temperature Drop Transient

Weighting	Model	Avg. Error	Max. Error	Error in highest power node
Unity	0-dimen.	16.5%	34.7%	-11.4%
Unity	1-dimen.	29.5%	213.6%	-21.2%
Galerkin	0-dimen.	21.0%	45.5%	-4.2%
Galerkin	1-dimen.	31.7%	222.7%	-20.8%
Unity	0-dimen. UL	16.7%	35.5%	-10.9%
Unity	1-dimen. UL	5.4%	12.0%	-5.3%
Galerkin	0-dimen. UL	21.2%	45.8%	-4.0%
Galerkin	1-dimen. UL	3.0%	-7.8%	-3.8%

The 1-D model, with unity or Galerkin weighting but restrained leakage is actually inferior to the 0-D model. Use of the unrestrained leakage approximation improves considerably the accuracy of both weighting schemes.

Conclusions

In this paper neither the range of simplified kinetics models examined nor the selection of transients used to test them is exhaustive. Hence it is difficult to know whether the behavior we have observed is truly characteristic of the approximation or an accident of the test cases examined. With that reservation, however, it does seem justified to offer several conclusions from the work.

First of all, the use of a stationary adjoint solution as a weight function improves accuracy consistently. For the point and one-dimensional (restrained leakage) models the improvement is striking.

Using the initial flux shape itself as a weight function (Galerkin weighting) rarely leads to improved accuracy and in most cases makes errors larger. For the point model this behavior can undoubtedly be improved by altering the group-to-group ratios of the regular flux so that they correspond more to those of the adjoint fluxes. However, for the 0-D and 1-D models examined in this paper, the individual energy group were allowed to vary independently so that the group-to-group ratios of both the flux and adjoint flux expansion matrices ($\overline{\psi}$ and $\overline{\psi}^*$ of Equa. 3) are of no consequence. We conclude that unity-weighting (although it may still be unacceptable for certain models) is generally to be preferred to Galerkin weighting.

For many cases, the unity-weighted zero-dimensional models are almost as accurate as the adjoint weighted schemes. Thus, if a solution of the adjoint nodal equations is unavailable, the unity-weighted, 0-D scheme, which is only slightly more complicated numerically than the point scheme, may be an attractive alternative.

One final conclusion that deserves attention is that the more general unity or Galerkin-weighted 1-D models can actually be far inferior to the corresponding 0-D model. Allowing the nodal leakages and node-average fluxes to vary independently (the 1-D UL approximation) seems to correct this defect. However, the equations embodying this approximation require more effort to solve, and probably the 1-D adjoint-weighted scheme provides a better procedure for avoiding the difficulty. In any event, the relationship of the 1-D and 1-D, UL models developed in this paper to others in current use ought to be examined to be sure that models now in use are closer to the 1-D, UL than to the 1-D scheme.

Acknowledgments

We wish to acknowledge the financial support received for this work from the Institute de Estudos Avancados/CTA (Brazil), the Comissao Nacional de Energia Nuclear (Brazil), the PSE and G Research Corporation and GPU Nuclear. Also, we are most grateful to Dr. Temitope Taiwo of Northeast Utilities for supplying us with adjoint weight functions for our test problems.

References

[A1] Argonne Code Center: Benchmark Problem Book, ANL-7416, Supplement 2 (1977).

[D1] Antonio F. V. Dias, "Systematic Derivation, From 3-D Nodal Equations, of Simpler Models for Describing Reactor Transients, " Ph.D. Thesis, Dept. of Nuclear Engineering, Massachusetts Institute of Technology, Cambridge, MA, May 1987.

[G1] G. Greenman, K. Smith, and A. F. Henry, "Recent Advances in An Analytic Nodal Method for Static and Transient Reactor Analysis," ANS Proceedings of the Topical Meeting on "Computational Methods in Nuclear Engineering," Williamsburg, Virginia, Vol. 1, p 3-49 (April 1979).

[K1] H. S. Khalil, P. J. Finck, and A. F. Henry, Reconstruction of Fuel Pin Powers from Nodal Results, Proceedings of a Topical Meeting, Advances in Reactor Computations, Salt Lake City, 28-31 March 1983, ANS Idaho Section, ISBN: 0.89448-111-8, 367.

[K2]   K. Koebke, "A New Approach to Homogenization and Group
       Condensation," paper presented at the IAEA Technical
       Committee Meeting on "Homogenization Methods in Reactor
       Physics," Lugano, Switzerland, 13-15 November 1978.

[K3]   K. Koebke, and L. Hetzelt, "On the Reconstruction of Local
       Homogeneous Flux and Current Distributions of Light Water
       Reactors from Nodal Schemes," Nucl. Sci. Eng. 91, p. 123,
       (1985).

[L1]   R. D. Lawrence, "Perturbation Theory Within the Framework
       of a Higher Order Nodal Method," Trans. American Nuclear
       Society, New Orleans, Louisiana, 3-7 June 1987, Vol. 45,
       p. 402-403.

[M1]   R. D. Mosteller, "Self-Adjointness of the Fast Flux in a
       Pressurized Water Reactor," Trans. Amer. Nuc. Soc. 50, 533
       (1985).

[S1]   K. S. Smith, "An Analytic Nodal Method for Solving the
       Two-Group, Multidimensional, Static and Transient Neutron
       Diffusion Equation," Nuclear Engineer's Thesis, Department
       of Nuclear Engineering, M.I.T., Cambridge, MA (March
       1979).

[S2]   K. Smith, A. F. Henry and R. Loretz, "Determination of
       Homogenization Diffusion Theory Parameters for Coarse Mesh
       Nodal Analysis," ANS Topical Meeting, Sun Valley, Idaho,
       September 1980.

[S3]   K. S. Smith and K. R. Rempe, "Testing and Applications of
       the QPANDA Nodal Model," paper presented at the
       International Topical Meeting on Advances in Reactor
       Physics, Mathematics and Computations, 27-30 April 1987,
       Paris, France, Vol 2, p. 861-873.

[T1]   T. A. Taiwo and A. F. Henry, "Perturbation Theory Based on a
       Nodal Model," Nuclear Science and Engineering, Vol. 92,
       34-41 (1986).

Computer Simulation of the Long-Term Stability of a Nuclear Waste Repository in a Salt Dome

M. WALLNER

Bundesanstalt für Geowissenschaften und Rohstoffe
Hannover, Federal Republic of Germany

Summary

Numerical simulation of repository-induced perturbations within the host rock is of particular significance because licensing procedure for a final repository requires a prior reliable and convincing demonstration of safety. Long-term assessment of the salt barrier integrity cannot be evaluated from experiments alone but only by computations. A proper geomechanical modeling is necessary to evaluate barrier efficiency. The necessity of validation of the geomechanical model is explained. Preliminary design calculations oriented towards problem identification and trend indication are presented.

INTRODUCTION

In the Federal Republic of Germany a geologic waste isolation system has been conceptualized consisting of parallel or inter-acting subsystems, that provide multiple natural and man-made barriers to prevent the release of radionuclides into the biosphere. The suitability of a particular geological formation for final waste disposal, however, has to be demonstrated by a comprehensive safety analysis showing that the interaction of the whole system "waste product / repository mine / overall geological situation" can maintain the pre-determined protection aims.

Although the radionuclide release, namely solubility, nuclide transport, barrier permeability, and retardation behavior, are first order characteristics in a safety analysis, it can be realized that the mechanical stability of the barrier is a prior necessary condition for its efficiency. Especially the temperature increase, resulting from the decay heat of high level waste, will cause geomechanical and geochemical reactions which may be of extremely important influence on the barrier integrity.

Geomechanical Modeling Methodology

Related to the different stages in the repository life, different geotechnical stability problems have to be considered:

- construction and operation phase:
  stability of the mine (rock bearing capacity, convergency,
  usability) to guarantee safe construction and operation,

- post-operation phase:
  long-term integrity of the salt formation and the man-made
  barriers (backfill and seals) to prevent or localize the
  release of radionuclides.

The above mentioned problems require time-dependent specifications because time sequence of thermal loading and time-
dependent creep deformations are considered to be important in
assessing detrimental effects on the serviceability of the
rock structure and the man-made barriers.

Common engineering methods are not sufficient or even inadequate for evaluating the entire geotechnical problem, due to
the complexity of geotechnical factors and processes which have
to be considered. The practical demonstration of the stability
of the final repository can only be carried out by a combination
of various investigations and computations. Engineering-geological and geotechnical investigations, rock-mechanical measurements, computations, in situ-monitoring, and mining experience
must receive equivalent consideration. [3]

Scope of computations

Numerical computations are of particular significance, because
the licensing procedure for a final repository requires a prior
reliable and convincing demonstration of safety. Thermally induced deformations, stresses, and resulting stability problems,
however, are neither covered by previous mining experience nor
have they been subject of practical applications. Therefore,

computations on the thermomechanical behavior have the following
objectives:

- analysing thermo-mechanical processes through computations
  shall lead to a proper assessment of consequences;
- experience-based conclusions can be extended by computations;
- rock mechanical criteria for a stable mine design can be
  developed from computational parametric studies;
- the long-term assessment of the salt barrier integrity
  cannot be evaluated from experiments alone, but only by
  utilizing computations.

Constitutive Modeling

The proper idealization of the repository mine in the salt
formation into a computation model is a fundamental basis for a
realistic computation. The geolocial environment has to be considered with complex properties, as internal structure, thermo-
mechanical behavior, and initial conditions. Thereby, the
correct description of the thermo-mechanical behavior of rock
salt within a constitutive model is of fundamental importance.

Based on numerous test results, it became evident that the
mechanical behavior of rock salt is highly nonlinear dependent
on e.g. stresses, temperature and time [5]. Beside an elastic
response, rock salt deformation behavior is characterized by
long-term creep deformation under any deviatoric loading. Under
constant test conditions (stresses, temperature, humidity)
these creep deformations converge against a steady state.

Although deviatoric stresses are reduced by creep deformations,
fracture phenomena also occur at relatively high loading rates
or under almost uniaxial stresses. The transition from failure
to creep may be defined as long-term strength [5].

The BGR Reference Constitutive Model for rock salt is summarized
in eq. 1-5, saying that the total strain rate is composed of an
elastic, a thermal, a creep, and a fracture strain rate part.

The elastic part (eq. 2) and the thermal part (eq. 3) are standard formulations. The steady state creep formulation (eq. 4) is based on a multi-mechanism creep model, suggested by Munson and Dawson [4]. Transient creep is not incorporated because the consistency of transient creep formulations is still discussed. As a first rough approximation, fracture deformation due to long-term strength is described by a viscoplastic model (eq. 5) using an extended (curved) Drucker/Prager criterion and an associated flow rule:

$$\dot{\varepsilon}_{ij} = \dot{\varepsilon}_{ij}^{el} + \dot{\varepsilon}_{ij}^{th} + \dot{\varepsilon}_{ij}^{cr} + \dot{\varepsilon}_{ij}^{f} \tag{1}$$

$$\dot{\varepsilon}_{ij}^{el} = -\frac{v}{E}\dot{\sigma}_{kk}\delta_{ij} + \frac{1+v}{E}\dot{\sigma}_{ij} \tag{2}$$

$$\dot{\varepsilon}_{ij}^{th} = \alpha_t \dot{T}\delta_{ij} \tag{3}$$

$$\dot{\varepsilon}_{ij}^{cr} = \frac{3}{2}\frac{\dot{\varepsilon}_{eff}^{cr}}{\sigma_{eff}}s_{ij}, \qquad \dot{\varepsilon}_{eff}^{cr} = \sum_{i=1}^{3}{}^{i}\dot{\varepsilon}_{eff}^{cr}(S,\sigma_{eff},T)$$

$${}^{1}\dot{\varepsilon}_{eff}^{cr} = A_1\exp(-Q_1/RT)(\sigma_{eff}/\sigma^{*})^{n1}$$

$${}^{2}\dot{\varepsilon}_{eff}^{cr} = A_2\exp(-Q_2/RT)(\sigma_{eff}/\sigma^{*})^{n2}$$

$${}^{3}\dot{\varepsilon}_{eff}^{cr} = 2[B_1\exp(-Q_1/RT) + B_2\exp(-Q_2/RT)] \times$$

$$\sinh(D < \frac{\sigma_{eff} - \sigma_{eff}^{0}}{\sigma^{*}} >) \tag{4}$$

$$\dot{\varepsilon}_{ij}^{f} = \frac{1}{\eta} < F > \frac{\partial F}{\partial \sigma_{ij}}$$

$$F = \alpha\left(\frac{|I_0|}{\sigma^{*}}\right)^{m-1}I_0 + \sqrt{II_s} - k \tag{5}$$

Numerical Modeling

In order to provide an efficient tool for the solution of thermomechanical response of rock salt in a nuclear waste repository, BGR in close cooperation with Control Data, Hamburg, developed the Special Purpose Code for Analysis of Nonlinear Thermomechanical Response of Rock Salt ANSALT [6].

The ANSALT code mainly consists of 4 modules with interactive data flow as schematically shown in fig. 1.

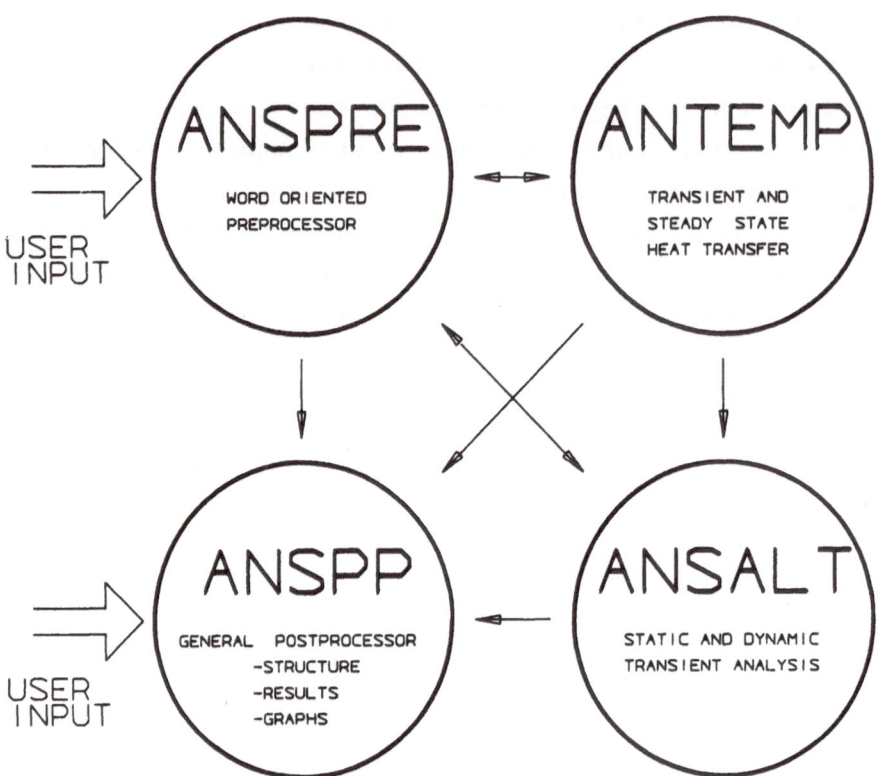

Fig. 1: ANSALT Data Flow

ANSPRE   is a word-oriented preprocessor for both finite element
         codes ANSALT and ANTEMP with simplified keyword -
         oriented input.

ANSALT   is a nonlinear finite-element-program to solve coupled
         thermomechanical problems.

ANTEMP   is a nonlinear finite-element-program to solve steady
         state or transient heat flow problems.

ANSPP   is a graphical display and result processor which
         provides capabilities to display or print selected
         results in user specified ways, e.g.: deformed struc-
         ture or parts of the structure, time history plots,
         line data as result display at user defined structure
         intersections.

A summary of the ANSALT features, material laws, geomechanical
model description, solution strategies, and special geomechanical conditions is shown in fig. 2.

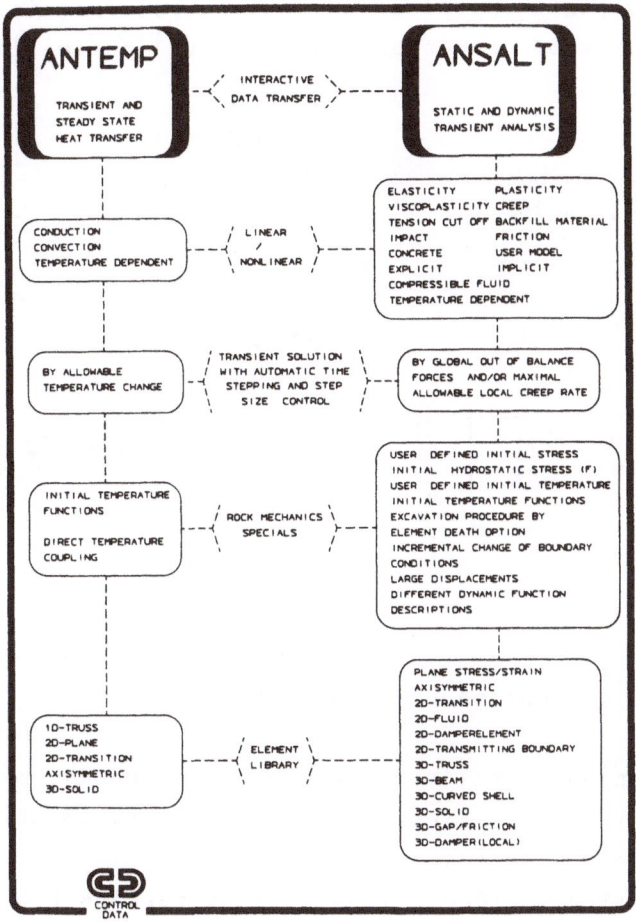

Fig. 2: ANSALT Code Capabilities

Model Validation

It is obvious that geomechanical modeling can only reach a certain level of accuracy, because the real thermo-mechanical
behavior of a complex geological structure will always remain
unknown up to a certain extend. The scientificly based
engineer`s approach, to overcome this general difficulty, is a
continuous improvement of the model, appropriate to the
improved knowledge of the input data. The main features of this
approach are the establishment of a consistent constitutive
relationship for the mechanical behavior, validation of this

model, and quantification of site relevant input data. The principle of geomechanical model validation is sketched in fig. 3.

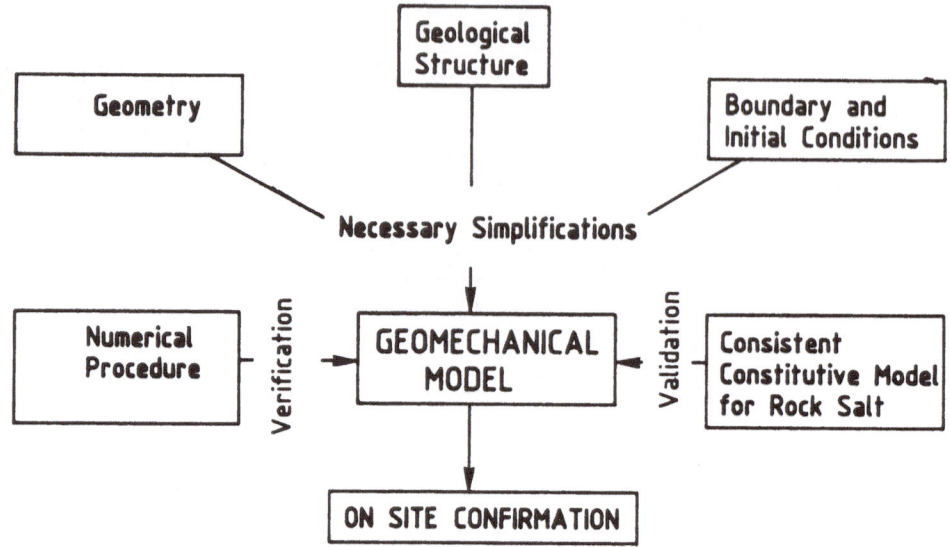

Fig. 3 Validation of Geomechanical Model

Developing a realistic geomechanical model, one has to bear in mind, that in a first step this already means necessary simplifications with respect to:
- geometry (determination of load-bearing behavior represented in spatial, plane, or axisymmetric models)
- geological structure (discretization of strata into zones of the same kind, namely the same geotechnical behavior)
- boundary and initial conditions (determination of stresses, temperatures, displacements, loads etc. at the boundary of the geometric model as well as their initital states or their changes through time).

Often these basic assumptions will not be clearly realized or at least will not be very well justified. Nevertheless, it is of crucial importance to represent those factors as realistically as possible. Model validation itself has to follow a strict scientific procedure:
- Prior to validation the numerical code, used for computation, has to be verified. This means, it has to be proved that the code gives mathematically correct answers.

Analytical solutions can verify portions of various numerical procedures. As a part of verification, benchmarks among various codes will be applied. However, benchmarks do not verify but only resolve why discrepancies occur.

- Model validation is achieved through successful predictions of laboratory test results or in situ results, taking into account a consistent constitutive model, proper boundary conditions, and initial conditions. Model validation in this sense is not curve fitting process by back-analysis, but a demonstration to what extend a particular consistent model is able to describe the thermomechanical response of the host rock, although the constitutive model perhaps does not take into account the entire mechanical behavior.

Model validation mainly means validating the consistency of the constitutive model. From this viewpoint laboratory testing can impose better well defined and controlled conditions than field testing because we can better isolate and validate aspects of the constitutive model by carefully designed laboratory test than we can in situ.

- A validated constitutive relationship for rock salt is than the proper basis for geomechanical modeling of the site specific geological situation. However, site specific units of equal mechanical behavior and related mechanical para- meters have to be determined and are to be confirmed by field tests.

SALT BARRIER INTEGRITY COMPUTATIONS

Gorleben Site

The Gorleben salt dome has been investigated from a geological point of view since 1980 in order to determine the suitability for the construction of a final repository for all kinds of radioactive wastes. In 1980 and 1981, four exploratory bore- holes were drilled from the surface down to a depth of 2,000 m as well as 31 boreholes down to the caprock. With respect to this information, two shaft pilot boreholes were drilled in the center of the Gorleben salt dome in 1982. Until 1985, 13

334

additional caprock boreholes have been drilled to investigate
the transition zone between salt dome and caprock. Moreover,
150 km of seismic reflection data have been obtained. With that
data, exploration from the surface was almost finished.

The underground exploration started in 1984 with extensive pre-
parations for shaft sinking. Sinking of the two shafts is
scheduled to be completed in 1989. Lateron, extensive underground exploration work will be carried out.

Complex structures have been recognized in all boreholes, thus
showing an intensive folding. Up to now, the structure of the
salt dome can only be roughly described. The Gorleben salt dome
has a length of almost 14 km. In the vertical direction it
extends from 250 into 3,100 - 3,300 m below surface. The width
of the salt dome reduces with depth and amounts to about 3 km
at the repository level. A representative geological profile of
a cross-section of the Gorleben salt dome is shown in fig. 4
[1].

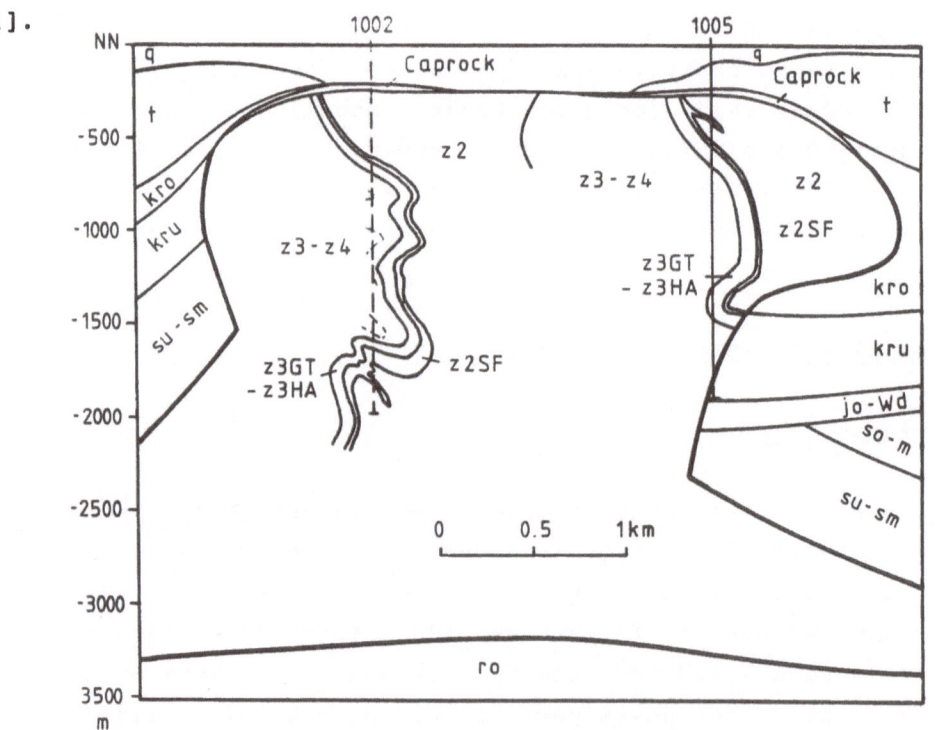

Fig. 4: Geological Profile of the Gorleben Salt Dome

The question, how good the salt formation is capable of
sustaining thermal loading, was rised early and has been subject of many investigations. From the geomechanical point of
view, thermally induced deformations and related stresses, as
well as their influence on the barrier integrity, have been
regarded.

Heat generation, due to radioactive decay, decreases with time.
Therefore, in the beginning temperatures are increased in the
vicinity of the repository, causing thermal expansion. This,
however, results in thermally induced stresses because the rock
mass builds up a transient resistance to deformation. Along
with decreasing temperatures in later times, the rock mass
tends to contract and stresses are reduced in the repository
area.

Assessment of the integrity of the geological barrier can only
be based on computations. Preliminary design calculations are
oriented towards problem identification and trend indication,
since final input data for the computational model are not sufficiently available at the present stage of exploration of the
Gorleben salt dome.

Table 1:

	Salt dome	Overburden	Adjacent rock
Thermal conductivity (W/mK)	$\dfrac{6.1}{0.0045T-0.229}$	2.1	2.6
Specific heat (Wd/m³K)	22.0	22.0	22.0
Coeff. of linear thermal expansion $\times 10^5$ (1/K)	4.0	1.0	1.0
Elastic constants: E (MPa) ν (-)	25 000 0.27	250 0.3	15 000 0.27
Steady state creep (s. eq. 4) A (d⁻¹) Q (kJ/mol) n (-)	$A_1 = 0.18$ $Q_1 = 54.0$ $n_1 = 5.0$	—	—

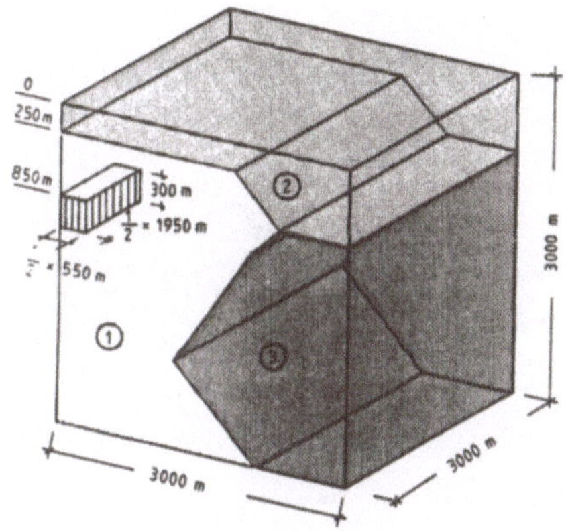

Fig. 5: 3-D Gorleben Repository Model

3 D-Model

From former thermal calculations it is known that for long-time ranges 2-D-model become invalid because they do not any more represent proper boundary conditions. Therefore, it is necessary to study this boundary effect from the three-dimensional behavior.

A simplified 3-D-model, representing only one quarter of the repository in the Gorleben salt dome, was considered. The geometric configuration is shown in fig. 5. Salt dome (1), overburden (2), and adjacent rock (3) were assumed to be homo-genous. Thermal properties and mechanical behavior are listed in table 1.

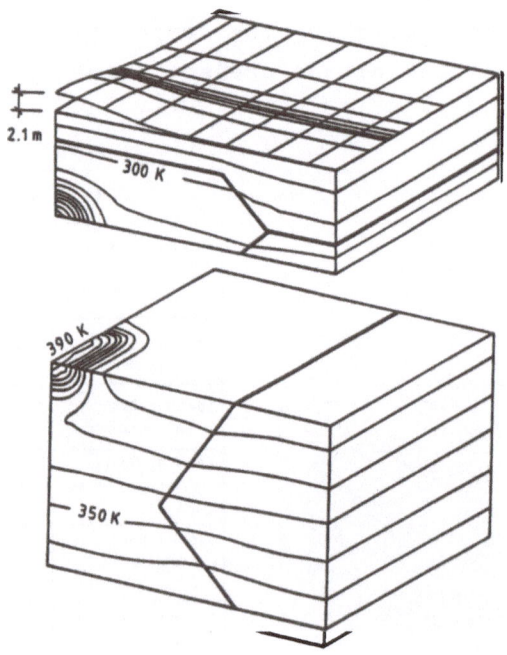

Fig. 6: Temperature Distribution and Surface Lifting after 1,000 Years

The initial heat generation in the repository corresponds to
0.24 W/m^3. The FE-Model consists of 592 spatial 20-node elements
with a total number of degree of freedom of 7,862. Temperatures,
deformations, and stresses are calculated over a time range of
10,000 years.

Characteristic results of the computation are plotted in
Fig. 6, showing the temperature distribution and the surface
lifting (scaled by a factor of 100) after 1,000 years. A
sequence of results was compared to corresponding 2 D-computa-
tions. In case of a repository elongated in one direction,
comparisons confirmed that two-dimensional computations give
a sufficiently appropriate thermo-mechanical response for the
central part of the repository which comes out to be the most
critical one with respect to maximal thermal loading.

2 D-Model

The geological situation can only be modeled at a somewhat
higher resolution at reasonable costs within a two-dimensional
computation. The speculative cross-section of the Gorleben salt
dome, as plotted in fig. 7, was assumed taking into account
15 separate homogenous layers.

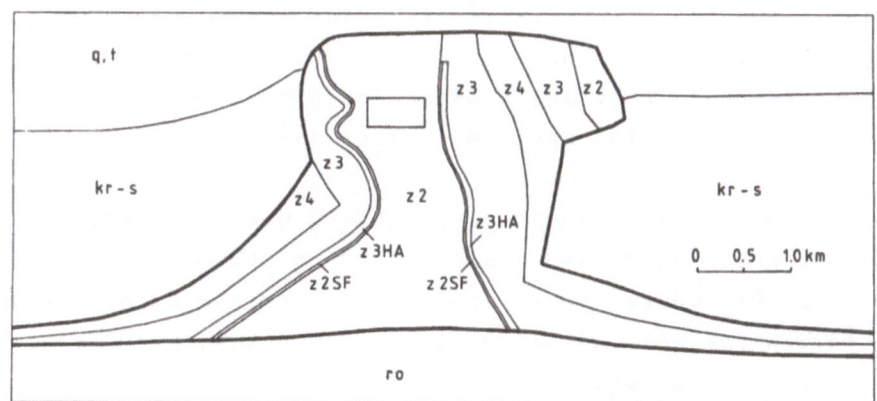

Fig. 7: Cross Section of the Gorleben Salt Dome as Assumed for
        Computation

	Zechstein salt			z2 SF (carnallite)	z3 HA (anhydrite)	Quaternary Tertiary q,t	Cretaceous Bunter kr-s	Rothliegen-des ro
	z2	z3	z4					
Thermal conductivity (W/mK)	90%*)	80%*)	70%*)	0.6	100%*)	2.1	2.4	2.7
Specific heat (Wd/m³K)	22.0			23.0	22.0	22.0	22.0	22.0
Coeff of linear thermal expansion × 10^5 (1/K)	4.0			2.5	1.6	1.0	1.0	1.0
Elastic constants E (MPa)	25.000			16.000	60.000	500	18.000	17.000
v (-)	0.25			0.27	0.23	0.33	0.25	0.25
Steady state creep (s eq 4) A (d^{-1}) Q (kJ/mol) n (-)	$A_1 = 0.05$ $Q_1 = 58.6$ $n_1 = 5.0$	$A_2 = 2.1 \times 10^6$ $Q_2 = 113.0$ $n_2 = 5.0$		$A_1 = 1.8$ $Q_1 = 54.0$ $n_1 = 5.0$	–	–	–	–

*) $\lambda = 6.1/(0.0045T - 0.229)$

Table II: Thermal and Mechanical Data, 2-D-Computation

Finite Element discretization consisting of 1.028 elements or 6,534 DOF, respectively, is shown in fig. 8. Due to the current state of site specific data evaluation, all thermal and mechanical parameters, as summarized in table II, are still assumed values. They are based on engineering-geological judgement. Especially the steady state creep behavior of all salt layers is uniformly considered to be at a lower bound. The heat loading in the repository corresponds to an initial heat generation of 0,219 W/m^3.

Temperature distribution after 1,000 years (fig. 9) and the related far field deformations (fig. 10, deformations are scaled by a factor of 100) illustrate the computed results. A maximum temperature of 182 C was calculated after 150 years. After 2,000 years, the maximum surface lifting amounts to 3,3 m.

The main objective of these computations was to ascertain whether fracturing due to thermal loading may occur within the salt dome. Of course, this cannot be determined by a single computation but only within a parametric study. Computations of this kind will be continued taking into account improved knowledge of the existing geological situation and the mechanical properties of the different layers. The computations carried out up to now indicate that thermally induced fractures will

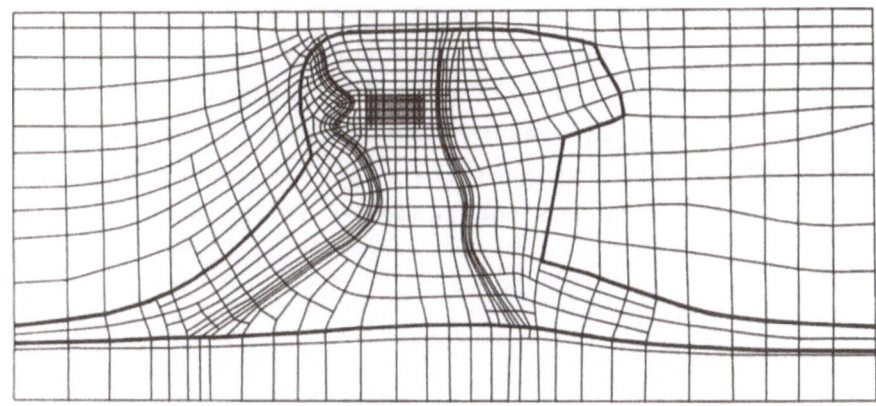

Fig. 8: Finite Element Mesh

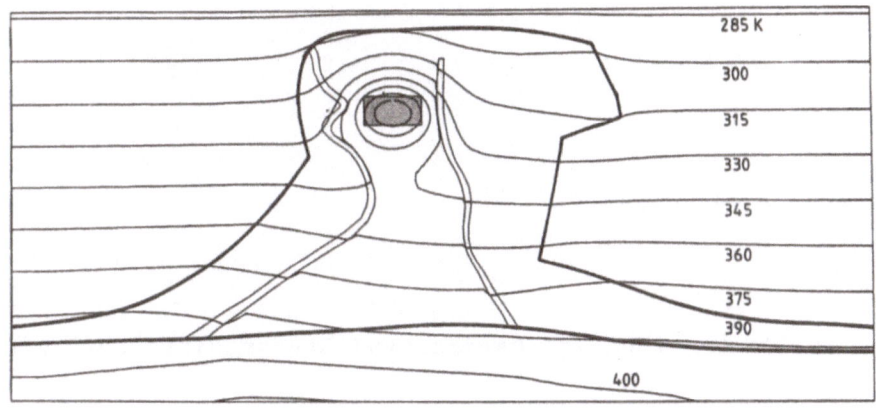

Fig. 9: Temperature Distibution 1,000 Years After Waste
        Implacement

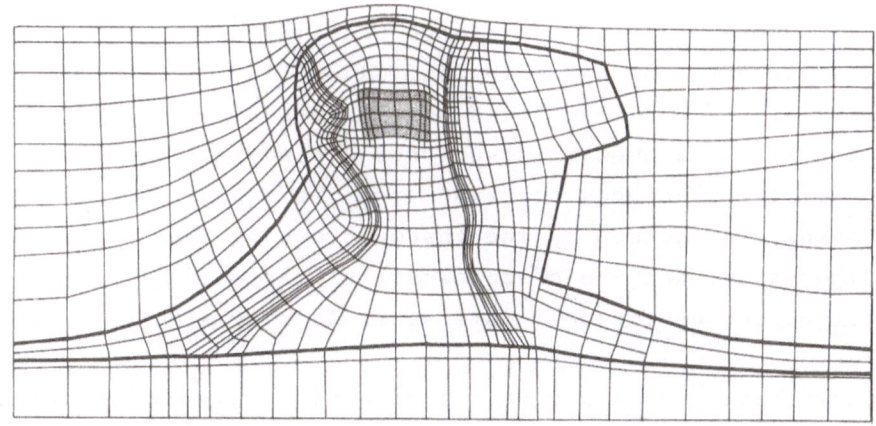

Fig. 10: Deformations 1,000 Years After Waste Implacement,
         Scaled by a Factor of 100

occur in the anhydrite layers. The salt layer between the repository mine and the anhydrite, however, will remain unfractured.
Moreover, the top of the salt dome is likely to be fractured.
The extend of these thermally induced tensile fractures depend,
self-evident, on the integral thermal loading but also on the
ductility of the salt and the geometry of the disposal field.

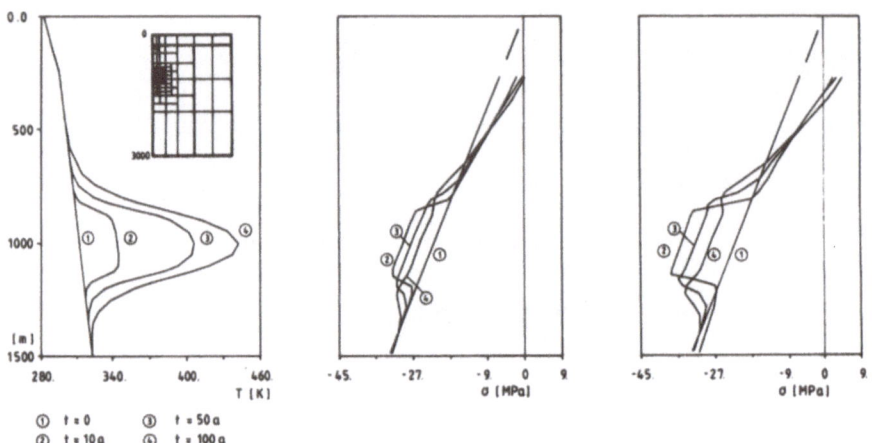

Fig. 11: Temperature and Horinzontal Stress Distribution with
Depth

Simplified Model

The occurence of tensile stresses or fractures at the top of
the salt dome was studied in more details applying a very
simplified model as shown in fig. 14. Justification for the
suitability of the simplified model was achieved through comparison with the complex structured 2 D-model. It turned out
that tensile fractures starting from the top of the salt dome
are mainly influenced by the thermal loading, the geometrical
configuration, and the creep capacity of rock salt. The internal structure of the salt dome seems to play a minor role.

Characteristic results from a parametric study with respect
to creep influence are shown in fig. 11. Tensile stresses or
fractures only occur within a relatively small time scale

(100 years). High ductility (high creep capability) enables rock salt to decrease the potential for fracturing while low ductility encourage the development of tensile fractures. Those thermally induced fractures, however, do not propagate too deep into the salt dome considering possible creep parameters of domal salt.

CONCLUDING REMARKS

This paper presents some fundamental ideas with regard to demonstrating structural stability of a final repository. Although tightness of the multi-barrier system is the first objective, mechanical stability of the barrier is a prior condition. Therefore, reliable predictions on the efficiency of the various barriers, based on numerical computations, are of particular interest.

In a salt formation, the rock mass itself naturally provides the main barrier function. Proper geomechanical modeling and validation of the geomechanical model is necessary in order to be able to assess the long-term geotechnical stability of the geological formation.

Some computed results on the thermo-mechanical response, resulting from the final disposal of high radioactive waste, are presented as application case examples. Although highly idealized, the simulation model for the Gorleben repository is able to predict thermo-mechanical effects, sufficiently. However, it is necessary to demonstrate that simplified models are suitable to study all essential thermomechanical effects.

The existing modeling technique provides a useful tool to compute barrier integrity calculations. Due to the present state of exploration, assumptions still have to be made, concerning the internal structure of the Gorleben salt dome and the definite repository configuration. However, the importance of various parameters, like thermal loading, ductility of rock salt, geometric configuration of the repository, etc., to the

integrity of the salt barrier can be evaluated in a sensitivity
study. Thereby, the non-occurance of tensile stresses within
the salt-barrier is considered to be a criterion for an
efficient and safe disposal mine layout.

Along with the underground exploration of the Gorleben site,
further development of the model must attempt to consider the
internal structure of the salt dome in more details. Further
computations will be carried out with the intention to adapt a
preliminary repository design to the explored internal structure of the Gorleben salt dome.

REFERENCES

[1]   BORNEMANN, O., GIESEL, W. & JARITZ, W.: Geoscientific
      Investigation of the Gorleben Site, Germany, IAEA Int.
      Symp. on the Siting, Design and Construction of Under-
      ground Repositories for Radioactive Wastes, Hannover,
      F.R. Germany, March 3-7 (1986)

[2]   LANGER, M. & VENZLAFF, H.: Sicherheitsnachweis und Stör-
      fallanalyse für ein Endlagerbergwerk im Salzgebirge.
      Geol. Jb. A 75, Hannover, p. 627-633 (1984)

[3]   LANGER, M., PAHL, A. & WALLNER, M.: Engineering-Geological
      Methods for Proving the Barrier Efficiency and Stabi-
      lity of the Host Rock of a Radioactive Waste Reposi-
      tory. IAEA Int. Symp. on the Siting, Design and
      Construction of Underground Repositories for Radio-
      active Wastes, Hannover, F.R. Germany, March 3-7 (1986).

[4]   MUNSON, D.E. & DAWSON, P.R.: Constitutive Model for the
      Low Temperature Creep of Salt (with Application to
      WIPP), SAND 79-1853, Albuquerque, N.M., Sandia
      National Laboratories (1979).

[5]   WALLNER, M.: Analysis of Thermo-Mechanical Problems
      Related to the Storage of Heat-Producing Radioactive
      Waste in Rock Salt. Proc. 1st. Conf. on the Mechanical
      Behavior of Salt, The Pennsylvania State University,
      University Park, Pennsylvania, Nov. 9-11, 1981,
      p. 739-763, Trans Tech. Publications (1984).

[6] WALLNER, M. & WULF, A.: Thermomechanical Calculations
      Concerning the Design of a Radioactive Waste Reposi-
      tory in Rock Salt, Int. Symposium Felsmechanik in
      Verbindung mit Kavernen- und Druckschächten, Aachen
      1983, Bd. 2, p. 1003-1012

[7] WALLNER, M.: Stability Demonstration Concept and Prelimi-
      nary Design Calculations for the Gorleben Repository.
      Proc. "Waste Management", Tuscon 1986, Vol II,
      p. 145-151

Response of Underground Openings to Dynamic Loadings

H.-J. ALHEID and K.-G. HINZEN
Bundesanstalt fuer Geowissenschaften und Rohstoffe Hannover, FRG

A. HONECKER
Control Data GmbH Hamburg, FRG

W. SARFELD
Gesellschaft fuer Strukturanalyse in
Forschung und Entwicklung, Berlin

Summary

A Finite-Element program for seismic analysis of underground structures is presented. The numerical formulation solves nonlinear dynamic problems including an infinite medium of rock or soil. As a general solution strategy a hybride numerical method is proposed.Its efficiency is demonstrated by examples. The combination of in-situ dynamic loading experiments and dynamic Finite-Element calculations gains the knowledge of dynamic response of underground structures and provides a basis for predictive numerical calculations.

Introduction

Finite-Element (FE) analysis of the dynamic response of underground structures gives an instructive insight into the effects of transient dynamic loading to the static prestressed vicinity of the structure. This knowledge is important in earthquake safety-analysis of underground terminal repositories for hazardous waste. If numerical dynamic analysis is combined with in-situ dynamic experimental results, the degree of damage of the rock mass surrounding the structure can be estimated. This is important even for static stability analysis.

In general, requirements on the spatial discretization and a complex geological characteristic as well as a complex geometry of the structure prohibit the inclusion of seismic source, travel path and structure in one single FE discretization.

To overcome the problems of limited storage capacities or computation time it is necessary to separate the signal flow from the source to the structure into steps and to combine the solutions of the single steps to get the dynamic response of the structure. This hybride concept has the great advantage, that different mathematical methods can be used to solve each partial problem and the most appropriate method can be applied (ALHEID et al.[2]). Three main steps in the signal flow are obvious:

- generation of the dynamic load in the source area

- transmission of the signal from the source region to the structure

- response of the structure to the dynamic excitation.

In the concept of submodells any source of dynamic excitation can be used as long as the farfield time-histories can be determined. This may be done analytically, numerically or even by measurements.

Concerning the transmission of the signal from the source to the structure, a vertically inhomogeneous medium is sufficient in most cases to model the geological characteristic. One dimensional FE analysis, reflectivity method or a ray theoretical approach can be applied to determine the development of the complete wavefield. The ray theoretical approach in time domain has the advantages of easy separation into up- and downgoing parts of the wavefield and small computational efforts.

If nonlinear material behaviour has to be considered the FE method in time domain is the appropriate tool to analyse the response of the structure to the dynamic excitation. The dimensions of the discretization should be as small as possible to reduce the numerical efforts. Thus only a part of the fullspace including the structure can be modelled. This requires suitable boundary conditions at all model boundaries such as transmitting boundaries in time domain (ALHEID et al.[1]) or at least viscous boundaries.

In addition the correct excitation has to be applied to the four
boundaries of the FE model depending on the direction of wave
propagation (ALHEID et al.[2]). This problem is essential in
the case of earthquake loading because the exciting wavefield is
composed of waves with different directions of propagation.
These waves act simultaneously on a structure.

Mathematical Model

The mathematical model used in this paper is based on the FE
technique and is incorporated in the computer-system ANSALT
developed cooperatively by Control Data GmbH and BGR
(Bundesanstalt fuer Geowissenschaften und Rohstoffe). Besides
quasistatic nonlinear analysis ANSALT allows nonlinear dynamic
analysis in time domain. The following features relate to
different highlights for the nonlinear transient dynamic
analysis:

- s o l i d e l e m e n t (4 – 8 nodes) for modelling rock
  and granular soil material. The nonlinear material
  behaviour is formulated for Mises and extended
  Drucker-Prager constitutive laws.

- v i s c o u s b o u n d a r y element (2 – 3 nodes) to
  simulate energy absorbtion by propagating shear- and
  compressional waves. Vicous boundaries are perfectly
  absorbing in time domain, if the angle of incidence is
  normal to the boundary.

- t r a n s m i t t i n g b o u n d a r i e s developed by
  theory of semi-infinite element can be used to propagate
  arbitrary incoming and outgoing waves at the FE boundary.
  This boundaries are defined as a stiffness and damping
  matrix and describe exactly any wave type in the frequency
  domain. In order to use transmitting boundaries in time
  domain a practical approximation was introduced, by taking
  the arithmetic mean-value matrices for a chosen number of
  frequencies (ALHEID et el.[1]). Assembling the described

element for a given dynamic problem leads to the well known equation of motion

$$m*a + c*v + k*u = p(t)$$

m: mass matrix   a: acceleration vector
c: damping matrix   v: velocity vector
k: stiffness matrix   u: displacement vector
        p(t): load-time history

which will be solved numerically using implicit time integration (Newmark algorithm).

This procedure for solving dynamic problems in the area of soil mechanics has been verified extensively using simple models and is now used in standard calculations.

Example 1

An earthquake focus is assumed to be situated at a depth of 15 km below an underground drift embedded in layered media at a depth of about 1 km. The depth of the upper boundary of the bedrock is 3 km.   The drift has a maximum height of 6 m and a maximum width of 7 m. Following the concept of hybride modelling the complete model is separated into three submodels:

1. The assumed earthquake source time-displacement history is a band limited box-car with a duration of two seconds. Band limiting to frequencies between 0.5 and 5.0 Hz was necessary to avoid alising effects in the FE calculation.   The resultant farfield velocity time history is shown in figure 1.

2. The normal-incidence wavefield in the layered medium was calculated using a time-domain discrete state-space model (Mendel et al.[4]; Ferber [3]) with a sampling rate of 3 ms. The resultant pulse responses of the layered medium at the depth of the embedded structure for the up- and downgoing waves, respectively, are shown in figure 1.

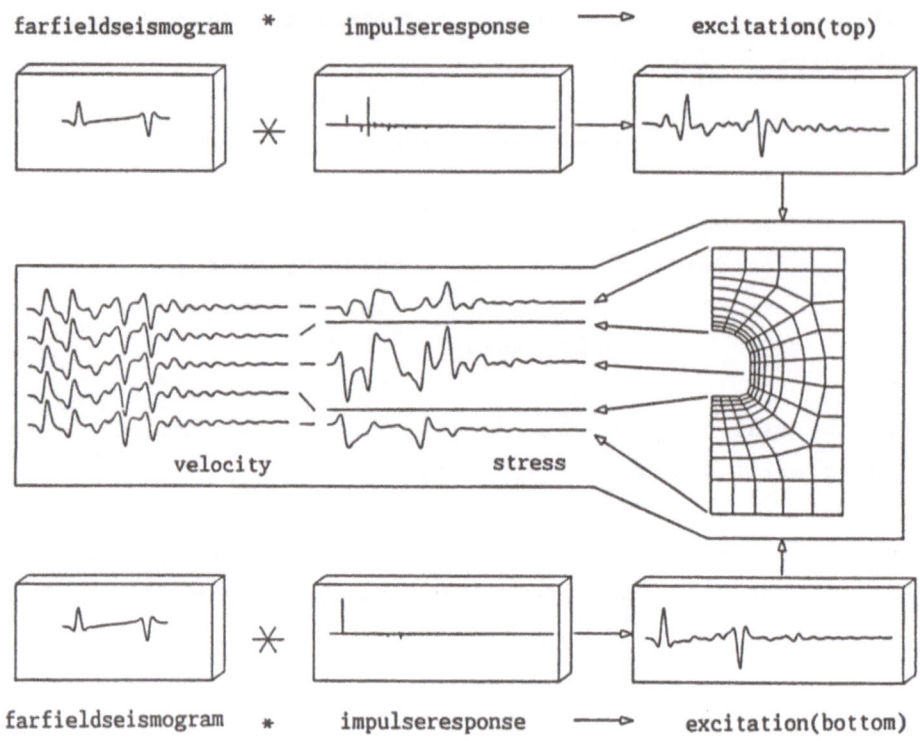

farfieldseismogram *   impulseresponse   ⟶   excitation(top)

velocity   stress

farfieldseismogram *   impulseresponse   ⟶   excitation(bottom)

Fig.1. Earthquake stability calculation. Excitations at bottom and top of the Finite-Element discretization are obtained by a convolution of numerically calculated impulseresponses and far-field seismogramms derived from analytical considerations.

3.   In the FE calculations we used the appropriate force-time histories derived from the velocity functions as input to the upper and lower boundaries of the FE model for down- and upgoing waves respectively. The time step used in the calculations is 12 ms. The response of the structure was calculated for a total time of 6.6 s, i.e. 550 time steps. The cross sectional two- dimensional discretization is shown in figure 1. The overall dimensions of the Finite- Element discretization are 12 m * 24 m. The system consists of 107 8-node isoparametric elements with 361 nodes and 684 degrees of freedom. In order to simulate the fullspace

surrounding of the model transmitting boundaries are used at the right hand side. At top and bottom of the FE region and of the transmitting boundaries viscous boundary elements are placed.

In figure 1 the resultant vertical velocity and vertical stress time histories are presented. The calculated velocity time histories describe the simple superposition of the two velocity input functions. Obviously the model acts nearly as a rigid body due to the low frequency content of the seismograms. Comparing the stress time histories considerable differences are obviuos. The large stress concentration due to the drift at the wall-face is evident.

Example 2

In-situ Experiment

In-situ experiments were performed in order to validate the output of the FE-code. For this purpose a special underground opening was constructed. This opening is a drift of nearly 100 m length. It is 3.5 m high, 3.0 m wide and the maximum overburden is about 55 m. In the construction of the in-situ settings care had to be taken, that the spatial dimensions, frequency content of the signals and the parameters describing the geology were in a range which could reasonably be modelled in our FE numeric experiments. The material is a rather homogeneous dolomite with a p-wave velocity of 3800 m/s. A two dimensional measuring array was constructed in a range with minor faults and joints at a depth of 66 m. This array consists of 6 boreholes. Figure 2 shows the geometry. In each borehole three 3D geophone-stations were planted. By shaking table testing the amplitude response of the geophones was found to be proportional to ground velocity within a frequency range from 30 Hz to 1500 Hz. The data were registrated with a computer controlled 64 channel digital recording system. The maximum sampling rate is 62.5 kHz per channel.

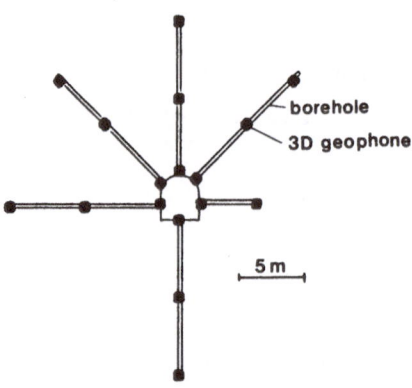

Fig.2. Geometry of in-situ measuring array

The dynamic load to the drift was exaggerated by the detonation of explosive charges of a weight from 170 g to 1360 g. The charges were fired in a borehole which was drilled from the top of the overburden within the two dimensional measuring array. The overburden of the drift is about 40 m in this range. The charges were fired at depths between 15 m and 20 m.

The seismograms are 12bit data with a sampling frequency of 31 kHz. In addition to the signals from 20 geophone stations with 60 seismic channels, the exact moment of the detonation and the pressure time-history in the blasthole were recorded. The pressure history was used to model the excitation function in the numeric experiments.

Figure 3 shows a typical result of an in-situ dynamic load experiment (IDLE). The vertical component velocity seismograms in this figure are normalized to their maximum amplitudes. The * indicates the position of the 1360 g charge. The positions of the geophone-stations are given by small circles. Only the right half of the symmetric measuring array is shown. The lines indicate the location of the corresponding geophone station for each seismogram. In case of two seismograms at one place, the corresponding ground movements in the left and right part of the

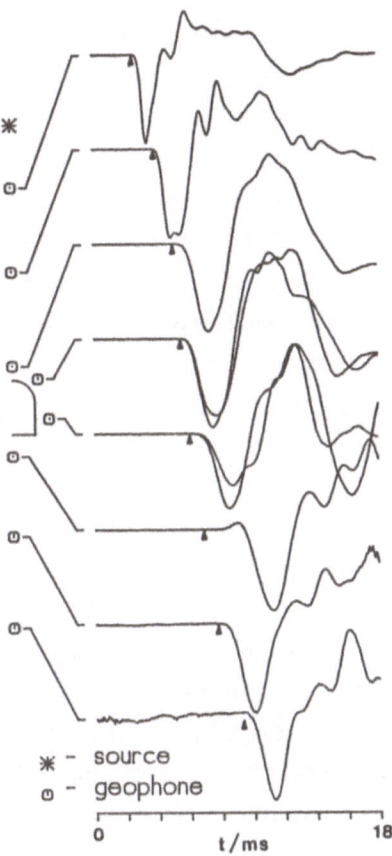

Fig.3. Velocity seismograms in the vicinity of a drift produced by the detonation of an explosive charge.

measuring array are given. In all IDELs the measured ground movements proved to be reproducible

The triangles at the seismic traces mark the arrival times, which would be expected, if the medium was a homogeneous halfspace with a p-wave velocity of 3800 m/s. While the first breaks for the two upper traces exactly meet the expected arrival times, a delay of p-arrivals is obvious at the measuring positions in the vicinity of the drift. The largest delay occures at the position close to the drift floor (6th trace in fig. 3). A zone of weakened material around the drift with a reduced p-wave velocity is a possible explanation for these observations. At an increased distance from the drift (lower trace) the influence of the weakened zone is no longer effective and the delay decreases.

The direction of the first motion is negative (away from the source) for all but one of the seismograms. The exeption is the trace which was recorded close to the drift floor. The small upward movement at the beginning of the signal is an effect of the dynamic drift response.

Numerical Calculations

To calculate the response of the underground drift and its vicinity a two-dimensional FE discretization was chosen. The overall dimensions of the discretization are 7 m * 14 m. The system consists of three element groups with different elastic constants, a total of 457 4-node isoparametric elements with 505 nodes and 987 degrees of freedom. The fullspace surrounding was simulated by means of transmitting boundaries (vertical boundaries) and viscous boundary elements (horizontal boundaries). According to the results of the measurements a disturbed zone with an extension of 4.5 m was assumed to surround the drift. This disturbed zone was simulated by a linear elastic material with a Young's modulus being one half of the Young's modulus of the intact rock.

As only downgoing waves had to be considered in this example, the excitation was only applied to the upper boundary of the discretization. The load-time history was found by a seperate FE calculation. In this analysis the stress-time history 5 m away from a borehole was calculated applying the measured pressure-time history as excitation to the borehole wall and using nonlinear material behaviour.

The delayed p-arrivals in the vicinity of the drift could be modelled easily using only one uniform disturbed zone surrounding the structure. However, with this model the small upward movement observed in the measurements at the bottom of

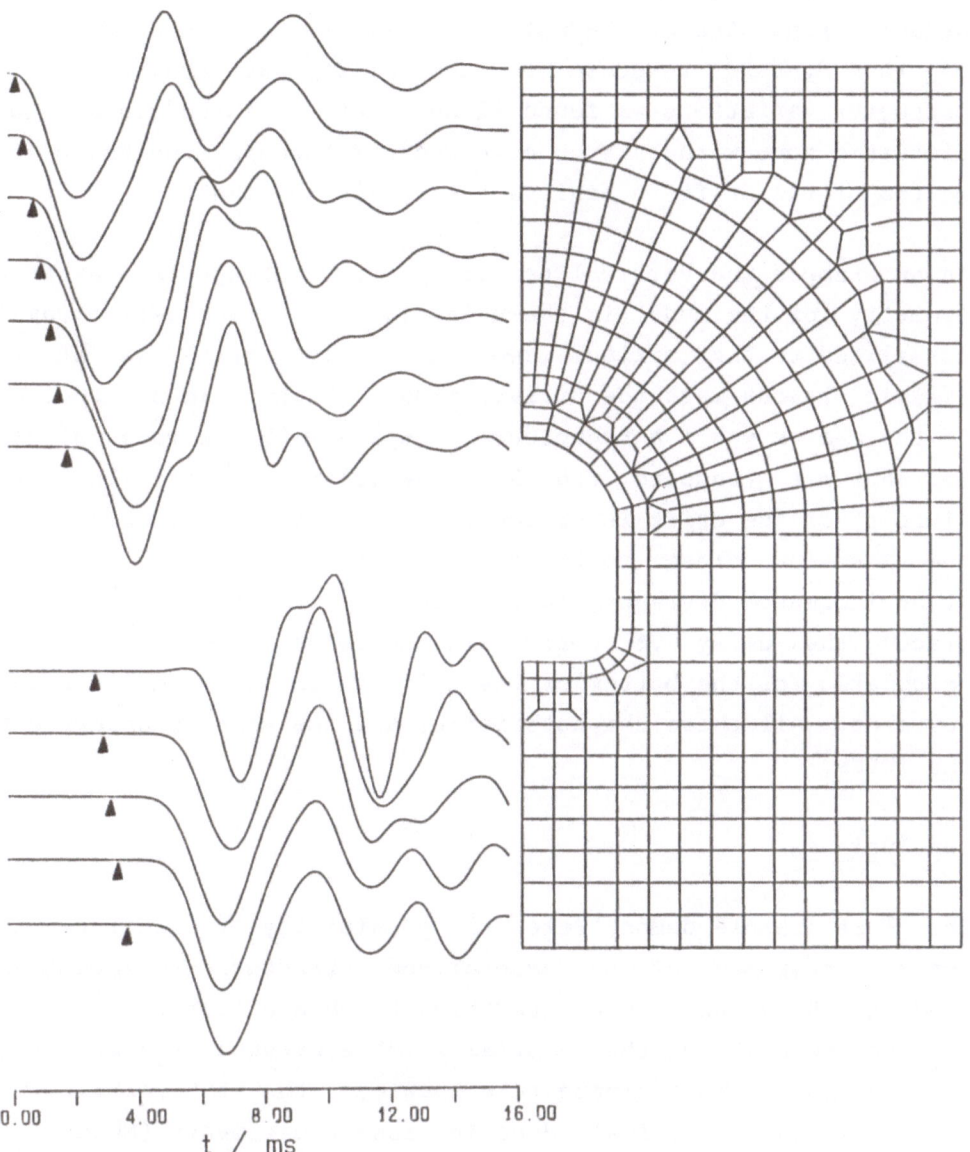

Fig.4. Velocity seismograms in the vicinity of a drift calculated by the Finite-Element method.

the drift could not be simulated. Thus the influence of the dynamic properties of the material below the bottom of the drift on the dynamic response of the drift was studied. From parameter variations we found it necessary to introduce a second disturbed zone with an even more reduced Young's modulus in the bottom of the drift to achieve the upward movement.

As an example the vertical seismic profile along the axis of symmetry of the model is shown in figure 4. Each seismogram is normalized to the maximum amplitude. The triangles at the seismic traces mark the arrival times, which would be expected, if the medium was a homogenious halfspace with a p-wave velocity of 3800 m/s (compare fig. 3). The delay of p-arrivals in the vivinity of the drift is obvious. The largest delay occures at the position close to the drift floor (8th trace in fig. 4). At an increased distance from the bottom of the drift (lower trace) the delay of p-arrivals decreases. The seismogram calculated for the bottom of the drift exhibits a small upward movement. All these characteristics are the same as observed by the IDLEs.

Conclusions

The first example demonstrates the possibility to analyse the dynamic response of an underground structure to earthquake loading. As a fundamental condition to obtain correct numerical results the state of the material which sourrounds the structure must be known. To evaluate this knowledge the interactive use of IDLEs and numerical dynamic load experiments (NUDLEs) is suggested. The increased knowledge of the state and extent of the disturbed zone around an underground structure is of general interest in many mining applications. The second example shows that numerical studies of physical effects help to understand measured data. After confidential assumptions about the material behaviour around an underground structure have been developed the FE method is a suitable tool to calculate the effects of earthquake loading to underground structures, which of course cannot always be simulated experimentally to their whole extent.

References

1. Alheid,H.-J., A. Honecker, W. Sarfeld and H. Zimmer, 'Transmitting boundaries in time domain for 2-D nonlinear analysis of deep underground structures', Proc. NUMETA 85, Swansea, ed. J. Middelton and G. Pande, 117-127, 1985

2. Alheid,H.-J., K.-G. Hinzen, A. Honecker and W. Sarfeld, 'Transient analysis in rock dynamics - some problems and solution strategies',Proc. ECONMIG 86, University of Stuttgart, 1986

3. Ferber,R.-G., 'Normal-incidence wavefield computation using vector-arithmetic', Geophys. Prosp. 33, 540-542, 1985

4. Mendel,J.M., N.E. Nahi and M. Chan, 'Synthetic seismograms using the state-space approach, Geophysics 44, 880-891, 1979

Rotated Title Index

Author Index